Engineering Materials and Processes

Series editor

Brian Derby, Manchester, UK

More information about this series at http://www.springer.com/series/4604

Ming Wang Fu

Design and Development of Metal-Forming Processes and Products Aided by Finite Element Simulation

Springer

Ming Wang Fu
Department of Mechanical Engineering
The Hong Kong Polytechnic University
Hong Kong
China

ISSN 1619-0181 ISSN 2365-0761 (electronic)
Engineering Materials and Processes
ISBN 978-3-319-83526-6 ISBN 978-3-319-46464-0 (eBook)
DOI 10.1007/978-3-319-46464-0

Printed on acid-free paper

This Springer imprint is published by Springer Nature
The registered company is Springer International Publishing AG
The registered company address is: Gewerbestrasse 11, 6330 Cham, Switzerland

To my parents, my wife and partner, Yan Lei, and my sons and daughter, Gabriel Yongquan Fu, Grace Yongan Fu, and Graham Yongji Fu.

Preface

In today's rapidly changing product marketplace, the critical requirements for product quality, productivity, and time-to-market have been becoming the most powerful driving force behind any new product design and development (PDD) paradigm for seeking the best design solution and quality product development. This would render the design and development of products full of challenges. From design perspective, how to come out with "design right the first time" and the best design solution is crucial for reducing "time-to-market" and ensuring "quality and defect-free" product development. From product realization point of view, it would reduce trial and error and shorten product development lead time.

In net-shape or near net-shape PDD, plastic deformation process is one of the important manufacturing processes for fabrication of metal-deformed parts and components. The parts or components produced by this traditional manufacturing process are widely used in many industry clusters, ranging from computer, home appliance, medical, consumer electronics, automobile to aerospace industries. The high demand for shorter design and development lead times, good dimensional accuracy, overall quality, and rapid design changes have become the bottleneck issues in metal-forming industry. For the companies which want to maintain their competitiveness and cutting edge in this industrial cluster, there is an urgent need to shorten design and development lead times, reduce production cost, improve product quality, and enhance productivity. The simulation-enabled PDD paradigm and technologies in metal-forming arena help address the above-mentioned issues in this traditional but now revitalizing and promising industrial cluster.

With the advent of finite element method (FEM), finite element (FE)-based simulation, and their wide applications in plastic deformation processes, the traditional metal-formed products development paradigm is shifted from the heuristic know-how and trial and error to in-depth scientific calculation, analysis and simulation to support metal-formed part design, forming process determination and configuration, product quality control and assurance, and the realization of mass production of defect-free metal-formed products. The simulation-enabled metal-formed PDD paradigm provides solutions to address these issues.

In the past decades, extensive researches on simulation-enabled metal-formed PDD have been thoroughly conducted in both academia and industry. The detailed methodologies and techniques to support the above-mentioned design activities have been well explored and developed. This book aims to report the state-of-the-art advances in these areas, which include the fundamentals of rigid-plastic FEM and the FEM-based simulation of metal-forming process, simulation-aided metal-formed part design, process determination and configuration, die design, and product quality assurance and control.

In this book, Chap. 1 first introduces the basis of plastic deformation of materials, plastic deformation or forming processes, and the forming system to realize plastic deformation process. With the forming process and system, the metal-formed parts are fabricated. In addition, the challenges of this traditional manufacturing process are summarized and how to address these issues is briefly summarized. In Chap. 2, the fundamental of rigid-plastic finite element method is articulated, which is the kernel of FEM-based simulation in metal-forming processes. The simulation of plastic deformation processes by FEM is then elucidated and the detailed case studies are given to show its applicability in metal-forming arena.

In Chap. 3, how the FE simulation helps metal-formed part design is introduced. For a given designed product, there can be many design alternatives and solutions generated from metal-formed part design perspective. How good they are and how the corresponding forming systems perform are assessed by using FE simulation. In addition, the forming process determination and process parameter configuration with the aid of FE simulation are presented. By using FE simulation, different process routes and process parameter configurations can be evaluated based on the formability of materials under the given designed metal-forming parts and the quality of the deformed parts. The focus of this chapter is on the evaluation of metal-formed part design, process route, process parameter configuration, and the designed metal-forming system by using FE simulation.

In metal-forming, die is an important tool to deform the workpiece in such a way the metal-formed part is fabricated. Die works under a very severe stress condition and the working stress is not uniformly distributed in the entire die structure and thus the working stress needs to be carefully and rationally designed and controlled to ensure the good performance and long service life. In Chap. 4, the simulation-enabled die and forming system design is presented, which includes different design approaches for die design aided by FE simulation and the methodology to evaluate the entire forming system design based on the proposed evaluation criteria.

In PDD, product quality, production cost, and time-to-market are three overriding issues. Product quality is the first and most important. In metal forming, many defects can be generated in forming processes and some of them can be classified into flow-induced and stress-induced defects. The former is caused by the irrational flow pattern, while the latter is generated by the working stress exceeding the strength limit of materials. The most common defect of the latter is the ductile fracture occurring in forming process. In this book, how to identify these defects and avoid them via FE simulation is presented in Chaps. 5 and 6. In Chap. 5, the flow-induced defects in multiscaled plastic deformation processes are introduced. The fundamental flow behaviors and formation mechanisms of defects in different scales are revealed and how to avoid the defects is also articulated. In Chap. 6, ductile fracture and stress-induced defects in multiscaled deformation are described. The defect formation mechanisms and evaluation criteria are summarized. The modeling, simulation, and prediction of the ductile fracture and stress-based defects are summarized, which provide a systematic basis for defect analysis in multiscaled, especially for microscaled plastic deformation.

Metal-forming has been becoming an important manufacturing process for long time due to its high productivity, good product quality, and low production cost. This conventional manufacturing process, however, has been traditionally linked to long years of apprenticeship and skilled craftsmanship. The traditional product design and development paradigm cannot meet more and more demanding requirements in the current competitive marketplace, and the simulation-enabled metal-formed PDD paradigm provides solutions to address these issues. This book is thus intended to give the comprehensive description and thorough articulation of the state-of-the-art simulation-enabled PDD paradigm in this arena for students, researchers, and engineers in industries, R&D organizations, and academia. In addition, this book also provides valuable information to this manufacturing field and can be used as a reference and textbook for a graduate course in learning this very useful and practical manufacturing process.

Hong Kong, China Ming Wang Fu

Acknowledgements

I am indebted to the following institutions and individuals who have helped make this book possible due to their generous supports and contributions to the research activities from which the research outcomes and findings summarized in this book arise.

- The Hong Kong Polytechnic University (HK PolyU), Faculty of Engineering, and the Department of Mechanical Engineering for providing various research scholarships, grants, and resources which are pertinent to establishing the research facilities and conducting the researches from which this book arises;
- The General Research Fund and the Innovation and Technology Fund from Hong Kong Government for funding the projects of B-Q08V, B-G33F (515012), ITS/028/07, ITS/313/09, and ITS228/11 to the support of the researches summarized in this book, and the project of No. 51575465 from the National Natural Science Foundation of China;
- Dr W.L. Chan, Dr. H. Li, Dr. J.G. Liu, Dr. J.H. Deng, Dr. Q. Hu, Dr. Y.Q. Ning, Dr. Y. Zhao, Dr J.Q. Ran, Dr. J.L. Wang, and Dr. Z.T. Xu for their contributions to the projects acknowledged here or contributing part of their research results when they worked under the above projects to this book.

Finally, I would like to express our sincere gratitude to my family for supporting me throughout conducting the researches and the publishing process. Their understanding and support are critical and invaluable for the completion of the above research projects and this book.

Ming Wang Fu

Contents

Chapter 1
Introduction

1.1 Introduction

Metal-forming is one of the most practical and efficient manufacturing processes to fabricate net-shape or near-net-shape parts via plastic deformation of materials. This traditional manufacturing process offers many attractive and unique advantages, including high productivity and efficiency, superior mechanical properties, excellent material utilization, low production cost, and the complex geometries of the deformed parts. With the ever-increasing costs of materials, energy, and manpower, more and more strict requirements from environment-friendly and sustainable development perspectives, and the marginal profit of this traditional industrial cluster, the design of metal-deformed parts, forming process, tooling and the entire forming system, as well as defect prediction and avoidance and the assurance and control of metal-formed product quality are becoming more critical. The new product design and development (PDD) paradigm thus needs to be developed for carrying out these design activities such that the design and development of metal-formed parts via plastic deformation can be realized with shorter design and development lead time, lower production cost, and higher product quality. This traditional industry can thus keep the cutting edge in the competitive marketplace.

Metal-forming refers to a set of plastic deformation approaches by which the given material in the form of blank or billet with different geometries and shapes including plate, sheet, bar, rod, block, wire, and tube is transferred into a desired final shape without change in the mass or composition of the material [1–3]. It can be classified into two simple categories, viz. bulk metal-forming and sheet

M.W. Fu, *Design and Development of Metal-Forming Processes and Products Aided by Finite Element Simulation*, Engineering Materials and Processes,
DOI 10.1007/978-3-319-46464-0_1

metal-forming [3–5]. The former refers to the deformation of materials with three dimensions in nature and includes the forming processes such as forging, extrusion, rolling, and drawing, while the latter refers to the metalworking process by which the sheet metal is deformed or stamped. In sheet metal-forming, the stresses of deformation are generally in the plane of sheet metal. The main operations in sheet metal-forming include shearing, drawing, bending, and forming. In detail, each operation can have its detailed subset operations. Shearing can have more detailed shearing-based operations including blanking, punching or piercing, notching, and trimming. Forming can also have its subset operations including stretch forming, flexible die forming, bulging, spinning, peen forming, and superplastic forming. But the nature behind the processes is the plastic deformation of materials. This chapter summarizes the basis of the plastic deformation of materials, metal-forming process, forming system, and the challenges arising nowadays in this widely used and unique industrial cluster. In addition, how finite element method (FEM) helps address the challenges is also presented.

1.2 Plastic Deformation of Materials

The deformation of materials can be defined as the change in the shape or size of materials driven by external force. The force could be in different forms such as tensile, compressive, and shear forces. The plastic deformation of materials is a permanent material flow, which transforms the solid materials from one shape to another. During this process, two main changes happen. One is the change of geometry of the deformation body, and the other is the change of microstructure of the material. For the former, the originally simple geometries of material are transferred into the final geometries of the deformed workpieces with complex and desirable features in such a way to meet the design specifications in terms of geometry and shape, surface topography structure, and dimensional accuracy. The latter is referring to the change of microstructure of materials driven by working temperature and plastic strain, especially in hot working process. The change of microstructure in plastic deformation process would be driven by phase transformation and change, movement of structural defects such as grain boundaries, point vacancies, line and screw dislocations, stacking faults and twins, and recrystallization in hot working process. All of these physical phenomena affect and determine the final microstructure of the hot-worked materials. Therefore, in design and development of metal-forming processes using plastic deformation of materials, it is necessary to consider the above-mentioned geometry and microstructure characteristics such that the mechanical properties and the final geometries of the deformed workpieces can be tailor-made and well controlled. In this new scenario of PDD of metal-forming products, there thus emerge two categories of design requirements, viz. "geometry accuracy-driven" and "desirable properties-driven" requirements, and thus, a new PDD paradigm is thus needed to be proposed to

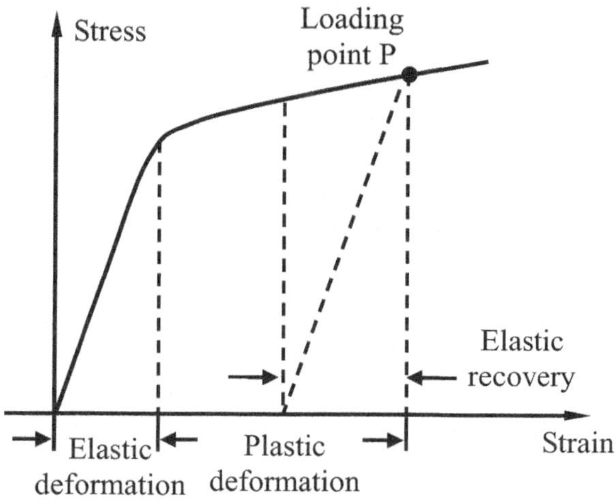

Fig. 1.1 Elastic and plastic deformation

realize these two categories of design requirements and specifications. The simulation-enabled PDD paradigm is the desired one to realize this goal.

From the mechanical behavior perspective, the deformation of materials can be classified into elastic and plastic deformation. Figure 1.1 shows the stages of elastic and plastic deformations by using the stress–strain curve of materials. In elastic deformation stage, the stress does not exceed the yield stress of materials and the relationship between stress and strain is linear. In addition, the elastic deformation is reversible and recovery. In metal-forming processes, it needs to consider this ubiquitous deformation phenomenon as it affects and deteriorates the dimensional accuracy, geometry precision of the deformed parts, and the performance of tooling. Therefore, in tooling design and process determination, the elastic deformation in tooling and the elastic recovery of the deformed parts need to be considered such that the dimension compensation in design of tooling and the deformed parts can be determined quantitatively in up-front design stage. In other words, the geometrical accuracy of the deformed part can be secured if the elastic deformations of both workpiece and tooling can be appropriately handled and addressed.

The plastic deformation of materials, on the other hand, is irreversible and permanent. Most materials in plastic deformation stage generally undergo elastic deformation and then plastic deformation when the effective stress of the deformation body exceeds its yield stress. In plastic deformation, the plasticity of materials is the propensity to undergo plastic deformation without fracture. In metal-forming arena, this capacity of materials is used to design and determine the forming process route and the sequence and number of forming operations. In addition, different forming processes including cold, warm, or hot forming are selected based on this capacity of materials.

For crystal structured materials, the crystal type of materials determines the plasticity of materials. The typical crystal types of materials include face-centered cubic (fcc), body-centered cubic (bcc), and hexagonal close-packed (hcp).

Generally, the fcc materials such as Au-, Ag-, Al-, Cu-, Ni-, and Fe (Austenite)-based alloys have good plasticity and a relatively low flow stress and thus high workability and followed by the bcc materials including Fe (Ferrite)-, Cr-, Mo- and W- and Nb-based ones with lower plasticity and further lower workability. The hcp materials such as Mg-, Zn-, and Zr-based ones have poor plasticity and further the bad workability. For most crystal-based materials, their plastic deformation is basically realized via slip movement and rarely by twining. The main approach to improve the plasticity of materials is thus to increase the number of slip systems in the deformation body. The most efficient way is to increase the working temperature. The increase of hydrostatic stress also enhances the plasticity of materials.

1.3 Forming Process

Forming process is one of the efficient manufacturing processes, which transfers the materials from one geometry configuration to the other and produces the required geometries and shapes of the deformed parts with the desirable mechanical properties. In this process, there is no material removal involved; instead, the material is deformed and displaced, and the volume of the deformed material is assumed to be constant. In addition, the process can be conducted at different working temperatures, and thus, it can be classified into cold, warm, and hot forming processes in the following.

1.3.1 Cold Forming

Metal-forming can be classified into cold, warm, and hot working according to its working temperature. Cold working usually refers to the metal-forming conducted at room temperature to about 30 % or less than its melting temperature of the materials on an absolute temperature scale, and this temperature is generally below the recrystallization temperature of the materials. At this working temperature range, the deformation phenomena and behaviors associated with cold working process occur. Dislocations increase, and their interaction and entanglement cause working hardening and the increase of flow stress, and thus, the strength of material increases with deformation. If the working stress exceeds the strength of material, ductile fracture would happen in the deformation process. To soften the material for further processing, thermal heat treatment is occasionally employed to facilitate the recovery and recrystallization of the material cold-worked such that the ductility of material can be increased and the flow stress of material be significantly reduced. From the quality perspective, ductile fracture, the so-called stress-induced defect in this book, is one of the main defects to be considered in cold working process and tooling design. In addition, the final surface finish and dimensional accuracy of the cold-formed parts are generally well controlled. Sometimes, annealing after the working process needs to be conducted to eliminate or reduce the residual stress generated by the unbalanced and non-uniform deformation in the deformation body. The residual stress in cold-worked parts can also be reduced or eliminated via annealing heat treatment.

In cold forming processes, there is no heating required. The deformed parts have a better surface finish and good dimensional accuracy compared with the warm- and hot-formed parts. In addition, the cold-deformed parts have a better reproducibility and interchangeability. The flowline-based directional properties can be imparted into the deformed parts, if there are no subsequent working processes such as heat treatment conducted. In addition, no heating is needed, and thus, the contamination problem can be minimized. Due to its good surface quality and dimensional accuracy, cold forming process is sometimes used after warm and hot working processes to improve the surface quality and dimensional accuracy of the deformed parts.

The disadvantages of cold forming process include higher deformation load and the greater capacity equipment needed. The process requires stronger tooling and good ductile materials. Meanwhile, the surfaces of billets need to be clean and scale-free. For multioperation cold forming process, intermediate annealing may be required to eliminate the strain hardening generated in the previous operations and to improve the ductility of materials for further processing. In this process, the imparted flowline and directional properties could be detrimental in some cases, and the undesirable residual stress induced in cold forming process could affect the mechanical properties and the quality of deformed parts. Furthermore, if the residual stress is undesirable in the deformed parts, it needs to be released or reduced by heat treatment. Otherwise, it would cause the early failure or fracture of the cold-formed parts in subsequent processing or service.

In cold forming process, lubricant plays an important role as it can improve significantly the homogeneous flow of materials, reduce friction and wear in between tooling and workpiece, decrease deformation load, and facilitate the ejection of the deformed part from die cavity. Therefore, selection of a suitable lubricant is important in cold forming process. The development of phosphating and saponification is a good example, which significantly enhanced the productivity of cold forming process in military industries a few decades ago when it was developed.

1.3.2 Warm Forming

Warm forming is usually conducted in the approximate temperature range of 30–60 % of the melting temperature of materials on an absolute scale. It is also believed that the forming process done below the recrystallization temperature of materials is considered as warm forming. In warm forming process, the deformation force needed is lower than that in cold working process, the final finish and dimension accuracy, however, are not as good as cold working, but better than hot working. Warm forming process has the drawbacks as those in hot working process, which will be discussed later. One of the major considerations for selection of warm working is economic, as there is no need to heat the workpiece up to hot working temperature, the working temperature in warm working process, on the other hand, can significantly reduce deformation load and improve the plasticity of materials. This could be the main driving force to choose this process.

Compared with cold forming processes, warm forming decreases the flow stress of material, and thus, the material is easier to work and the energy consumption is less. In addition, it also increases the ductility of materials, and thus, the medium complex geometries of the metal-deformed parts can be made by this category of processes. In addition, as the working temperature is not so high compared with hot working, the surface quality and dimension accuracy of the deformed parts are better than those produced by hot forming processes. Furthermore, there is no recrystallization in the process, the severe deformation in warm forming processes can also break the large microstructure and the original texture down to small ones, and thus, grain refinement happens in this process.

1.3.3 Hot Forming

Hot working is usually conducted at the temperature of 61 % or above of the melting temperature of the materials on an absolute scale, and the working temperature is generally above the recrystallization temperature of materials. At elevated temperature, materials have low flow stress, and thus, the deformation load needed is reduced and the capacity of forming equipment can be small. This, on the other hand, compensates a little for the energy consumption in heating up the workpiece to hot working temperature. In addition, recrystallization occurs in hot working process, which could cause new grains to continually form and grow in the process. The continual formation of new grains causes the ductility of materials to remain high and allows a large amount of deformation to be imparted without fracture, which is caused by the deformation exceeding the ductility of materials. In hot working process, control of the final dimensions of the deformed parts is more difficult due to scale formation and volumetric change in the subsequent cooling of workpiece.

In hot working, the flow stress of materials is decreased and it is thus easier to deform and consumes less energy. In addition, the ductility of materials is increased and the diffusion, which can remove or reduce chemical inhomogeneity, increased too at the elevated temperature. The pores and defects in the materials may be reduced in size or close completely during the deformation process, and impurities can be broken down. For some materials, the easy-to-deform phase is deformed at elevated temperature in such a way to facilitate deformation and enhance deformation compatibility. Taking carbon steel as an instance, the austenite, which is weak, is ductile, and has the crystal structure of face-centered cubic, is deformed instead of the strong body-centered cubic ferrite at lower temperature. Considering recrystallization, it can happen in some materials in hot working process, which refines the microstructure of materials and improves the mechanical properties of the deformed parts.

The main disadvantages of the hot working process of materials include the undesirable reactions between the materials and the surrounding atmosphere, which are driven to happen in high-temperature environment. In addition, less precise tolerances and dimensional accuracy due to thermal contraction, warping from uneven cooling, and distortion caused by the ejection die cavity exist in the deformed parts. Furthermore, the microstructures of the workpiece may vary throughout for different reasons, which affect the properties of the deformed parts. For the tailored–customized mechanical properties of the deformed parts requested by unique application scenarios, how to design the hot forming process and process parameters, and further to ensure the desirable properties as well as the requested geometry and dimensional accuracies of the deformed parts, is a critical issue in the current metal-formed PDD paradigm.

1.4 Metal-Forming Process and System

In manufacturing industries, many different forming processes are used. But they can be simply classified into two categories, viz. bulk metal-forming and sheet metal-forming, in addition to the classification presented in Sect. 1.3, which is done based on the working temperature of deformation. This classification, however, is based on the geometry and shape of billet materials. For the former, the billet material generally takes the form of block, bar, or cylinder with a specific volume and three dimensions in nature, and the forming process includes forging, extrusion, rolling, and drawing. For the latter, on the other hand, it covers shearing, bending, and drawing. The billet material usually has the form of sheet. In sheet metal-forming, the surfaces of sheet metals play an important role and the deformation causes the changes of geometry and shape of the sheet metal surfaces in such a way to produce the desirable geometries of the sheet metal parts. In this process, the thickness of sheet also changes, but it is usually not desirable in most cases. The surfaces of sheet metals further serve as one part in the interface in between the sheet material and tooling. For bulk forming, however, the geometry and shape of the volumetric billet change and the materials flow from one to another desired form such that the parts with specific geometry and shape are made.

In realization of deformation, billet material is placed into a die set and the force applied by forming equipment through die components deforms the material. The material thus flows along the predefined geometries of die cavity. When the space of die cavity is fully filled up, the deformation is completed and the desired part is produced. In this process, the interaction and interplay among different factors, including die assembly and components, billet material and its microstructure and mechanical properties, and the designed process route and process parameter configuration including working temperature, deformation amount, and strain rate,

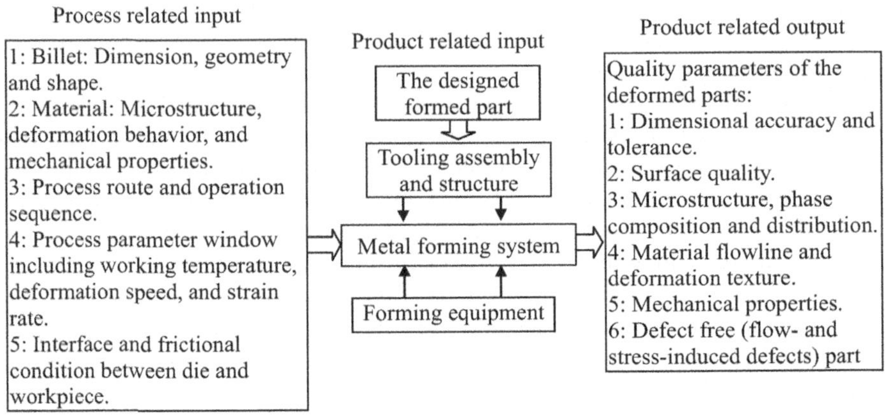

Fig. 1.2 Input and output relationship in a forming system

all affect the quality of the final product in terms of dimensional accuracy, surface quality, microstructure, material flowline and deformation texture, and finally the mechanical properties of the deformed parts. Therefore, the entire forming process can be considered as the working process of a forming system in which all the output variables are affected by the input variables and determined by the performance of the forming system. Figure 1.2 shows the input and output relationship of a forming system.

A metal-forming system can be considered to include tangible and intangible elements. The tangible ones include billet and its geometry and shape, material microstructure and properties, tooling assembly, structure and its detailed components, interface and lubrication condition between tooling and billet material, and the forming equipment. The intangible ones include process, process chain, process parameter configuration, etc. Each element in the system contributes to the entire system performance and its final product-related outcome.

Design of a forming system is to figure out all the tangible and intangible elements described above and to optimize the system performance with the best output via figuring out the relationship between input and output variables. In tandem with this, design of each element, understanding of the interaction and interplay among different elements, and relationship between input and output variables are all critical issues to be addressed.

1.5 Challenges in Metal-Formed Product Development

In metal-forming industries, there are three overriding issues, which are shown in Fig. 1.3. They are productivity, production cost, and product quality. Due to the marginal profit and the severe competition in global and local marketplaces in this

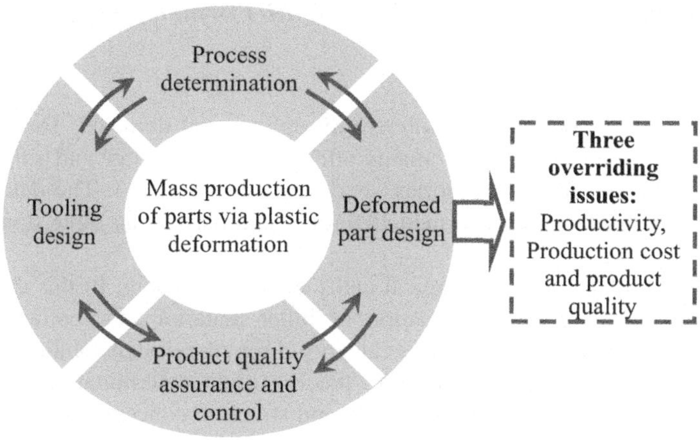

Fig. 1.3 Design activities and the overriding issues in metal-forming PDD

industrial sector, how to improve productivity, cut production cost, shorten time to market, and enhance product quality are all the bottleneck issues in this traditional manufacturing cluster. Therefore, a new product design and development paradigm needs to be developed to address these issues and to ensure "design right the first time" or reduce the tryout times of PDD in workshop before mass production can be implemented and realized. The new PDD paradigm thus needs to support decision-making and solution generation in different design stages including design of metal-formed parts based on the geometry, dimension, and quality requirements of the final product, process route determination and process parameter configuration, design of tooling assembly, structure and the detailed components, and the quality control and assurance of the deformed part. From this perspective, the finite element simulation-enabled design and development of metal-formed products is a promising approach to realizing this goal.

To develop a simulation-enabled PPD paradigm to support the metal-forming process and product development, the main design activities and the overriding issues involved in the metal-forming PDD are articulated here. Figure 1.3 presents the above-mentioned three overriding issues that are associated with four categories of design activities. They are deformed part design, forming process determination, including the detailed process parameter configuration, tooling design, and product quality assurance and control. The simulation-enabled design and development paradigm efficiently supports the decision-making in these design activities, helps figure out the interaction and interplay of the designs in different categories, and optimizes the whole forming system design. In detail, the following are systematically described and presented from the perspectives of simulation-enabled solution generation and decision-making.

1.5.1 Multidomains Involved in Metal-Formed Product Development

Metal-forming process involves different disciplines and domains. The interaction and interplay among different domains affect decision-making and solution generation in different design categories, as described in Fig. 1.3. The detailed multidomains involved in metal-formed PDD are summarized as three domains and shown in Fig. 1.4.

The first domain is the mechanical and mechanics domain. In this domain, the phenomena including the deformation of billet material and tooling and their interactions in between belong to this category. From the design solution generation perspective, the deformed part design, process route determination and process parameter configuration, tooling assembly and structure conceptualization and final design, and product quality assurance and control are all under this domain, viz. domain 1, as shown in Fig. 1.4. The above-mentioned "geometry accuracy-driven" requirements are generally considered in this domain.

In addition to the deformation of materials, the evolution and development of microstructure, including recovery, recrystallization, grain growth, and phase transformation of workpiece material, may simultaneously occur in the deformation process, especially in hot working process. All these phenomena are considered to belong to the materials science domain, i.e., domain 2, as shown in Fig. 1.4. In design of metal-deformed parts, forming process, and finally the product quality and properties, the issues related to these phenomena need to be fully considered. In addition, the previously defined "desirable properties-driven" requirements need to be considered in this domain and how these requirements can be met needs to be evaluated in this domain.

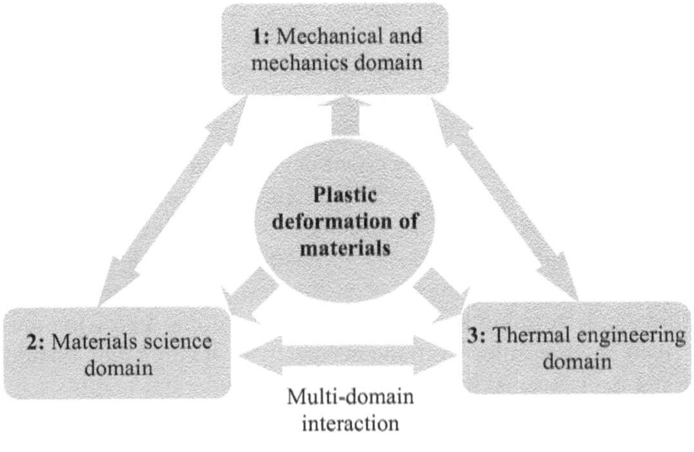

Fig. 1.4 Different domains involved in metal-deformed product development

The last one is the thermal engineering domain or domain 3, also shown in Fig. 1.4. In the plastic deformation of metallic materials, about 90 % of mechanical energy or plastic deformation energy is converted to thermal energy, which contributes to the temperature increase inside the deformation body. Since the generated deformation is not uniformly distributed in the deformation material, the temperature increase induced by deformation is thus also not uniform in the workpiece, which would further lead to different deformation behaviors, microstructure evolutions, and local properties of the deformed parts. Therefore, this phenomenon also needs to be well analyzed quantitatively such that the mechanical properties and dimensional quality of the deformed parts in up-front design stage can be predicted.

In metal-forming process, the above-mentioned three domains ubiquitously and simultaneously exist. The interaction and interplay among these domains thereupon prevail. Taking domains 1 and 2 as an instance, the plastic deformation behaviors and phenomena in domain 1 would affect or facilitate the microstructure evolution in domain 2, such as grain refinement, dynamic recovery, recrystallization, and even phase transformation. One of the typical examples between deformation and microstructure evolution is severe plastic deformation (SPD) which generates severe stress state or shear stress to break the coarse grains of materials to ultrafined or even nanostructured microstructure in such a way to improve the properties of materials. In addition, deformation can induce phase transformation in plastic deformation process of materials such as the phase transformation from austenite to martensite in plastic deformation of austenitic stainless steel and the phase transformation from β to α phase in hot working of Ti alloys. From the grain refinement point of view, the plastic deformation facilitates the evolution of dislocation substructure, consisting of the generation of dislocations, their self-organization into structures inside the grains, evolution of the structure into low-angle graining boundaries (LABs), characterized by subgrains, and transformation of LABs into high-angle grain boundaries (HABs) with the aid of annealing or dynamic recovery in hot working process. All of these would lead to grain refinement in the microstructure of the materials. Therefore, the effect of deformation in domain 1 is closely associated with the microstructure evolution of materials in domain 2 in the material working process.

On the other hand, the microstructure of materials and the evolution of microstructure in working process affect the properties and deformation behaviors of materials. The grain size of materials has a close relationship with the flow stress of materials in plastic deformation, which is well represented by Hall–Petch equation. In addition, the microstructure evolution of materials such as recovery, recrystallization, grain growth, phase transformation, reorientation, and texturing of grains also affects the material deformation and flow behaviors in working process, which further influence the final microstructure and the mechanical properties of the deformed parts. If the mechanical properties of the deformed parts need to be well controlled and the "desirable properties-driven" requirements to be met, the prediction and analysis of the evolution of microstructure need to be done in such a way to help process determination and process parameter configuration in up-front design process.

Furthermore, domains 2 and 3 are also fully interlinked. Their interaction in between is also critical in the deformation process of materials. The properties and microstructure of the deformation materials and their variation in the course of deformation in domain 2 are affected by the temperature and temperature gradient specified in domain 3. However, the changes in domain 2 including phase transformation and microstructure evolution in turn affect the thermal behaviors and phenomena in domain 3. The interplays in between the two domains occur simultaneously and concurrently. The revealing of these interactions and interplays is important as they affect process determination, process parameter specification, and tooling design in up-front design stage.

There is also reciprocal action between domains 1 and 3. In domain 1, the mechanical and mechanics behaviors such as deformation behaviors, flow rheology of materials at warm or hot working temperature, and deformation mechanisms are fully associated with thermal phenomena in domain 3 such as temperature magnitude, distribution, and gradient in the deformation body, which further affect those behaviors in domain 1. In addition, the thermal behaviors in domain 3 are also affected by the mechanical phenomena in domain 1. Taking the plastic work done in domain 1 as an instance, 90 % of this work is converted to the thermal energy in domain 3 and further increases the temperature level, distribution, and gradient in the deformation body. The temperature change in domain 3 in turn affects the flow stress and deformation behaviors and mechanisms in domain 1. This mutual interaction and interplay in between the two domains is also simultaneous and bidirectional.

In the above statement, the interactions and interplays in between two domains are articulated. Among these three domains, there are reciprocal actions. First of all, the material flow rheology, deformation behaviors and mechanisms, friction, and interfacial phenomena in domain 1 are determined by the microstructure and metallurgical properties of materials and their change and variation in domain 2, which are in turn affected by the thermal phenomena in domain 3. On the other hand, thermal-related phenomena such as temperature level, distribution, and gradient I domain 3 affect the microstructure of materials and its evolution including dynamic recovery, recrystallization, grain growth, and phase in the course of warm or hot working process in domain 2. Microstructure and its evolution in domain 2 in turn affect the deformation and flow properties of materials, which further influence the flow and deformation behaviors in domain 1.

From modeling and simulation perspectives, simultaneously coupled simulation of these physical phenomena and behaviors in different domains would be helpful for fully revealing and understanding the panorama of interaction and interplay among different domains. However, it is actually quite difficult, if not impossible. The feasible and possible practice is to simulate the interaction among them by sequential coupling in between. In addition, assumptions need to be made to simplify the real forming systems by considering the major behaviors or performances of the systems to be revealed, while ignoring some low-priority factors. However, trade-off needs to be made to ensure the efficient, viable, accurate, and feasible simulations can be conducted via considering sufficient factors and phenomena of the system and without loss of the accuracy and authenticity of the simulation. Therefore, how to efficiently and accurately model and simulate the simultaneous behaviors and phenomena of different domains and their interactions is still an eluded and tantalized issue to be explored and addressed.

1.5.2 Design of the Deformed Parts

Metal forming is a net-shape or near-net-shape manufacturing process. In the near-net-shape forming scenario, secondary operations are generally needed. In other words, the metal-deformed parts or components are usually different from the final products from geometry and dimension perspectives. Therefore, the metal-deformed parts need to be designed based on the geometry and dimensional requirements of the final products. Since the design can have infinite design space for the given design parameters, a good design is not good enough if there is a better one. The better design is always not the final goal if there is the best one. In searching for the best design solution, the trial-and-error approach based on the heuristic know-how is not efficient to identify the best solution as it is difficult to fully assess the performance of different designs and reveal the needed information related to all design parameters for best solution generation. In this way, a plethora of tryouts in workshop for design change and fine-tuning are needed and the design and development lead times would be longer, which affects time to market. Therefore, simulation-enabled design paradigm is a feasible, viable, and efficient approach to seeking the best design solution from all the potential design alternatives in the whole design space.

Design of metal-formed parts significantly affects the subsequent forming process determination, tooling structure and component design, tooling design and service life, material utilization, energy consumption, and product quality and assurance. Taking the example as shown in Fig. 1.5 as an instance, the product shown in Fig. 1.5a is a flanged part with a central through hole. From the forming process perspective, the central through hole cannot be fabricated directly by just one forming operation and a patch-up is needed to patch it. The patch-up will be trimmed away in the second operation. For design of metal-deformed parts based on this designed product, three design variables can be introduced and considered here. The first one is the location of patch-up. Figure 1.5b shows three different locations to be considered, which are at the bottom, the middle, and the top of the deformed part. The second design variable is the thickness of patch-up, which may have different values, as shown in Fig. 1.5c. The last one is the shape of patch-up. The different shapes of the patch-up are shown in Fig. 1.5d.

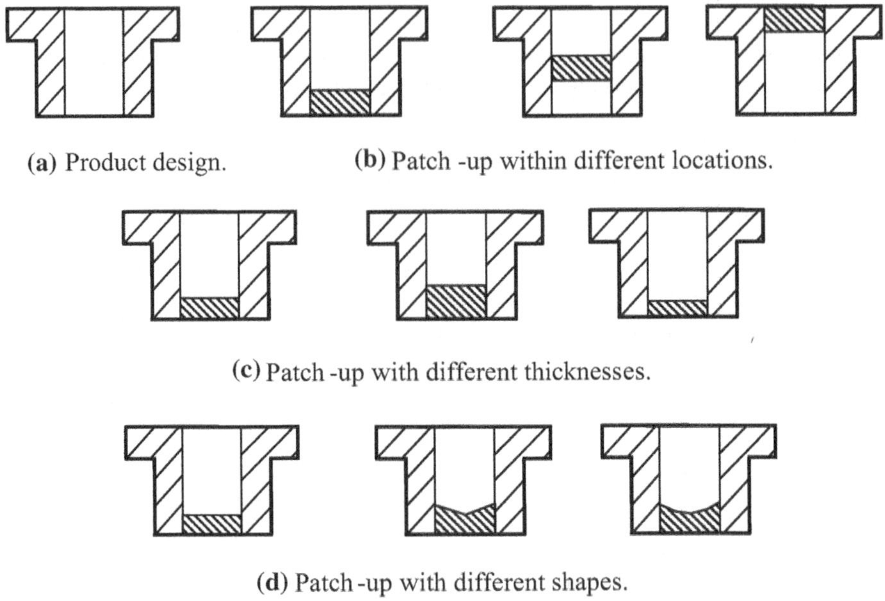

Fig. 1.5 Design of deformed part based on the given product design

From the material flow perspective, different designs of the deformed part would lead to different shapes and geometries of die punch and die cavity. The corresponding different shapes and geometries of die punch and die cavity in different design scenarios would lead to different material flow sequence, amount, and pattern, which would generate different strain magnitudes, distributions, and gradients inside the deformation body, different flowline, and deformation textures, or even generate flow-induced defects. In addition, for different deformed part designs, different deformation loads are thus needed, which require different energy consumptions and different capacities of forming equipment. From the die stress and die service life angle perspective, different designs can have different working stress levels, stress distributions, stress amplitudes, and stress concentration scenarios in die structure and die components. Die service life can thus be different. From material utilization consideration, it is known that material cost takes up about 60 % of the product cost in metal-forming industries and different designs would have different material costs. A little bit reduction of raw material cost would be critical in industrial cluster as the profit is quite marginal. Therefore, the design of deformed part is the first step in metal-formed product development and affects all other designs and the performance of the subsequent designs.

1.5.3 Process and Process Parameter Configuration

The design of forming process and process route is generally determined based on the geometry, shape, and dimensional tolerance and accuracy requirements of metal-deformed parts and the workability of the materials to be used. For a given or designed process and process route, the process parameters can have different configurations. Therefore, there are two issues to be addressed; one is process and process route determination, while the other is process parameter configuration.

For forming process determination, the manufacturability of the deformed parts and the workability of the material to be used are the first two factors to be considered. Whether the cold, warm, or hot forming process is selected needs to consider these two factors. For determination of the detailed process, such as bulk metal-forming or sheet metal-forming, it needs to take into account the geometry and shape of the deformed part. The volumetric deformed parts are fabricated by bulk metal-forming processes, while the sheet metal parts are made by sheet metal-forming processes. The detailed forming operations such as drawing, bending, or shearing for sheet metal-forming or extrusion, forging, rolling, and drawing for bulk metal-forming and the operation sequence are usually determined based on the know-how and experiences of designers and the related data, information, and knowledge available from different sources such as metal-forming handbooks, monographs, and the archives of similar product development. The design solution is thus generated in such a way based on those heuristic know-hows, knowledge, and experiences from prior arts. How finite element (FE) simulation can help design solution generation in these design activities? The answer is that it is used to evaluate and assess the designs and identify which one is good, better, or best from all the design alternatives based on the given criteria and the performance of each design. In addition to design solution evaluation, FE simulation can help designers generate designs in up-front design process via revealing the manufacturability, deformation, and deformation behaviors and mechanisms of the designed metal-deformed parts and the deformability and workability of the material to be used under the pregiven processing conditions. Therefore, FE simulation cannot directly generate design solution, but it helps decision-making and design solution generation in design process by providing the needed information through simulation and evaluation of the designs.

For process parameter configuration, it conducts the detailed value setting for a set of process parameters in metal-forming process. The process parameters for metal-forming include forming temperature, deformation velocity, deformation amount, and lubrication condition. The parameter configuration is to determine the detailed values of these parameters. Since each parameter has its own feasible value range, configuration of these parameters is thus to specify the detailed value for each parameter from the whole value range, and thus, there are different combinations and configurations. On the other hand, there are interactions and interplays among different process parameters, and different configuration can have totally different process performance, deformation behavior and mechanism, and the quality of the deformed parts. Process parameter configuration is thus critical in deformation process. Although the final deformed part fabricated by different process parameter configurations may have almost the same geometries, however, the internal microstructure, deformation texture, material flowline distribution, defect formation scenario, and dimensional accuracy of the deformed parts could be quite different. Therefore, how to conduct parameter configuration is also a non-trivial issue.

How can FE simulation help process parameter configuration for metal-deformed product development? The role FE simulation plays is still more focused on the evaluation of process parameter configurations and identifies the good, better, and best configurations. Similar to process and route generation, FE simulation evaluates process configuration in tandem with the predefined evaluation criteria via revealing the needed information specified by evaluation criteria. On the other hand, how FE simulation helps generate different configurations is still more focused on providing information to support decision-making by process designer in configuration generation. Since process parameter configuration needs domain-specific knowledge, know-how and experience, and the configuration space is generally infinite, how to come out with the promising configurations which cover the best configuration is critical. In addition, identification of the best configuration should be a viable process and the configuration space cannot be too big for facilitating the identification of the best one via FEsimulation.

1.5.4 Die Design and Its Service Life Analysis

Metal-forming die is a set of tooling to realize the forming of materials with the determined process and process route under a given process parameter configuration. Die design is a geometry mapping process from the geometry and shape of the deformed part to the geometry and shape of punch and die cavity via strictly considering the dimensional accuracy and tolerance requirements of the deformed parts. Die design thus needs to ensure this mapping accurate, and the determination of die geometries follows the requirements of the geometry and dimensional accuracy of the deformed parts. The domain-specific information and knowledge related to the elastic deformation of tooling and the plastic deformation of workpiece material are thus needed to be provided to accurately determine the dimensions and geometry of tooling. How to accurately provide this category of information for decision-making via revealing the elastic deformation of tooling and the plastic deformation of workpiece material is critical. On the other hand, the design of die affects material flow, deformation behavior and mechanism, dimensional accuracy, and the flow-related texture of the deformed parts in cold forming, or even the microstructure and microstructure evolution in hot working processes. All of these, on the other hand, affect the dimensional accuracy and mechanical properties of the deformed part. Furthermore, die design correlates the quality of the deformed part.

From the die service life perspective, die design significantly affects the service life of the designed die. For die component level, the feature, shape, and dimension of die component all affect the working stress and stress distribution within the component, which further affect die service life. It is believed that 10–15 % decrease of working stress can prolong the die service life for a few folds. From the die structure perspective, die structure also significantly affects the working stress level and distribution inside the die. Taking a multilayer extrusion die as an instance, the change of two-layer die structure to three-layer one, the die service life can be tremendously increased.

Traditionally, die design is an experienced and know-how based practice. Long time of apprenticeship and craftsmanship is a must for a skillful master of die design techniques. With the increase of product structural integrity, complexity, and miniaturization of metal-deformed parts, knowledge-based design is now a popular practice with the aid of FE simulation. From the geometry representation perspective, the three-dimensional (3D) geometry and shape representation has become a trivial issue due to the increasing capacity of computer-aided design (CAD) technologies and off-the-shelf CAD systems [4]. Design solution generation, however, is still a non-trivial issue. Although knowledge-based systems provide an efficient approach for retrieving the needed data, information, knowledge, guidelines, formulas, and prior designs to support decision-making, the decision-making in die design, however, is still done by designers. Their know-how and experience play a big role. With the aid of FE simulation, the traditional heuristic- and experience-based decision-making is now becoming more scientific-based analysis and quantitative calculation. The FE simulation plays this role and helps design solution generation from this perspective.

In addition to design generation aided by FE simulation, FE simulation helps evaluation and optimization of die design. From the component design perspective, it helps evaluate the critical dimension determination such as the radius and taper angle of punch, the clearance between die insert and punch, and the fillet radius of die insert and cavity based on the strength, function, and performance of die structure. The evaluation can be conducted via simulation of different designs to reveal the working performance of the designs based on the predefined criteria. In this process, how to interpret the simulation results and extract the useful information to validate and verify the designs also needs domain-specific knowledge, viz. metal-forming and die design knowledge.

Furthermore, die service life is one of the key performance indicators in die design and development, and thus, die service life analysis and prediction in die design is also an important design activity. FE simulation is of critical help in providing the useful information, such as cyclic stress, peak stress, and amplitude stress, for analysis and calculation. Currently, there are two traditional approaches including stress- and strain-based approaches for die service life analysis. For the former, FE simulation identifies the maximum loading stage in deformation process and the maximum stress location in the die structure at this stage. Based on the maximum stress value and S-N curve generated by stress-life governing equation, the potential service life of die can be predicted and determined. For the strain-based approach, the procedure is almost the same, but the die service life is assessed and determined based on the maximum accumulated strain and its variation and distribution in the die structure in a cyclic service period. Since FEM simulation cannot provide absolute solution but does give comparable results, the comparison and decision in design process are made via comparing different die service lives predicted by simulation, and thus, the optimal design with the longest service life can be identified.

1.5.5 Defect Formation, Prediction, and Avoidance

As mentioned before, different designs of metal-deformed parts can have different deformation behaviors, material flow scenarios, deformation amounts, working stresses, and stress distributions in the workpiece. From the material flow perspective, unreasonable flow mechanism and pattern result in the formation of the so-called flow-induced defects in this book. From the stress point of view, when the stress exceeds the load-carrying capacity or the strength limit of materials, fracture may happen in the materials and stress-induced defect such as ductile fracture could occur. To predict and avoid the occurrence of these two types of defects, viz. flow- and stress-induced defects, it is necessary to reveal the deformation mechanism, stress and stress distribution, and flow behavior of the workpiece material and further the defect formation mechanism, if any, in the deformation process such that defect avoidance methods can be developed and the defect-free deformed parts and forming process can be figured out.

Defect occurrence cycle can be considered to cover defect initiation, formation, and growth stages, which are closely related to the deformation mechanism and deformation behavior of the material for the designed deformed part. To fabricate defect-free products, it is necessary to explore and identify the conditions under which defect initiates and the mechanisms of initiation, formation, and growth of defect. FE simulation helps reveal the detailed process of defect occurrence and figure out the defect occurrence condition through virtually and numerically representing and modeling the real forming process from time and space domains. In terms of the representation of physical variables from the both domains, FE simulation virtually realizes and demonstrates the deformation process and reveals the needed physical variables and process performance instantaneously in the deformation time domain and the three-dimensional space domain defined in the 3D space of forming system. To reveal defect initiation, it is necessary to identify the defect-related physical variables and their variations in the entire deformation process. Based on the variation of the related physical variables such as the instantaneous velocity of up-front material flow, the formation of flow-induced defect such as folding or unfilling can be identified and revealed. Similarly, the occurrence and formation of stress-induced defect can be determined according to the maximum stress level, distribution, variation, and location in the workpiece in the deformation process. Of course, the conditions in which defect occurs need to be systematically investigated, identified, and established first.

For defect analysis and prediction using FE simulation, the process covers the simulation of the entire deformation process, identification of the defect-related physical variables and their variations, and further prediction of the defect formation based on the predefined defect formation criteria and conditions. From the defect avoidance perspective, defect avoidance methods need to be excogitated based on the defect formation mechanisms and conditions. The developed defect avoidance methods can be implemented via redesign of the forming process and the deformed part.

1.5.6 Optimization of Metal-Forming System

As described in Sect. 1.4, a forming system includes (a) forming material, billet and billet geometry and shape; (b) tooling assembly, structure and its detailed components, and working geometries of the tooling; and (c) the embedded forming process, and process chain and process parameter configuration. In the previous sections, the design of the deformed part, forming process, process chain and process parameter configuration, and tooling aided by FE simulation, and further defect prediction and assurance are detailed and summarized. As a whole, a forming system can be evaluated with the aid of FE simulation. Since the best design of individual system elements may not be able to ensure the best performance of the whole forming system, evaluation of the performance of the forming system as a whole and further identification of the optimal one is of importance.

In the evaluation of a forming system, a set of evaluation criteria need to be developed from different perspectives such as small deformation load, low energy consumption, using less material, good deformation uniformity and flow rheology, small working stress and uniform stress distribution, long tooling service life, and defect-free deformation. For different forming system designs, evaluation of the different forming systems and further optimization of them can be conducted based on the FE simulation results against the developed criteria. Although FE simulation cannot provide absolutely accurate results, it does provide the wonderful and comparable solution for evaluation of forming system design and identify the better and eventually the best one from all the potential designs [6, 7].

1.6 Summary

This chapter presents the basic concepts of plastic deformation of materials and metal-forming process and system. Different domains involved in metal-forming process are elaborated and their interaction and interplay, which exist ubiquitously in any metal-forming processes, are presented. A panorama of metal-forming system, its input and output variables, and their relationship are articulated. An epistemological understanding of a plethora of challenges in metal-forming PDD is then given, and the evaluation of forming systems is elucidated. The FE simulation-enabled metal-forming PDD paradigm to support metal-formed part design, forming process determination, process parameter configuration, tooling design, defect prediction, and avoidance is delineated. This chapter presents a framework in which how the FE simulation supports metal-forming PDD is elaborated. All of these will be systematically presented in the subsequent chapters in this book.

References

1. Valberg HS (2010) Applied metal forming: including FEM analysis. Cambridge University Press, Cambridge
2. Altan T, Ngaile G, Shen G (2005) Cold and hot forgings: fundamentals and applications. ASM International, Materials Park
3. Kobayashi S, Oh SI, Altan T (1989) Metal forming and the finite-element method. Oxford University Press, New York
4. Bolianovic V (2014) Sheet metal forming processes and die design, 2nd edn. Industrial Press, South Norwalk
5. Hosford WF, Caddell RM (2011) Metal forming: mechanics and metallurgy, 4th edn. Cambridge University Press, Cambridge
6. Fuh JYH, Zhang YF, Nee AYC, Fu MW (2004) Computer-aided injection mould design and manufacture. Marcel Dekker Inc, New Year
7. Fu MW, Chan WL (2014) Micro-scaled products development via microforming: deformation behaviors, processes, tooling and its realization. Springer, Berlin

Chapter 2
Rigid-Plastic Finite Element Method and FE Simulation

2.1 Introduction

The finite element method (FEM) is one of the most powerful and efficient numerical methods for solving governing equations, which are generally partial differential equations, of a complex system over a continuous physical domain. The complex system over the continuous physical domain is discretized into simple geometric shapes called elements for solving the governing equations. This discretization process is called meshing. For a FEM mesh, the points in which the elements are connected are called nodes. A FEM mesh thus consists of all the elements and nodes. FEM can be used for solving field physical variables of concern in different domains including physical, mechanical, and thermal domains represented by governing equations, which can be simplified as a set of simultaneous algebraic linear equations after linearization in the following:

$$[K]\{u\} = \{F\} \tag{2.1}$$

where $[K]$ is the property-related matrix; $\{u\}$ is the vector of physical variables to be solved and represents the working behaviors of the system of concern; and $\{F\}$ is an external loading. The whole system is thus represented by the relationship of $[K]$, $\{u\}$, and $\{F\}$ under a specific physical domain with a detailed physical meaning. Therefore, Eq. (2.1) is a general equation, which represents the working relationship of a complex system in any physical domain. Taking an object deformation as an instance, Eq. (2.1) represents the property and behavior relationship of the system under an external loading $\{F\}$. For this deformation case, $[K]$ is called stiffness matrix, representing the material and geometry properties of the system, and $\{u\}$ is the displacement or velocity matrix, representing the

M.W. Fu, *Design and Development of Metal-Forming Processes and Products Aided by Finite Element Simulation*, Engineering Materials and Processes,
DOI 10.1007/978-3-319-46464-0_2

displacements or deformation velocities of nodes. $\{F\}$ is the external loading, which causes the deformation represented by the displacements or deformation velocities of the nodes, but determined by the properties of the system. If Eq. (2.1) models a thermal system, then $[K]$ represents the conductivity matrix of the system, while $\{u\}$ denotes the temperatures at the nodes; and $\{F\}$ is the heat source. If the equation represents a fluid system, $[K]$ is the viscosity matrix of the system, while $\{u\}$ is the velocity vector of the nodes; and $\{F\}$ represents the body force of the liquid system. Therefore, for any domain of concern, the nonlinear governing equations of a complex working system can be linearized by mathematical methods such as Newton–Raphson method to get the general linear formulation designated by Eq. (2.1).

Regarding the FEM applications in metal-forming PDD, various types of material models for representing different deformation behaviors of materials have been developed and used for the analysis and simulation of different deformation phenomena and behaviors of a single deformation body or the whole forming system. There thus exist different types of FEM by using different material models for solving different problems and meeting different needs. Elastic FEM uses the elastic material model, which assumes the constitutive relationship between strain and stress is linear. The elastic FEM is used to analyze the elastic deformation of deformation bodies in such a way to address the elastic deformation of tooling and further the dimensional accuracy of the deformed parts. For tooling design and service life prediction, the elastic FEM is usually used for strength-based design and estimation of tooling service life via determining the working stress and the accumulated deformation strain in deformation process. In addition, if the material model is elastic–plastic or rigid-plastic, the FEM with this type of material model would be the elastic–plastic or rigid-plastic FEM. The elastic–plastic FEM treats the workpiece as an elastic–plastic body. Under this material model, the material undergoes elastic deformation first and then plastic deformation once the effective stress of the material exceeds its flow stress or yield stress. For this kind of deformation body, the elastic recovery or springback of deformation body can be determined by using this material model such that the elastic deformation of tooling and workpiece can be considered in the design of metal-deformed parts and tooling to meet the geometry accuracy driven requirement. If the elastic deformation of workpiece can be ignored, the rigid-plastic FEM can be used. For the rigid-plastic FEM, the material model is rigid-plastic and the large-scale plastic deformation of the deformation body can be analyzed by the rigid-plastic FEM. In hot working process or superplastic deformation of materials, the material deformation is very sensitive to strain rate in the deformation process. The viscoplastic FEM can thus be used. For this type of FEM, the material model is viscoplastic model by which the strain rate sensitive behavior of materials can be modeled and analyzed. For most hot working processes in which the strain rate must be considered, the viscoplastic FEM should be used.

From application perspective, FEM is widely used in different deformation scenarios for analysis of different physical phenomena and behaviors, addressing design and engineering issues, and eventually supporting design and production activities. This chapter aims at summarizing the concepts of modeling and simulation, the rigid-plastic FEM formulation, and the FEM applications-related issues in metal-forming PDD.

2.2 Modeling and Simulation

Modeling and simulation are two imitating approaches to representing the real systems with the concerned physical behaviors and phenomena by virtual or physical models on the basis of the associated physical principles. Modeling and simulation display or demonstrate the system to be investigated virtually or physically to reveal the phenomena, behaviors, functions, or performances of the systems in such a way to obtain the needed information for decision-making in research, design, and production or to explore the unknown phenomena and behaviors of the systems. The goal of modeling and simulation is to reduce the physical experiment, time, and cost in realization of the above-mentioned activities and explorations. Most importantly, modeling and simulation can reveal the hidden phenomena and behaviors or the instantaneous phenomena and performances of the system, which is difficult, if not impossible, to be identified and revealed by the traditionally physical experiments.

In modeling and simulation, models are generally developed first. A model of a real system represents the system physically or virtually from the perspective of the phenomena or behaviors of the real system to be explored and investigated. A model is similar to but generally simpler than the real system it represents, while abstracting and approximating most of the same salient features of the real system as close as possible. A good model is a judicious trade-off between realism and simplicity [1]. A key feature of a model is manipulability [1]. A model can be a physical model, which should tangibly represent the system to be modeled by scaling down or up the real geometry and dimension of the system using the same or replaceable materials with similar properties and can demonstrate the functions and performances of the system. A virtual model is a digital mock-up which represents the real system mathematically and physically. Similarly, it can represent and demonstrate the functions or performances of the system of concern. Modeling is an action process to build model. The models can be classified static and dynamic models. The most important difference in-between the static and dynamic models is the run-time representation. The static modeling is a time-independent representation of the system, while dynamic modeling is the run-time model of the system. Therefore, the dynamic models keep changing with reference to time, whereas the static models are at equilibrium in a steady state.

Simulation is a process of using models to reveal the behaviors, phenomena, functions, or performance of a real or theoretical system via identification of the values and variations of the related physical variables of concern. In simulation process, models are used to represent the characteristics of a system to be simulated. The purpose of a simulation is to represent the characteristics of the real system by manipulating variables that cannot be controlled in a real system. Simulation evaluates the model of the system such that the design of the system or the performance of the system can be optimized. In addition, simulations are useful to study the real-life systems that would be difficult to access such as too complex, too large or small, too fast or slow, not accessible, too dangerous or unacceptable to engage. While a model aims to be true to the system it represents, a simulation can use a model to explore the states that would not be possible in the original system [1].

2.3 Rigid-Plastic Finite Element Method

As mentioned before, FEM uses different material models to simulate different deformation behaviors and phenomena of materials, and thus different types of FEM have been developed and used in metal-forming PDD. From the deformation perspective of billet materials, the elastic deformation is generally ignored as it is quite small compared with the large plastic deformation the materials undergo and thus the rigid-plastic FEM is widely used to handle this type of deformation scenario. To conveniently represent the formulation of FEM, some basic concepts of tensor are introduced in the following.

2.3.1 Cartesian Tensor Representation

There is the following relationship if a Cartesian coordinate (x, y, z) is transformed into the other (x', y', z'):

	x	y	z
x'	l_1	m_1	n_1
y'	l_2	m_2	n_2
z'	l_3	m_3	n_3

$$\left.\begin{aligned} x' &= l_1x + m_1y + n_1z \\ y' &= l_2x + m_2y + n_2z \\ z' &= l_3x + m_3y + n_3z \end{aligned}\right\} \qquad (2.2)$$

where $l_1 = \cos(x', x), m_1 = \cos(x', y)\ldots l_2 = \cos(y', x)\ldots$

Using x_j and x'_i (i, j = 1, 2, 3) to represent x, y, z and x', y', z', respectively, the transformation formula of Eq. (2.2) can be designated as:

$$x'_i = \sum_{j=1}^{3} l_{ij}x_j \quad (i = 1, 2, 3) \qquad (2.3)$$

In the above, there are two types of suffixes:

(1) One appears precisely once, e.g., i, which is known as free suffix;
(2) The other appears precisely twice, e.g., j, which is known as dummy suffix.

In tensor representation and calculation, the dummy suffix can be used to represent the summation in the following:

$$x'_i = \sum_{j=1}^{3} l_{ij}x_j = l_{ij}x_j \tag{2.4}$$

where j equals to 1, 2, and 3. The dummy suffix j represents the summation of the three terms from j = 1 to 3. This is also called the rules of summation convention.

On the other hand, u_i and v_j are two vectors in a Cartesian coordinate (x, y, z). After they are transformed to (x', y', z') coordinate, they are:

$$u'_k = l_{ki}u_i \quad v'_k = l_{kj}v_j \tag{2.5}$$

If the following matrix T_{ij} is defined:

$$T_{ij} = \begin{bmatrix} u_1v_1 & u_1v_2 & u_1v_3 \\ u_2v_1 & u_2v_2 & u_2v_3 \\ u_3v_1 & u_3v_2 & u_3v_3 \end{bmatrix} \tag{2.6}$$

The tensor format can thus be designated as:

$$T_{ij} = u_iv_j \tag{2.7}$$

After coordinate transformation:

$$T'_{kh} = u'_kv'_h = l_{ki}u_il_{hj}v_j = l_{ki}l_{hj}T_{ij} \tag{2.8}$$

The tensor-like T_{ij} with nine components is called second-order Cartesian tensor. Obviously, the stress and strain are both the second-order tensors. Similarly, a tensor with 27 components is called third-order tensor. To transform a three-order tensor, following equation can be derived which is similar to the vector's and the second-order tensor's transformation.

$$S'_{mnp} = l_{mi}l_{nj}l_{pk}S_{ijk} \tag{2.9}$$

In addition to the above basic concepts of Cartesian tensor representation, it is necessary to introduce the following basic operations of the second-order tensor as they are needed in therepresentation of the formulations of rigid-plastic FEM and the computation and derivation of those equations.

(A) Addition: Each element of one tensor is added to the corresponding element of the other one, respectively.

$$C_{ij} = A_{ij} \pm B_{ij} \tag{2.10}$$

(B) Multiplication: There are two kinds of multiplication of tensors, viz. outer and inner product.

- *Outer product* If the subscripts of the two second-order tensors are different, their outer product is a fourth-order tensor:

$$C_{ijkh} = A_{ij}B_{kh} \tag{2.11}$$

- *Inner product:* If there are the subscripts of the two second-order tensors, a second-order tensor will be obtained due to the rules of summation convention.

$$C_{ik} = A_{ij}B_{jk} \tag{2.12}$$

(C) Differentiation: The differentiation of a tensor is represented in the following.

$$\frac{\partial u_i}{\partial x_j} = u_{i,j} \, \frac{\partial T_{ij}}{\partial x_i} = T_{ij,i} \tag{2.13}$$

(D) Substitution: The Kronecker Delta is introduced as follows:

$$\delta_{ij} = \begin{cases} 1, & i = j \\ 0, & i \neq j \end{cases} \tag{2.14}$$

With the Kronecker Delta, there is the following relationship:

$$A_i = \delta_{ij}A_j = \delta_{i1}A_1 + \delta_{i2}A_2 + \delta_{i3}A_3 \tag{2.15}$$

In fact, the Kronecker Delta is:

$$\delta_{ij} = \begin{pmatrix} 1 & 0 & 0 \\ 0 & 1 & 0 \\ 0 & 0 & 1 \end{pmatrix} \tag{2.16}$$

In addition, the deviatoric stress is denoted as:

$$\begin{aligned} S_x &= \sigma_x - \sigma_m & S_{xy} &= \tau_{xy} \\ S_y &= \sigma_y - \sigma_m & S_{yz} &= \tau_{yz} \\ S_z &= \sigma_z - \sigma_m & S_{xz} &= \tau_{xz} \end{aligned} \tag{2.17}$$

Using the Kronecker Delta, Eq. (2.17) can be represented as:

$$S_{ij} = \sigma_{ij} - \sigma_m\delta_{ij} = \sigma_{ij} - \frac{1}{3}\sigma_{kk}\delta_{ij} \tag{2.18}$$

(E) Symmetric and anti-symmetric tensor:

If $T_{ij} = T_{ji}$, T_{ij} is a symmetric tensor.
If $T_{ij} = -T_{ji}$, T_{ij} is an *anti-symmetric* tensor, the diagonal elements of the *anti-symmetric* tensor are equal to zero.

Any tensor can be written as the sum of symmetric and anti-symmetric tensors in the following.

$$T_{ij} = \overbrace{\frac{1}{2}\left(T_{ij} + T_{ji}\right)}^{\text{Symmetric}} + \overbrace{\frac{1}{2}\left(T_{ij} - T_{ji}\right)}^{\text{Anti-symmetric}} \tag{2.19}$$

For concise representation of the rigid-plastic FEM, the following basic equations used in the rigid-plastic FEM can be represented by tensor and its representation.

(1) Equilibrium equation:

$$\sigma_{ij.j} = 0 \tag{2.20}$$

which can have the detailed formulations in the following:

$$\begin{aligned}
\frac{\partial \sigma_x}{\partial x} + \frac{\partial \tau_{xy}}{\partial y} + \frac{\partial \tau_{xz}}{\partial z} = 0 \\
\frac{\partial \tau_{xy}}{\partial x} + \frac{\partial \sigma_y}{\partial y} + \frac{\partial \tau_{yz}}{\partial z} = 0 \\
\frac{\partial \tau_{xz}}{\partial x} + \frac{\partial \tau_{yz}}{\partial y} + \frac{\partial \sigma_z}{\partial z} = 0
\end{aligned}$$

(2) Flow rule:
The Levy–Mises equations can be denoted in tensor form as:

$$\dot{\varepsilon}_{ij} = \dot{\lambda} S_{ij} \tag{2.21}$$

where $\dot{\lambda} = \frac{3\dot{\bar{\varepsilon}}}{2\bar{\sigma}}$. The full formulations of the above Levy–Mises rule are:

$$\begin{aligned}
\dot{\varepsilon}_x &= \frac{3\dot{\bar{\varepsilon}}}{2\bar{\sigma}} S_x & \dot{\varepsilon}_{xy} &= \frac{3\dot{\bar{\varepsilon}}}{2\bar{\sigma}} \tau_{xy} \\
\dot{\varepsilon}_y &= \frac{3\dot{\bar{\varepsilon}}}{2\bar{\sigma}} S_y & \dot{\varepsilon}_{yz} &= \frac{3\dot{\bar{\varepsilon}}}{2\bar{\sigma}} \tau_{yz} \\
\dot{\varepsilon}_z &= \frac{3\dot{\bar{\varepsilon}}}{2\bar{\sigma}} S_z & \dot{\varepsilon}_{xz} &= \frac{3\dot{\bar{\varepsilon}}}{2\bar{\sigma}} \tau_{xz}
\end{aligned}$$

(3) Geometric equations:
The relationship between deformation velocity and strain rate can be designated as:

$$\dot{\varepsilon}_{ij} = \frac{1}{2}\left(v_{i,j} + v_{j,i}\right) \tag{2.22}$$

This is the so-called geometry equation and can be extendedly formulated as:

$$\begin{aligned} \dot{\varepsilon}_x &= \frac{\partial v_x}{\partial x} & \dot{\varepsilon}_{xy} &= \frac{1}{2}\left(\frac{\partial v_x}{\partial y} + \frac{\partial v_y}{\partial x}\right) \\ \dot{\varepsilon}_y &= \frac{\partial v_y}{\partial y} & \dot{\varepsilon}_{yz} &= \frac{1}{2}\left(\frac{\partial v_z}{\partial y} + \frac{\partial v_y}{\partial z}\right) \\ \dot{\varepsilon}_z &= \frac{\partial v_z}{\partial z} & \dot{\varepsilon}_{xz} &= \frac{1}{2}\left(\frac{\partial v_x}{\partial z} + \frac{\partial v_z}{\partial x}\right) \end{aligned}$$

(4) Yield criterion:
The Mises yield criterion is:

$$J_2 = -\left(S_x S_y + S_y S_z + S_x S_z\right) + \left(S_{xy}^2 + S_{yz}^2 + S_{xz}^2\right) \tag{2.23}$$

When J_2 reaches the critical value k^2 of materials (k is the shear strength limit of materials), the yielding of material starts. The deviatoric stresses thus follow the following equations:

$$(S_x + S_y + S_z)^2 = S_x^2 + S_y^2 + S_z^2 + 2S_x S_y + 2S_y S_z + 2S_x S_z = 0$$

Hence:

$$J_2 = \frac{1}{2}\left(S_x^2 + S_y^2 + S_z^2\right) + S_{xy}^2 + S_{yz}^2 + S_{xz}^2 = \frac{1}{2} S_{ij} S_{ij}$$

The yield condition can be formulated as:

$$\frac{1}{2} S_{ij} S_{ij} = k^2 \tag{2.24}$$

Furthermore, according to the flow rule $\dot{\varepsilon}_{ij} = \dot{\lambda} S_{ij}$, there is the following relationship:

$$\dot{\varepsilon}_{ij}\dot{\varepsilon}_{ij} = (\dot{\lambda})^2 S_{ij} S_{ij} \Rightarrow \dot{\lambda} = \frac{\sqrt{\dot{\varepsilon}_{ij}\dot{\varepsilon}_{ij}}}{\sqrt{2}k}$$

Thus:

$$S_{ij} = \frac{\sqrt{2}k}{\sqrt{\dot{\varepsilon}_{ij}\dot{\varepsilon}_{ij}}}\dot{\varepsilon}_{ij} \tag{2.25}$$

(5) Incompressibility:
In deformation of materials, if the elastic deformation is not considered, the whole deformation can then be considered as a rigid-plastic deformation. The volume of materials thus remains unchanged in the deformation process. This is the so-called volume constancy condition of materials in plastic deformation process and can be designated as:

$$\dot{\varepsilon}_x + \dot{\varepsilon}_y + \dot{\varepsilon}_z = 0$$

It can further be denoted as:

$$\dot{\varepsilon}_{ij}\delta_{ij} = 0 \tag{2.26}$$

(6) Boundary condition:
The loading at the boundaries of deformation body must satisfy the following conditions:

$$\begin{aligned} \sigma_x l + \tau_{xy} m + \tau_{xz} n &= \bar{p}_x \\ \tau_{xy} l + \sigma_y m + \tau_{yz} n &= \bar{p}_y \\ \tau_{xz} l + \tau_{yz} m + \sigma_z n &= \bar{p}_z \end{aligned}$$

which can be concisely designated as:

$$\sigma_{ij} n_j = \bar{p}_i \tag{2.27}$$

2.3.2 Basic of Rigid-Plastic Finite Element Method

The Markov's variational principle is the fundamental of rigid-plastic FEM, which can be stated as follows:

Among all the kinematically admissible and incompressible fields $\dot{\varepsilon}_{ij}$ and v_i, which satisfy the stationary point of the following variational function will be the actual solutions:

$$\phi^* = \sqrt{2}k \iiint_V \sqrt{\dot{\varepsilon}_{ij}\dot{\varepsilon}_{ij}} \mathrm{d}V - \iint_{S_p} \bar{p}_i v_i \mathrm{d}S - \iiint_V \alpha_{ij}\left[\dot{\varepsilon}_{ij} - \frac{1}{2}(v_{i,j} + v_{j,i})\right]\mathrm{d}V + \iiint_V \lambda\dot{\varepsilon}_{ij}\delta_{ij}\mathrm{d}V - \iiint_{S_V} \mu_i(v_i - \bar{v}_i)\mathrm{d}S \tag{2.28}$$

where a_{ij}, μ_i, and λ are the Lagrangian multipliers, which introduce the constraints of velocity–strain rate relation, incompressibility or volume constancy of materials and velocity boundary condition into the variational equation of Eq. (2.28) in such a way to ensure they are satisfied.

The variation in Eq. (2.28) can further be formulated as:

$$\delta\phi^* = \sqrt{2}k \iiint \frac{2\dot{\varepsilon}_{ij}}{v2\sqrt{\dot{\varepsilon}_{ij}\dot{\varepsilon}_{ij}}} \delta\dot{\varepsilon}_{ij}\mathrm{d}V - \iint_{S_p} \bar{p}_i\delta v_i \mathrm{d}S - \iiint_V \delta\alpha_{ij}\left[\dot{\varepsilon}_{ij} - \frac{1}{2}(v_{i,j} + v_{j,i})\right]\mathrm{d}V - \iiint_V \alpha_{ij}\left[\delta\dot{\varepsilon}_{ij} - \frac{1}{2}(\delta v_{i,j} + \delta v_{j,i})\right]\mathrm{d}V + \iiint_V \delta\lambda\dot{\varepsilon}_{ij}\delta_{ij}\mathrm{d}V + \iiint_V \lambda\delta\dot{\varepsilon}_{ij}\delta_{ij}\mathrm{d}V - \iint_{S_v} \delta\mu_i(v_i - \bar{v}_i)\mathrm{d}S - \iint_{S_v} \mu_i\delta v_i\mathrm{d}S \tag{2.29}$$

Since $S_{ij} = \frac{\sqrt{2}k}{\sqrt{\dot{\varepsilon}_{ij}\dot{\varepsilon}_{ij}}}\dot{\varepsilon}_{ij}$, $S_{ij} = \sigma_{ij} - \sigma_m\delta_{ij}$ and $\frac{1}{2}(\delta v_{i,j} + \delta v_{j,i}) = \delta v_{i,j}$, the following thus exists:

$$\delta\phi^* = \iiint_V (\sigma_{ij} - \sigma_m\delta_{ij})\delta\dot{\varepsilon}_{ij}\mathrm{d}V - \iint_{S_p} \bar{p}_i\delta v_i\mathrm{d}S - \iiint_V \left[\dot{\varepsilon}_{ij} - \frac{1}{2}(v_{i,j} + v_{j,i})\right]\delta\alpha_{ij}\mathrm{d}V - \iiint_V \alpha_{ij}\delta\dot{\varepsilon}_{ij}\mathrm{d}V + \iiint_V \alpha_{ij}\delta v_{ij}\mathrm{d}V + \iiint_V \delta\lambda\dot{\varepsilon}_{ij}\delta_{ij}\mathrm{d}V + \iiint_V \lambda\delta\dot{\varepsilon}_{ij}\delta_{i,j}\mathrm{d}V - \iint_{S_V} \delta\mu_i(v_i - \bar{v}_i)\mathrm{d}S - \iint_{S_V} v_i\delta v_i\mathrm{d}S \tag{2.30}$$

By using integration by parts and Gauss formula, there is the following equation

$$\iiint\limits_V \alpha_{ij}\delta v_{i,j}\mathrm{d}v = \iiint\limits_V \left(\alpha_{ij}\delta v_i\right)_{,j}\mathrm{d}V - \iiint\limits_V \alpha_{ij,j}\delta v_i\mathrm{d}V$$
$$= \iint\limits_{S_p+S_v} \alpha_{ij}\delta v_i n_j\mathrm{d}S - \iiint\limits_V \alpha_{ij,j}\delta v_i\mathrm{d}V$$

Substitute the above equation into Eq. (2.30), the following is obtained:

$$\begin{aligned}\delta\phi^* = &\iiint\limits_V \left(\sigma_{ij}-\alpha_{ij}\right)\delta\dot{\varepsilon}_{ij}\mathrm{d}V + \iiint\limits_V (\lambda-\sigma_m)\delta_{ij}\delta\dot{\varepsilon}_{ij}\mathrm{d}V + \iiint\limits_{S_p} \left(\alpha_{ij}n_j-\bar{p}_i\right)\delta v_i\mathrm{d}S \\ &+ \iiint\limits_V \left[\dot{\varepsilon}_{ij}-\frac{1}{2}\left(v_{i,j}+v_{j,i}\right)\right]\delta\alpha_{ij}\mathrm{d}V + \iint\limits_{S_v} \left(\alpha_{ij}n_j-\mu_i\right)\delta v_i\mathrm{d}S - \iiint\limits_V \sigma_{ij,j}\delta v_i\mathrm{d}v \\ &+ \iiint\limits_V \dot{\varepsilon}_{ij}\delta_{i,j}\delta\lambda\mathrm{d}V - \iint\limits_{S_v} (v_i-\bar{v}_i)\delta\mu_i\mathrm{d}S\end{aligned} \tag{2.31}$$

Let $\delta\phi^*=0$ and taking into account the randomness of the variations in the above equation, the following can be obtained in the whole deformation body V:

$$\begin{aligned}&\sigma_{ij}-\alpha_{ij}=0 \Rightarrow \sigma_{ij}=\alpha_{ij} \\ &\sigma_{ij,j}=0, \quad \dot{\varepsilon}_{ij}-\frac{1}{2}\left(v_{i,j}+v_{j,i}\right)=0 \\ &\lambda-\sigma_m=0, \quad \dot{\varepsilon}_{ij}\delta_{ij}=0 \\ &\text{In } S_p\text{: } \alpha_{ij}n_j-\bar{p}_i=0 \\ &\text{In } S_v\text{: } v_i-\bar{v}_i=0, \alpha_{ij}n_j-\mu_i=0\end{aligned}$$

The above equations show as long as the deformation velocity v_i and strain rate $\dot{\varepsilon}_{ij}$ are the solutions of $\delta\phi^*=0$, they will be the true solutions as they satisfy all the basic equations listed above. Since the velocity–strain rate relation and velocity boundary condition are easily satisfied, the constraint of incompressibility or the volume constancy of materials is kept in the variational function and the following Eq. (2.32) is obtained.

$$\phi=\sqrt{2}k\iiint\limits_V \sqrt{\{\dot{\varepsilon}\}^{\mathrm{T}}\{\dot{\varepsilon}\}}\mathrm{d}V - \iint\limits_{S_p} \{v\}^{\mathrm{T}}\{p\}\mathrm{d}S + \iiint\limits_V \lambda\{\dot{\varepsilon}\}^{\mathrm{T}}\{C\}\mathrm{d}V \tag{2.32}$$

where $\{\dot{\varepsilon}\}$ is the strain rate matrix:

$$\{\dot{\varepsilon}\} = \left[\dot{\varepsilon}_x \quad \dot{\varepsilon}_y \quad \dot{\varepsilon}_z \quad \frac{1}{\sqrt{2}}\dot{\gamma}_{xy} \quad \frac{1}{\sqrt{2}}\dot{\gamma}_{yz} \quad \frac{1}{\sqrt{2}}\dot{\gamma}_{zx}\right]^{\mathrm{T}} \tag{2.33}$$

$\{p\}$ is the surface force matrix applied on the boundary S_{p}:

$$\{c\} = [1 \quad 1 \quad 1 \quad 0 \quad 0 \quad 0]^{\mathrm{T}} \tag{2.34}$$

Let the deformation body be discretized into m elements with n nodes. For a given shape function [N], we have

$$\{v\}^e = [N] \cdot \{\mu\}^e \tag{2.35}$$

where $\{v\}^e$ is the velocity at any arbitrary point inside the element and $\{\mu\}^e$ is the nodal velocity of the element. Therefore, there is

$$\{\dot{\varepsilon}\}^e = [B]\{\mu\}^e \tag{2.36}$$

Thus:

$$\dot{\bar{\varepsilon}}^e = \left(\frac{2}{3}\dot{\varepsilon}^e_{ij}\dot{\varepsilon}^e_{ij}\right)^{\frac{1}{2}} = \left(\frac{2}{3}\{\mu\}^{e\mathrm{T}}[B]^{\mathrm{T}}[B]\{\mu\}^e\right)^{\frac{1}{2}} = \left(\frac{2}{3}\{\mu\}^{e\mathrm{T}}[K]\{\mu\}^e\right)^{\frac{1}{2}} \tag{2.37}$$

Hence, we have the following variational function of element e:

$$\begin{aligned}\phi^e = \sqrt{3}k \iiint\limits_{V_e} \left(\frac{2}{3}\{\mu\}^{e\mathrm{T}}[K]\{\mu\}^e\right)^{\frac{1}{2}} \mathrm{d}V \\ - \iint\limits_{S_{\mathrm{p}}} \{\mu\}^{e\mathrm{T}}[N]^{\mathrm{T}}\{p\}\mathrm{d}s + \iiint\limits_{V_e} \{\mu\}^{e\mathrm{T}}\lambda^e[B]^{\mathrm{T}}\{C\}\mathrm{d}V\end{aligned}$$

$$\phi^e = \phi^e(\{\mu\}^e, \lambda^e)$$

$$\Phi \approx \sum_{e=1}^{m} \phi^e(\{\mu\}^e, \lambda^e) \approx \Phi(\mu_1, \mu_2, \ldots, \mu_{3n}, \lambda_1, \lambda_2, \ldots, \lambda_m) \tag{2.38}$$

As a result, the variation in Φ can be written as:

$$\delta\Phi = \sum_{e=1}^{m} \left(\frac{\partial\phi^e}{\partial\mu_i}\delta\mu_i + \frac{\partial\phi^e}{\partial\lambda_j}\delta\lambda_i\right) = 0 \tag{2.39}$$

Hence:

$$\begin{aligned}\sum_{e=1}^{m}\frac{\partial\phi^e}{\partial\mu_i}&=0(i=1,2,\ldots,3n)\\ \sum_{e=1}^{m}\frac{\partial\phi^e}{\partial\lambda_j}&=0(j=1,2,\ldots,3n)\end{aligned}\tag{2.40}$$

Just considering element e, we know:

$$\frac{\partial\phi^e}{\partial\{\mu\}^e}=\sqrt{3}k\iiint\limits_{V_e}\frac{\frac{2}{3}[K]\{\mu\}^e}{\dot{\bar{\varepsilon}}^e}\mathrm{d}V-\iint\limits_{S_\mathrm{p}}[N]^\mathrm{T}\{p\}\mathrm{d}S+\lambda^e\iiint\limits_{V_e}[B]^\mathrm{T}\{C\}\mathrm{d}V=0\tag{2.41}$$

$$\frac{\partial\phi^e}{\partial\lambda^e}=\{\mu\}^{e\mathrm{T}}\iiint\limits_{V_e}[B]^\mathrm{T}\{C\}\mathrm{d}V=\iiint\limits_{V_e}\{C\}^\mathrm{T}[B]\{\mu\}^e\mathrm{d}V=0\tag{2.42}$$

Let $\varphi_1(\{\mu\}^e,\lambda^e)=\frac{\partial\phi^e}{\partial\{\mu\}^e}$ and $\varphi_2(\{\mu\}^e)=\frac{\partial\phi^e}{\partial\lambda^e}$. By using Newton–Raphson approach and Taylor series, we have the following equation by neglecting the terms with the order of two or above. The linear equations in the following are thus obtained. This process is the so-called linearization.

$$\varphi_1(\{\mu\}^e,\lambda^e)=\varphi_1(\{\mu\}^e_{n-1},\lambda^e)+\left(\frac{\partial\varphi_1}{\partial\{\mu\}}\right)_{\{\mu\}^e_{n-1}}\cdot\Delta\{\mu\}^e_n=0\tag{2.43}$$

$$\varphi_2(\{\mu\}^e)=\varphi_2(\{\mu\}^e_{n-1})+\left(\frac{\partial\varphi_2}{\partial\{\mu\}^e}\right)_{\{\mu\}^e_{n-1}}\cdot\Delta\{\mu\}^e_n=0\tag{2.44}$$

The above equations can be further designated as:

$$\begin{aligned}&\sqrt{3}k\iiint\limits_{V_e}\frac{\frac{2}{3}[K]\{\mu\}^e_{n-1}}{\dot{\bar{\varepsilon}}^e_{n-1}}\mathrm{d}V-\iint\limits_{S_\mathrm{p}}[N]^\mathrm{T}\{p\}\mathrm{d}S+\lambda^e\iiint\limits_{V_e}[B]^\mathrm{T}\{C\}\mathrm{d}V\\&+\left(\sqrt{3}k\iiint\limits_{V_e}\left[\frac{2}{3}[K]\dot{\bar{\varepsilon}}^e_{n-1}-\frac{\frac{2}{3}[K]\{\mu\}^{e\mathrm{T}}_{n-1}\cdot\frac{2}{3}[K]^\mathrm{T}}{\dot{\bar{\varepsilon}}^e_{n-1}}\right]/\dot{\bar{\varepsilon}}^{e2}_{n-1}\cdot\mathrm{d}V\right)\cdot\Delta\{\mu\}^e_n=0\end{aligned}\tag{2.45}$$

$$\left(\iiint\limits_{V_e}\{C\}^\mathrm{T}\cdot[B]\mathrm{d}V\right)\cdot\{\mu\}^e_{n-1}+\left(\iiint\limits_{V_e}\{C\}^\mathrm{T}\cdot[B]\mathrm{d}V\right)\cdot\Delta\{\mu\}^e_n=0\tag{2.46}$$

Thus:

$$\begin{bmatrix} \frac{2}{\sqrt{3}}k \iiint\limits_{V_e} \left(\frac{[K]}{\dot{\bar{\varepsilon}}^e_{n-1}} - \frac{2}{3} \cdot \frac{[K]\{\mu\}^e_{n-1} \cdot \{\mu\}^{eT}_{n-1}[K]^T}{\dot{\bar{\varepsilon}}^{e3}_{n-1}} \right) \mathrm{d}V & \iiint\limits_{V_e} [B]^T\{C\}\mathrm{d}V \\ \iiint\limits_{V_e} \{C\}^T[B]\mathrm{d}V & 0 \end{bmatrix} \cdot \begin{Bmatrix} \Delta\{\mu\}^e_n \\ \lambda^e \end{Bmatrix}$$
$$= \begin{Bmatrix} \iint\limits_{S_p} [N]^T\{p\}\mathrm{d}s - \frac{2}{\sqrt{3}}k \iiint\limits_{V_e} \frac{[K]\{\mu\}^e_{n-1}}{\dot{\bar{\varepsilon}}_{n-1}}\mathrm{d}V \\ -\left(\iiint\limits_{V_e} \{C\}^T[B]\mathrm{d}V \right) \cdot \{\mu\}^e_{n-1} \end{Bmatrix} \tag{2.47}$$

Equation (2.47) can be simply represented as follows:

$$[S]_{n-1} \cdot \begin{Bmatrix} \Delta\{\mu\} \\ \{\lambda\} \end{Bmatrix}_n = \{R\}_{n-1} \tag{2.48}$$

To solve Eq. (2.48), the iterative approach is employed. The nth iteration is conducted based on the solution of $(n - 1)$th iteration. If following condition is satisfied, it is assumed the iterative solution is the desirable one.

$$\|\Delta\{\mu\}_n\| / \|\{\mu\}_{n-1}\| < \delta \tag{2.49}$$

where δ is a very small positive number and could be 10^{-5} or even smaller. In addition, the detailed representation of the norms in Eq. (2.49) is:

$$\|\Delta\{\mu\}_n\| = \sqrt{\sum_{i=1}^{3n} \Delta\mu^2_{in}} \quad \|\{\mu\}_{n-1}\| = \sqrt{\sum_{i=1}^{3n} \mu^2_{in-1}} \tag{2.50}$$

In the iteration process of solution generation, the initial velocity for the next step iteration is determined based on the solution of the previously iterative step. In detail, it can be determined based on the following equation:

$$\{\mu\}_n = \{\mu\}_{n-1} + \beta \cdot \Delta\{\mu\}_n \tag{2.51}$$

where β is chosen between 0 and 1, and β is called the acceleration parameter for convergence.

To handle the material properties and working hardening phenomenon in iteration process, it is generally assumed that the flow stress of materials is constant in each iteration step. As shown in Fig. 2.1, the flow stress of materials is supposed to

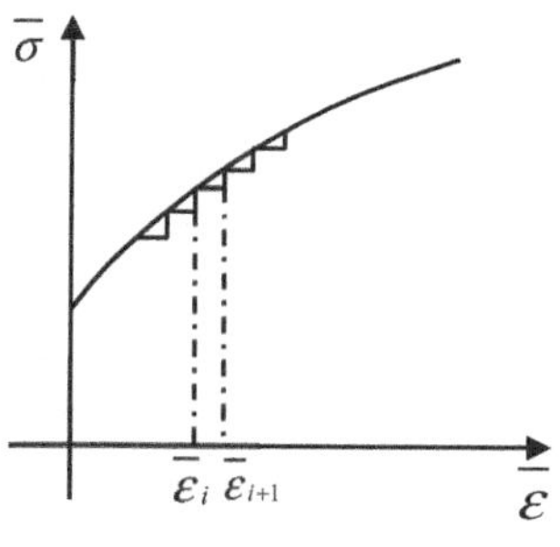

Fig. 2.1 Processing of material property and working hardening in FE simulation process

be constant in iteration step *i*, and a small flat line is used to substitute the small slope curve segment in the stress–strain curve with the working hardening effect. In such a way, the flow stress-related parameter *K* in Eq. (2.47) can be considered as a constant in all the calculations in the iteration step *i*; therefore, it can be shifted out of the integration sign in the above equations to simplify and facilitate the calculation.

Similarly, in processing of time-dependent deformation scenarios, such as viscoplastic deformation in hot working or superplastic forming, and in each time step *i* with a very small time increment interval, all the time-dependent parameters are considered to be constant in the given time step in such a way to realize the time-dependent simulation.

2.3.3 Finite Element Simulation

Finite Element (FE) simulation is an efficient approach to analyzing the behavior and performance of complex objects or systems in working conditions using the models generated based on the FEM theory. Generally, the models cover physical, mathematical, and computational models. The physical model abstracts and idealizes the real systems of concern to comply with certain physical theory with assumptions. It simplifies the real systems to be feasible enough to model and simulate. The mathematical model specifies the mathematical equations such as the so-called governing equations the physical model should follow. It also details the boundary and initial conditions and constraints. The computational model describes the element type, mesh density, and solution parameters. The solution parameters further provide the detailed incremental interval of iteration, calculation tolerance, error bound, iteration control, and convergence criterion.

In FE simulation, it involves the numerical discretization of the system to be analyzed into a number of small units, which are termed as elements, often a very large number, but limited or the so-called finite to calculate the physical variables,

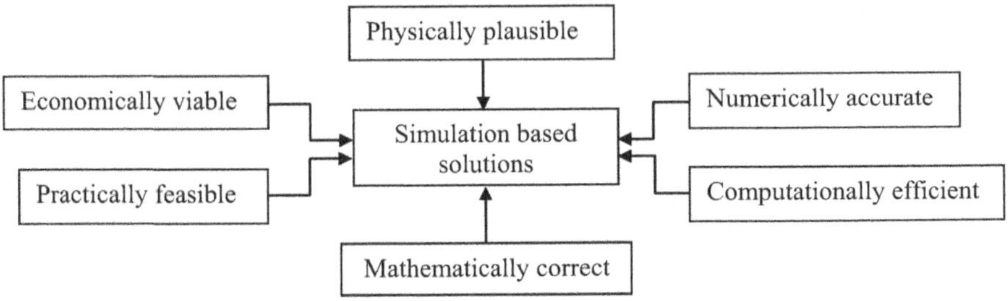

Fig. 2.2 Requirements for generation of simulation-based design solution

such as temperature, stress, strain, and velocity of each individual element such that an approximate governing equation can be generated for each individual element. The solution for the overall system can thus be obtained via solving the assembled governing equations from each individual element. The assembled governing equations represent the physical behaviors of the whole system from the perspective of the interest concerned.

FE simulation can be used in different physical domains to get the needed solutions of the field variables of real systems. The domains can cover diverse areas such as mechanical, thermal, electrical, electronic, environment, civil, aeronautical, aerospace, and manufacturing. The real systems to be analyzed could be difficult to access by human beings for different reasons such as too complex, dangerous, small, fast, or difficult to be physically realized or non-accessible. FE simulation can thus be used to physically and mathematically model and represent these systems in computer and virtually realize the phenomena and behaviors of the systems to be explored and studied. The FEM-based simulation is not only for cutting experimental cost, shortening research and development, effort and time, the most important is that it makes the impossible possible and the unrealistic realistic.

From engineering perspective, the final goal of FE simulation is to help generate feasible and practical solutions for the design and development of products and systems or provide useful information for decision-making and solution generation in different application scenarios. It is also used to reveal the instantaneous behaviors, phenomena, and performances of the system. The requirements for simulation-based solution generation are shown in Fig. 2.2. The requirements can be classified into two categories. The first category of requirements is more on how the FE simulation is established and then implemented. First of all, the simulation must be physically plausible from its physical model construction perspective. The physical model should be reasonably abstract and idealize the real systems with reasonable assumptions and comply with certain physical principles and rationales. From mathematical point of view, the mathematical model of FE simulation such as governing equations should be fully correct. From numerical implementation point of view, the computational model, which deals with meshing, element type, mesh size, calculation error, iteration, and convergence control, should be reasonable and accurate. Regarding the computation efficiency, it should be efficient and not too much time-consuming as designers and engineers usually need to patiently wait for the completion of simulation and then they can extract the useful information to support decision-making and solution generation in design stage.

The second category of requirements is more focused on solution generation based on the simulation results. The generated solution must be economically viable. In other words, the solution generation and solution implementation should not increase the design and development lead time and cost. Eventually, it should reduce tryout and streamline the design process. On the other hand, the generated solution should be practically feasible such that it can be implemented in industries. Therefore, how to translate the FE simulation results into design solution generation is a non-trivial issue as it needs sufficient domain knowledge to interpret simulation results first and then come out with scientific- and domain-specific data, information, and knowledge to support design.

The applications of FE simulation in metal forming include metal-formed part design, forming process determination, die design, defect prediction, product quality control, and assurance, covering dimensional accuracy, surface quality control, and mechanical properties assurance. In addition, prediction and analysis of microstructure evolution of deforming materials is also a promising application. These applications are based on the revealed physical variables and the identified phenomena and behaviors in the deformation process of a forming system under a certain working condition by FE simulation. In detail, it reveals and discloses flow pattern, material flow velocity, and its distribution, stress, strain, and temperature and their distribution in the entire deformation body. Furthermore, flowline, filling status, deformation load, and upfront material flow can also be revealed. With all of these different categories of information, the rationality of the deformed part design can be explored, verified, and validated in design stage. From forming process design perspective, FE simulation helps figure out the design of forming process, operation step, and sequence. It aids the design of the shape, geometry, volume, and dimension of billet and preform. In addition, it also helps process route determination and process parameter configuration, and further verification and optimization of the designs.

From die design point of view, the elastic deformation of die components is usually considered. FE simulation is used to reveal the deformation and deflection of die components during the deformation process and further to determine the stress and strain of die components and their distribution within the components. By employing certain criteria such as the minimum stress and the least stress concentration and further the information of physical variables identified and revealed by simulation, the die structure and assembly relationship and components are determined. This design paradigm is aimed at reducing the tryout in die design and shortening the design and development lead times via offline FE simulation. The final goal is to realize the development paradigm of "design right the first time." On the other hand, when the die design is aided by FE simulation, the coupling simulation of the plastic deformation of workpiece and the elastic or elastic–plastic deformation of die needs to be done simultaneously. The simultaneous simulation of the deformation of workpiece and die is termed as "coupled simulation," while separated simulation of the deformation of workpiece and die sequentially is called "uncoupled simulation" in which the deformation of workpiece is simulated first and the stage with the maximum deformation load is identified. The boundary condition between the workpiece and die in the maximum deformation load stage is then extracted out and transferred to the boundary condition in the simulation of die deformation under the maximum deformation loading. The details will be discussed in Chap. 4.

From die service life design perspective, FE simulation can be used to design die fatigue life under certain forming process and conditions. First of all, FE simulation is used to determine the cyclic working stress of die under a certain forming process configuration, die structure and assembly, and the materials' property configuration of die and workpiece. It then identifies the deformation stage at which the maximum deformation load occurs in the whole deformation cycle, the die component with the most potential fatigue failure and fatigue location in the die component, the maximum working stress at the peak of deformation load. With all of these information and knowledge, die designers can identify, reveal, and analyze the potential die fatigue mode and the possible root causes of fatigue. Based on the revealed working stress, fatigue mode, and root causes, the methods for die service life improvement can be proposed and implemented eventually in die design to prolong die service life. Similar to die design, the coupled or uncoupled simulation of the deformations of workpiece and die structure can be conducted for die service life enhancement.

From product quality assurance and control perspectives, FE simulation can be used to help reveal the formation mechanisms and root causes of defects generated in different forming processes. The defects can be classified into flow-induced, stress-induced, dimension-related, microstructure-related, and property-related defects, and their formation mechanisms and root causes can also be totally different. Through figuring out the mechanisms and root causes, the methods to avoid the defects can be proposed and developed, and further used in defect-free metal-forming PDD.

2.4 FE Simulation of Metal-Forming Systems

As described in Chap. 1, the elements in a metal-forming system can be classified into tangible and intangible ones. The tangible one refers to the physical items, while the intangible one covers the virtual items imbedded in the forming system. The physical items further refer to the materials used in the system, billet, and preform including their shape, geometry and dimensions, feeding mechanism, forming and ejection structures, and the lubrication condition in-between the workpiece and tooling. The virtual items include process and process chain, process parameters and process parameter configuration, which are generally imbedded in the design of the related physical elements in the forming systems. From FE simulation perspective, all the needed information related to the above physical and virtual items needs to be input into the FE simulation system via different models or predefined in the simulation system.

2.4.1 Modeling of Die and Workpiece

From modeling point of view, all the die components of concern need to be modeled from geometry perspective. The geometry models of die components are usually created in a computer-aided design (CAD) system and then exported into a FE simulation system for simulation as the latter generally does not have a strong geometry modeling capacity to model the complex and nonlinear geometries of die components. For the CAD models of die components generated in a CAD system, they are generally converted into a data exchange format such as initial graphics exchange specification (IGES), standard for the exchange of product model data (STEP) or stereo lithography (STL), which can be recognized by FE simulation systems and exported into a FE simulation system for preprocessing of simulation. In addition, the CAD model of workpiece or preform is also needed to be created in a CAD system and then followed by exporting to a FE simulation system via data exchange conversion. From deformation point of view, the die components with different materials can be treated as different deformation models including rigid body, elastic, and elastic-plastic models. For the workpiece, on the other hand, the deformation behaviors of workpiece materials can be modeled as an elastic-plastic, rigid-plastic, or visco-plastic model. In detail, they are described in the following.

When a die component is treated as a rigid body, it means that the component is assumed to not have any deformation in the simulation process. The real life, however, is not so, and this is an idealization as elastic deformation, no matter how small it is, exists ubiquitously in any deformation process. The function of this model is only the representation of component geometries. In other words, an object under the rigid body model is non-deformable, but the model represents the geometry profile of the object in a three-dimensional space and the geometry constraint imposed by the model in the 3D space of die assembly. When the rigid body model is used to represent a die component, the elastic deformation of die component is not critical and can thus be neglected. For this model, the movement speed of die component is specified initially, and there is no information related to the displacement, stress, strain, and thermal expansion of the die component.

For the elastic deformation model, it is usually used for modeling of die component for strength design, fatigue analysis, design optimization, and dimensional accuracy analysis. From material properties perspective, the main elastic material properties include Young's modulus and Poisson's ratio. In the scenarios of using this model, the information of stress, strain, and deflection of die components is available and can be used in the above-mentioned design activities. In simulation process, when the working stress of die component exceeds the yield stress of die materials, the stress, strain, and displacement determined by simulation will be incorrect as the real deformation of die component is no longer in elastic deformation scope. A good practice in elastic deformation simulation is to check the effective stress in simulation iteration process frequently to ensure this situation would not happen.

Regarding the plastic deformation model, it generally includes rigid-plastic and rigid-viscoplastic models, which are basically used for modeling of the deformation of workpiece materials. For the rigid-plastic deformation model, it assumes the deformation of workpiece has only plastic deformation and does not consider any elastic deformation involved. In addition, if the deformation is rate-independent, viz. the deformation behavior and flow stress of materials are independent of deformation velocity; there is thus no strain rate included in the constitutive model of materials. In these cases, where the strain rate sensitivity is not important and can thus be ignored, the rigid-plastic deformation model is used. The real-life example is cold forming processes in which the elastic deformation of workpiece can be ignored and the effect of strain rate can also be neglected. For the rigid-viscoplastic deformation model, on the other hand, the elastic deformation of materials is also not considered. The deformation is rate-dependent and the strain rate sensitivity plays an important role as it affects the deformation behavior and flow stress of materials in deformation process. The rigid-viscoplastic deformation model is generally used for analysis and modeling of warm and hot forming processes. The superplastic forming and isothermal forging are also good examples, which can be accurately simulated by using the rigid-viscoplastic deformation model.

For elastic-plastic deformation model, the material represented by this model is treated as an elastic deformation body when the effective stress is less than the yield stress of material, while as a plastic deformation object when the yielding condition of material is met. In elastic-plastic deformation of materials, the total strain is the combination of elastic and plastic strains. This model can be used for modeling of the elastic and plastic deformations of workpiece and die components. It can also be used for analysis of elastic recovery such as the springback in sheet metal forming and the strain induced by thermal expansion in warm and hot working processes. The model can further be used for analysis of residual stress in bulk cold forming. Since the strain rate sensitivity is not considered in this model, the hot working processes with large plastic deformation cannot be handled efficiently by this model. Furthermore, this model takes a longer calculation time compared with the time needed by using the rigid-plastic deformation model. And sometimes, it may have difficulty with convergence by using the model.

For viscoplastic deformation model, it represents the rate-dependent plastic deformation well. For viscoplastic deformation scenario, the deformation of materials depends on the rate at which the external loads are applied, or more accurately, the deformation velocity applied. The viscoplastic deformation materials exhibit not only the permanent deformation after the application of loads, but also the creep flow with time under the applied loads [2]. In metal forming, the creep flow is generally ignored, but the rate-dependent plastic deformation is the point to be considered.

2.4.2 Modeling of Frictional Behaviors

In metal forming, the friction behaviors and conditions in-between the die and workpiece interfaces significantly affect deformation and flow behaviors, deformation texture and flowlines, formation of new surfaces, and the surface quality of the deformed parts. In addition, friction also affects deformation load, energy consumption, occurrence of surface and internal defects, die stress and die service life, and process parameters configuration. In FE simulation, friction behaviors need to be quantitatively represented and modeled such that it can be fully embedded into the FE simulation as an input.

To quantitatively model and represent the frictional behaviors, the relationship between the relative movement of workpiece and die and the material properties of workpiece need to be quantitatively determined. There are three widely used models in metal-forming arena listed in following [3, 9].

A: Coulomb model
The Coulomb model represents the frictional stress and the compressive normal stress in the interface in-between the die and workpiece in the following equation.

$$f_s = up \tag{2.52}$$

where u is the friction coefficient and p is the compressive normal stress.

B: Shear stress model
Frictional stress in the interface of the die and workpiece is designated as follows:

$$f_s = mk \tag{2.53}$$

where m is the friction factor, which has the value from 0 to 1, and k is the shear strength of workpiece material.

C: Modified shear stress model
In addition, the frictional stress can also be represented as the following formulation:

$$f_s = mk[(2/\pi)\tan(v_r/a|v_d|)]t \tag{2.54}$$

where m is the friction factor, k is the shear strength of workpiece material. v_r is the magnitude of relative velocity between die and workpiece. a is a small positive constant such as 10^{-5}. $|v_d|$ is the absolute value of die velocity and t is a unit tangential vector. In this model, the relative velocity in between the die and workpiece is considered and the unit tangential vector t can be used to specify the direction.

For various forming conditions, the values of m vary, and the following provide a reference for different materials in different working processes [3]:

m = 0.05–0.15 in cold forming of steels, aluminum alloys, copper, using conventional phosphate–soap lubricants or oils.
m = 0.2–0.4 in hot forming of steel, copper, and aluminum alloys with graphite-based (graphite–water or graphite–oil) lubricants.
m = 0.1–0.3 in hot forming of titanium and high-temperature alloys with glass lubricants.
m = 0.7–1.0 when no lubricant is used, e.g., in hot rolling of plates or slabs and in non-lubricated extrusion of aluminum alloys.

2.5 Geometric Symmetry in FE Simulation

Symmetry is an attribute of a shape or relation. It refers to the correspondence in size, form, and arrangement of parts on opposite sides of a plane, line, or point. It can also deal with the regularity of form or arrangement in terms of like, reciprocal, or corresponding parts. Geometry symmetry deals with the symmetry of geometric shapes or objects. An object is geometry symmetric if it can be divided into two or more identical pieces that are arranged in an organized fashion [4]. This means that an object is symmetric from geometry perspective if there is a transformation that moves individual pieces of the object but does not change the overall shape. The type of symmetry is determined by the way the pieces are organized, or by the type of transformation.

Translation shifts a pattern some distance from its original in a specific direction and leaves the pattern appearing unchanged. In this process, the orientation is kept the same. Figure 2.3 shows the translational symmetry, and the symmetry is referring to the symmetry plane.

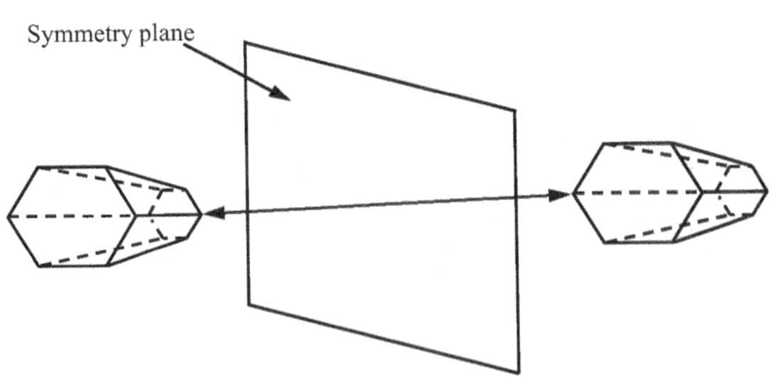

Fig. 2.3 Translation symmetry of an object

Reflection is a transformation in which the direction of one axis is reversed. An object has reflectional symmetry if there is a symmetry plane which divides it into two mirror images of each other, as shown in Fig. 2.4.

An object has rotational symmetry if it remains the same after a certain amount of rotation and without changing the overall shape. In this rotation process, the relative relationship of the geometry entities of the object keeps the same, but the orientation of the whole object changes with the rotation, as shown in Fig. 2.5.

In addition, scale symmetry is also important in FE simulation. In multiscale modeling and simulation, there is a size effect phenomenon from macroscale to mesoscale and further to microscale. In the subsequent chapters, these different size scales are often referred in different size-scaled metal-forming domains and handling the size effects in different size-scaled domains. Scale symmetry can be employed for scale-based modeling and simulation. An object has scale symmetry if it does not change shape when it is expanded or contracted or scaled up or down. From geometry point of view, a fractal is a shape made up of parts that are the same shape as itself and are of smaller and smaller sizes. Fractals also exhibit a form of scale symmetry, where small portions of the fractal are similar in shape to large portions. In FE simulation, scale symmetry is more referring to the geometrical similarity, ratio, and proportion of the system to be modeled. From the physical behavior of real-life perspective, there are two scenarios: One is the same physical behaviors and performances of the size-scaled system to be modeled, and the other

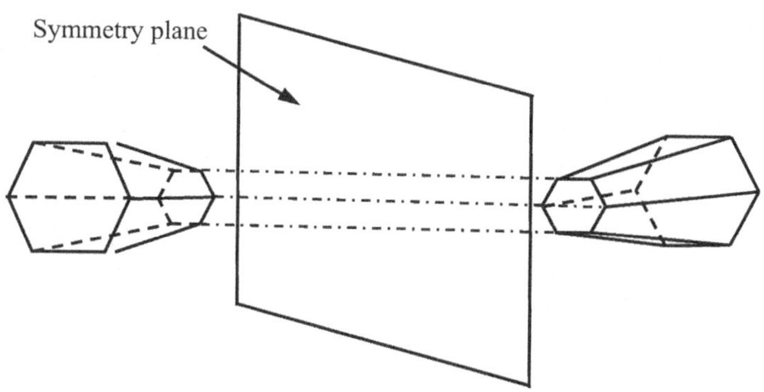

Fig. 2.4 Reflectional symmetry of an object

Fig. 2.5 Rotational symmetry of an object

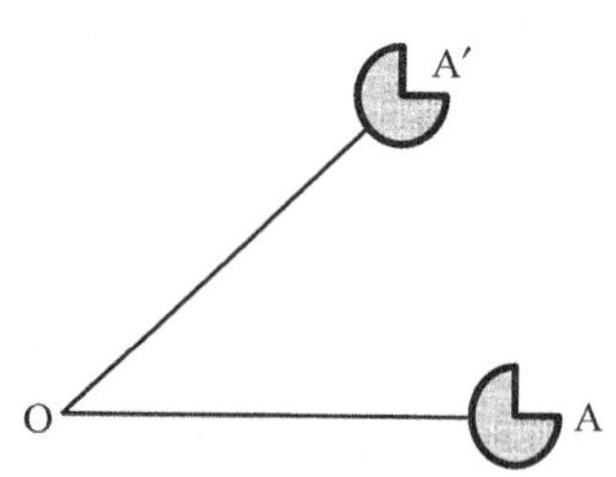

is the different physical behaviors and performances of the size-scaled systems due to size effect. In reality and practice, the scale-based physical behaviors and phenomena are of importance to be explored and investigated by using the scaled-down size model to reduce the sample size and further the capacity of experimental devices in such a way to cut experimental cost or even to make the impossible possible due to the downsizing of the scale, which becomes feasible to be implemented and realized. From FE simulation perspective, the scale geometries of the real and complex systems can also be employed to downsize the systems to be modeled and simulated in such a way to reduce the meshing size, the memory, and computational capacities of computers and further to enhance the computational efficiency. Figure 2.6 shows the objects with different scale factors from geometry proportional perspective.

In addition, axial symmetry is of importance in simulation as the computation can be significantly reduced by using the axisymmetric characteristic of the systems to be simulated. The axial symmetry refers to the symmetry of systems or objects around an axis. In other words, the appearance of the axisymmetric systems or objects would not be changed if they are rotated around an axis. For any cross section of an axisymmetric object with the symmetrical axis passing through, the cross section is divided into two reflectional geometries by the symmetrical axis. Figure 2.7 shows a cylinder with axial symmetry. In the figure, the cross section of the cylinder is divided into two reflectional parts around the symmetrical axis. From geometry perspective, one reflectional geometry can represent the other. In simulation, the axisymmetric systems or objects can be simplified as a 2D problem or can be represented by part of the axisymmetric systems. In engineering, there are many axisymmetric products or systems, which can be analyzed and simulated by using the characteristics of the axisymmetric characteristics.

In FE simulation, geometric symmetry of the systems to be modeled or simulated should fully be exploited through simulating part of the systems such as a half or one-quarter of the systems to reduce preprocessing job, shorten computation time, save storage and memory of computer, and increase the efficiency and cut the cost of simulation. By using the geometric symmetry of the systems to be modeled, the following conditions must be met:

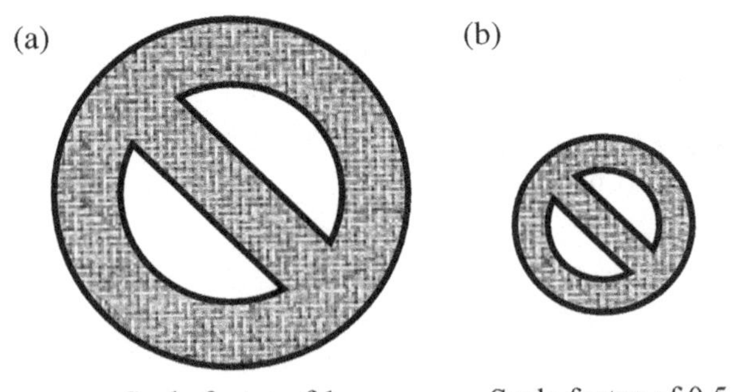

Fig. 2.6 Scale symmetry of an object

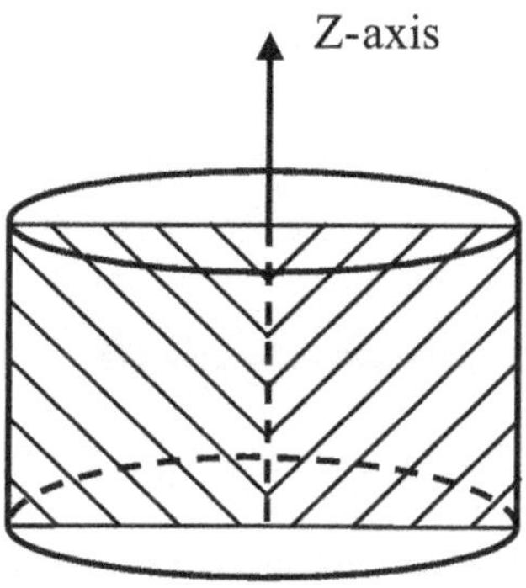

Fig. 2.7 Axial symmetry of an object

- The system to be simulated is of geometrical symmetry.
- The boundary conditions of the system are also symmetry from both physics and geometry perspectives.
- The constraints of the system are also symmetrical and can be applied symmetrically.
- The physical behaviors and performance of the systems must be the same symmetrically.

In this way, part of the system to be modeled can fully represent the whole system to be modeled.

In addition, there are some products or systems, which have pattern-based symmetry. Figure 2.8a shows an example of this type of symmetry case with pumpkin shape. From geometry modeling perspective, it can be generated via copying the basic unit, such as Fig. 2.8b, around the symmetrical axis. Assuming that the object undergoes compression deformation and the loading is uniformly applied and distributed on the top surface of the object in *Z*-direction as shown in the figure. The loading, boundary condition and constraints are symmetrically the same referring to the basic pattern unit. The pumpkin-shaped part can thus be approximately simulated based on the characteristics of pattern-based symmetry if the effect of the friction on the top surface is not so significant. In this way, one-eighth of the part is simulated as it can approximately represent the deformation of the whole system, as shown in Fig. 2.8b. In addition, Fig. 2.8c shows its boundary condition. On the symmetrical planes OS_1 and OS_2, which pass through the symmetrical axis OZ, the deformation is assumed to only happen on the symmetry planes, viz. the materials cannot move away from the symmetry planes OS_1 and OS_2. In other words, the original OS_1 and OS_2 are planes, and they are also plane after deformation. Therefore, the constraints on the two symmetry planes are defined and shown in Fig. 2.8c.

In this case, the deformation load determined based on the one-eighth of the part just represents the deformation force needed for deformation of this one-eighth of the part. For the whole system, the deformation load is the eight times the deformation load of the one-eighth of the part. For other field variables such as stress,

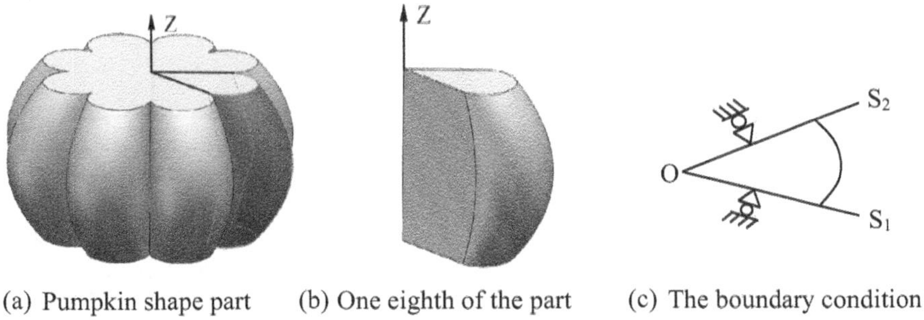

Fig. 2.8 Example of axial symmetry part

strain, and displacement, they can represent the whole part from symmetry point of view. In other words, the values of physical field variables determined based on the one-eighth of the part at a specific location are the same as those in the symmetrical location in the corresponding reflectional place in the part or the whole system. In the real application scenarios, the simulation can be done in this way to shorten the time for execution of simulation, reduce the data storage space for storing the simulation results, and increase the simulation efficiency.

2.6 Validation and Verification of FE Simulation

FE simulation has been widely used in many areas in engineering and science and is being heralded by many as one of the most important developments in the advance of engineering and science in modern history. FE simulation has become an enormously important tool to support decision-making via prediction and revealing of the behaviors and performances of engineering systems or physical events. However, the reliability of such predictions has been an issue with growing concern for a long while. Therefore, validation and verification of FE simulation has become a non-trivial issue, which perplexes people in simulation arena for decision-making by using the data and information provided via FE simulation as the results of FE simulation have been rendered as a legitimate basis in decision-making process. This is also understandable as there are a plethora of factors affecting the validity of simulation models and the reliability and accuracy of simulation results [5–8].

The simulation models defined in this book include physical, mathematical, and computational models generated for modeling and simulation of real systems or physical events from the perspective of physical behavior and functional performance of systems to be simulated. The physical model idealizes the problems or behaviors of interest of the system to be studied and abstracts them to comply with certain physical theory and assumptions. Taking the deformation of a beam as an instance, the physical model of the beam needs to comply with the continuum mechanics and assumptions describing the physical characteristics of the beam, such as static, three-dimensional, elastic type of material. The mathematical model specifies the mathematical equations such as the differential equations in FEM analysis the physical model should follow. The detailed formulation includes the governing equation of the systems and provides abstract representation of the systems consistent with a scientific theory. It also details the boundary and initial conditions and constraints. The computational model describes the element types, mesh density, and solution parameters. The solution parameters further provide detailed calculation tolerances, error bounds, iteration specifications, and convergence criteria. For the tailor-made commercial simulation systems for domain-specific applications, the common parts of these models are generally built into the FEM systems. Other model-related data and information of the systems to be modeled need to be input by users using different approaches.

Simulation model is an approximate imitation of real-world systems, but never exactly emulates the real-world ones. The validity of the models thus needs to be assessed. If the simulation models are not valid and accurate, the simulation and further the results arising from the simulation are of virtually no value. On the other hand, if the models are fully valid and reasonable, the simulation and the results may not be fully correct and accurate as the correct implementation of the simulation models is another issue. Therefore, the simulation models and the implementation of the models need to be validated and verified, respectively, to ensure the models are correct, and the results arising from simulation can be used as a legitimate basis for decision-making.

Validation and verification of simulation models and results are two important steps in FE simulation. It is generally believed that validation is the process to ascertain the generation of "right model." It is the process of determining whether the simulation models can accurately represent the real-world systems from physics perspective. Model validation determines the physical theories and assumptions underlying the models are correct, and the model's representation of the real system's behaviors being modeled is reasonable for the intended purpose of the model [5]. Verification, on the other hand, is concerned with building the "model right." It is the process by which it is ascertained that the simulation models can be accurately implemented and the implementation of simulation models can accurately represent the intended purposes of the models from mathematic point of view [6]. Figure 2.9 shows the positions of validation and verification in modeling and simulation.

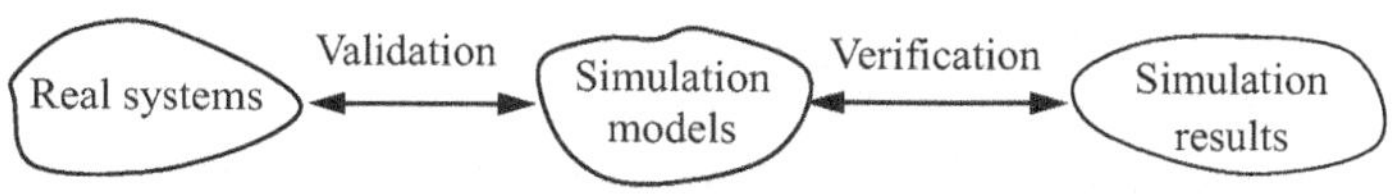

Fig. 2.9 Validation and verification in modeling and simulation

In building of simulation models, assumptions are generally made in establishing the physical, mathematical, and computational models. The assumptions simplify, idealize, and abstract the real systems such that the systems being simulated can be accurately represented by the simulation models. Similarly, assumptions need to be correct to ensure the validity of simulation models. However, it is difficult to validate whether the assumptions are correct alone. Therefore, the assumptions and the built simulation models should be validated together. Following are some practices to ensure the model generation is valid or "right model" and further the simulation results are accurate.

(A) Cross-checking the correctness of the established models. It is a good practice to discuss with experts on the correctness of the assumptions and further the established simulation models since different people have different understandings of the physical essence of the systems being modeled and cross-checking of the assumptions and the idealized models can figure out the loopholes in assumption making and further in establishment of the models.
(B) Comparison with the available analytical solutions. If there are available analytical solutions for simple systems, which are similar to the systems being simulated, the assumptions and the established models can be used to simulate the simple systems and then compare the simulation results with the available analytical solutions of the simple systems to validate the validity and correctness of the assumptions and the established models. The deviation in-between can be used to fine-tune the assumptions and the established models.
(C) Corroboration with the known behaviors or performances of the systems. Simulation is used to explore or identify the concerned behaviors or performances of real systems. The assumptions and the built models can also be used to simulate the similar systems with the known behaviors or performances such that the revealed behaviors or performances can be compared with the known ones of the similar systems.
(D) Validation by experimental results. The experiments on the systems being simulated in terms of the concerned behaviors can be conducted, and the experimental results can be used to validate the assumptions and the built models. In addition, if the system being simulated is not yet available, the experiments can be done on the scaled-down prototype system with the similar behaviors or performances. The prototype system can be simulated for comparison with the experimental results.

In tandem with validation, verification is the process to ascertain the accuracy of model implementation, as shown in Fig. 2.9. In FE simulation of metal-forming processes or forming systems, the implementation of simulation models is generally realized by coding of the models into the simulation platform or system through the application programming interfaces (APIs) provided by the system or by inputting the model-related data into the simulation system via the input mechanism provided by the system. The verification of implementation can be done using the following approaches:

(A) Structured walkthrough: Cross-checking of the implemented codes among peers to evaluate and determine the correctness of the codes if the implementation of the simulation models needs coding. If there is no coding required, input of the model-related data is generally needed. In this way, checking of the correctness of the generated data and its input is usually conducted. This can be considered as a static verification as it does not involve the execution of the coded program to examine its validity.

(B) Execution of program: Upon implementation of the simulation models by using the coding or direct input of the model-related data, the simulation can then be executed in simulation systems. In the course of execution, the intermediate output can be designed to monitor the simulation and ensure the correct simulation is conducted based on the evaluation and assessment of the intermediate output. In addition, the input and output relationship can be used to examine the correctness of implementation and simulation in the execution process.

(C) Outcome verification: The outcomes from simulation can be compared with (a) the available analytic solution of the forming processes or systems; (b) the detailed physical variables which are easily measured, such as the deformation load of a forming system; (c) geometry and dimension data such as the diameter and height of the experimental parts as they are easy to be measured or originally designed.

It is impossible to have total and absolute validation and verification of models and model implementation. Validation and verification need to be conducted feasibly and viably. Over validation and verification are not necessary and time-consuming and costly.

2.7 Summary

In this chapter, the basic concepts of modeling and simulation are briefly introduced and the notions of FEM and the FE simulation are presented. Considering the unique characteristics of metal-forming process, the rigid-plastic FEM is extensively described and the formulations are presented. By using the FEM, how the metal-forming system is simulated is articulated via considering the modeling of different elements in the system. Since many metal-forming systems are geometric symmetry, simulation of the whole system can be done by using the symmetry of systems in such a way to cut simulation cost and shorten lead time. Therefore, how to conduct symmetry-based simulation is summarized. Since simulation results provide a legitimate basis for decision-making, how to ensure the model generation and implementation are both correct is a critical issue with growing concern in simulation-based decision-making. The validation and verification of FE simulation models and model implementation are finally articulated in the end of chapter. All of these lay down theoretical basis for the subsequent chapters on FEM applications metal-forming process and product development.

References

1. Schwendimann B (2010) What is the difference between a simulation and a model. Proto-Knowledge. http://proto-knowledge.blogspot.hk/2010/12/what-is-difference-between-simulation.html
2. Perzyna P (1966) Fundamental problems in viscoplasticity. Adv Appl Mech 9(2):244–368
3. Altan T, Oh S, Gegel H (1983) Metal forming fundamentals and applications. American Society for Metals, Materials Park
4. Lockwood EH, Macmillan RH (1978) Geometric symmetry. Cambridge Press, Cambridge
5. Pace Dale K (2004) Modeling and simulation verification and validation challenges. Johns Hopkins APL Techn Digest 25(2):163–172
6. Sargent RG (2013) Verification and validation of simulation models. J Simul 7:12–24
7. Babuska I, Oden JT (2004) Verification and validation in computational engineering and science: basic concepts. Comput Methods Appl Mech Eng 193:4057–4066
8. Szabo B, Babuska I (2011) Introduction to finite element analysis: formulation, verification and validation. Wiley, London
9. Scientific forming Technologies Corporation (2014) DEFORM™ 3D Manual, Version 8.11

Chapter 3
Evaluation of Forming System Design

3.1 Introduction

In the current metal-forming product design and development paradigm, product cost, time-to-market, and product quality are three overriding issues, which determine the competitiveness of the developed products. In up-front design process, the first 20 % of design activities commits to more than 75 % of product development cost and product quality issues. How to conduct "design right the first time" is critical to ensure low product development cost, high product quality, and short time-to-market. To address these issues, state-of-the-art technologies are needed to support design solution generation, evaluation, and optimization in metal-forming product design and development. Traditionally, computer-aided design (CAD) and computer-aided manufacturing (CAM) technologies provide approaches for representation and realization of design solutions physically [1]. However, how to generate design solution and conduct design solution evaluation and optimization is a non-trivial issue. In metal-forming product design and development, it is difficult to simultaneously address the design issues related to metal-formed part design, forming process determination, process parameter configuration, tooling structure design, material selection, prediction of the properties of deformed part, and finally the product quality control and assurance. In addition, how to reveal, assess, and evaluate the interaction and interplay of different design variables or factors in the above-mentioned different stages and areas is another critical issue. Computer-aided engineering (CAE) technology fills this gap as it helps practitioners generate, evaluate, and optimize design solutions before the best design solution is feasibly and uniquely identified and practically and physically implemented. In the current

M.W. Fu, *Design and Development of Metal-Forming Processes and Products Aided by Finite Element Simulation*, Engineering Materials and Processes,
DOI 10.1007/978-3-319-46464-0_3

metal-forming product development paradigm, CAE simulation technology is one of the state-of-the-art technologies, which has been widely used in addressing the above-mentioned technical issues and will be used to solve the emerging bottleneck problems in the next upward trend of technology and product development.

CAE simulation technology uses numerical techniques to quantitatively represent the working behaviors of a physical system and imitates the dynamic and physical behaviors of the system in working process and condition. In metal-forming processes, the plastic flow of materials and the deformation of tooling are the main physical behaviors of a forming system, which can be simulated by numerical approaches through modeling of the plastic flow of materials and the elastic or elastic–plastic deformation of tooling. The numerical results from simulation are correspondingly related to the physical content of the forming system being simulated. Currently, most CAE simulations employ commercial finite element (FE) simulation packages, in which the numerical technique, such as the finite element method, is the enabling and kernel technology.

The finite element method (FEM), whose engineering birth and growth in the 1960s was due to the advent and wide application of high-speed computer to structural analysis, has spread into a variety of engineering and science disciplines in the past a few decades. It has become a standard design and analysis tool to support decision-making and solution generation and evolution in many areas in engineering and science, help reveal and explore different behaviors and phenomena in scientific research, and verify and validate findings in different explorations. The applications of FEM to the large plastic deformation of materials and metal-forming engineering can be traced back to the pioneer work done by Lee and Kobayashi in 1973, which laid down the foundation for simulation of large plastic deformation [2]. With the advance and mature of the computational algorithms of FEM for large plastic deformation, the development of commercial FE simulation systems became feasible and off-the-shell FEM systems were then available in marketplace. FEM was thus widely used in metal-forming industries for analysis of material flow behavior, determination of forming processes, configuration of forming parameters and process window, design and optimization of die structure, and further the entire forming system. FE simulation, as a subset of modeling and simulation technology, plays an important role in the development of new and innovative products in modern industries in the past and future.

Regarding the detailed applications of FE simulation in metal-forming product development, the focus is more on some individual issues such as revealing of deformation behavior, deformation loading, and flow scenario for process route determination and forming machine selection, stress and strain analysis for tooling design, and defect prediction and avoidance for product quality control and assurance [3–18]. However, how to use this technology to help design an entire metal-forming system, assess and evaluate the system performance as a whole, and eventually optimize the system design via taking into account all the potential affecting factors has not yet been fully addressed. On the other hand, it is well known that a metal-forming system comprises of all the input variables related to the billet geometry and material, the tooling, the conditions at tool–material interface, the equipment used, the forming process and process parameter configuration, and the characteristics and the requirements of the final product. Evaluation of the performance of a forming system thus needs to deal with these input variables. In addition, the interplay and interaction of these inputs are quite complex and the physical experiments to reveal them at design stage are not possible. Even if it is feasible by using physical modeling experiment, it is also quite time-consuming, expensive, and not efficient. FE simulation, however, provides an efficient approach for evaluation of a forming system designed as a whole and reveals the instantaneous behaviors and performances of the system in such a way the whole panorama of concern of the designed system can be systematically understood and evaluated.

In this chapter, a simulation-enabled methodology to evaluate the performance of metal-forming systems is presented and how it is implemented is also articulated. First of all, FE simulation-aided design of deformed parts is presented and the designs are evaluated and assessed based on the identified values or performances of the proposed physical variables revealed by FE simulation. Moreover, the procedure for simulation-based forming process and die design is given. By using the developed quantitative evaluation criteria, the forming system performance is assessed and the best system design can thus be determined.

3.2 Evaluation of Metal-Forming Systems

3.2.1 Factors Affecting the Design of Metal-Forming Systems

FE simulation of a metal-forming system models and imitates the dynamic behavior and performance of the system in working conditions. The kernel of this simulation is the numerical method of FEM, which can quantitatively represent the input–output relationship of the forming system with the determined process and process parameter configuration in forming of a designed deformation part. The "input" of the system here refers to the input variables defined previously and tailor-abstracted as the input for simulation. The "output", on the other hand, is the desirable information related to material flow and deformation velocity, flowline distribution, advance of material flow front in forming process, and the magnitude and distribution of stress and strain in the deformation body. The outcome information is used to help design solution generation, evaluation, and optimization in terms of metal-deformed part design, process determination, process parameter configuration, tooling design and service life assessment, prediction and avoidance of flow- and stress-induced defects, and dimensional accuracy and property-related product quality control and assurance. The input–output relationship is affected by the characteristics and capacities of the designed tooling structure, feeding and ejection mechanisms, forming equipment, the properties of the materials used, and the efficiency of the designed process and process parameter configuration, etc.

Through FE simulation, the design of forming system can be evaluated and optimized in up-front design process, which is an important design activity in the simulation-enabled design paradigm for metal-forming product design and development. It shifts the traditional design and development paradigm from trial-and-error and heuristic know-how to more scientific calculation, analysis, and simulation. By using this simulation-based design and development paradigm, the evaluation and verification of design solutions thus needs less trial-and-error experiments in workshop in such a way to shorten product development lead time, reduce development cost, and improve product quality and productivity.

In the traditional design practice, however, design conceptualization and solution generation are generally a trial-and-error process based on experiences and know-hows. The kinds of experiences and know-hows are usually acquired through long years of apprenticeship and skilled craftsmanship [1]. Therefore, the traditional design paradigm is time-consuming and error-prone and needs more experimental tryouts in workshop to verify the design, and further, it may involve more design changes. Moreover, if the designs of product, process, and the entire forming system are not satisfactory, it is also difficult to identify the reasons and pinpoint the root causes by the traditional approaches due to the complex interaction and interplay among a plethora of affecting factors, which are generally difficult to be revealed, explored, and quantitatively represented.

In a metal-forming system, the geometries and shapes of the deformed parts are generated via mapping of the geometries and shapes of the die into the deformed parts. The geometries and shapes of the die, on the other hand, are determined based on the geometries, dimensions, and tolerance requirements of the deformed parts, which are further decided by the design of final products. In net-shape forming process, there is no subsequent machining or working process needed to further shaping the geometries of the deformed parts. The difference of the geometries and shapes between the deformed parts and the final products lies in the geometry compensation taken in design of the deformed parts via considering the elastic recovery and volume contraction in deformation process. The dimensional accuracy and tolerance of the deformed parts, however, are also controlled by the machining quality of the die. In terms of forming process, it is actually embedded in the built-in forming procedure, the designs of billet, preform and die structure, and further the configured forming process parameters. The physical forming system covers feeding system, ejection mechanism, die structure, lubrication, and the selected forming equipment with auxiliary subsystems. The design of these elements and the entire forming system is affected by various related factors and their interplay and interaction, as shown in Fig. 3.1. These factors are listed as follows:

(a) Geometry, shape, and dimension of the metal-formed part;
(b) Material, material microstructure, and the properties prepared by previous working processes;

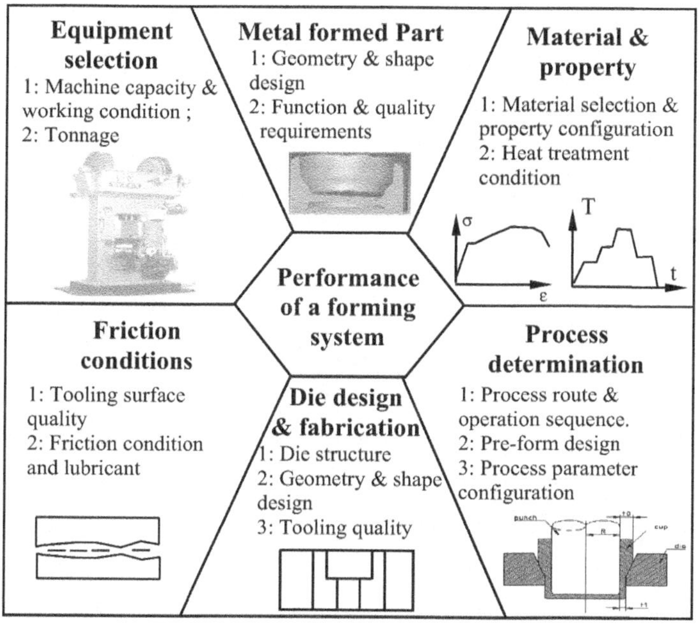

Fig. 3.1 Factors affecting the design of a forming system [7]

(c) Process route, process parameters configuration, and design of billet and preform;
(d) Die geometry, structure, and quality;
(e) Friction and lubrication condition at the interface between workpiece and tooling; and
(f) Equipment capacity and working condition setting.

All of these factors and their interactions complicate the design of forming system and make it difficult to evaluate and optimize the designs. It is thus believed that the optimal conceptualization of a forming system design is a non-trivial issue due to the difficulty to establish and explore the entire panorama of the affecting factors and their interactions. In conceptual design stage, prioritization of different factors and translation of them into design variables is of importance to be addressed.

To have an in-depth understanding of the eluded and tantalized interactions and interplays among different factors involved in forming system design, Fig. 3.2 shows an instance for the extra cost induced due to the uncertainty in determination of the maximum deformation load for screw forming process [3]. According to the figure, a 30 % uncertainty in determination of the maximum deformation load could lead to the extra capital cost of equipment close to 90 % and the extra direct cost of the forged parts between 5 and 9 %, depending on the cost of materials. In addition, it further confirms that the determination of deformation load, which is only one of

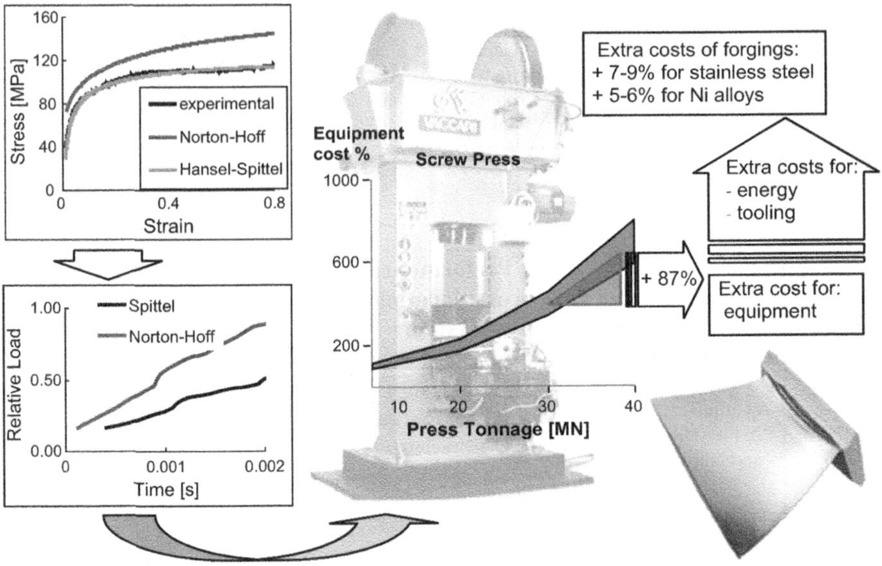

Fig. 3.2 Extra cost due to uncertainty in the determination of the maximum deformation load [3]

the intermediate performances of the input variables, could significantly affect forming equipment selection and product development cost. In the traditional forming product development, a 30 % uncertainty in loading estimation is a common issue as the deformation load is generally estimated by empirical formula via simplifying or ignoring some low-priority factors. It is thus very difficult to give an accurate evaluation and prediction of these performance and outcome of the designed forming system.

The CAE simulation, however, provides an efficient, robust, feasible, viable, and pragmatic approach to support the design and development of metal-forming product, forming process, die set, and the entire whole forming system [4–6]. It has become a standard tool to support metal-formed part design, process determination, process parameter configuration, tooling design, and product quality control and assurance [2–17]. By using FE simulation technology, the systematic evaluation and verification of design solutions via considering the intended performance of the forming system and the deformation behaviors of the deforming workpiece based on the predefined evaluation criteria thus becomes possible, feasible, and practical. The evaluation of design solutions needs to take into account the interactions and interplays of different design variables and further to identify the good, better, and best design solutions.

Although there is no panacea to fully address all the issues and finally come out with the perfect design solutions, quantitative analysis of design solutions and comparison of their performances aided by FE simulation would ensure that the best design solutions could be generated and identified from all the design alternatives. Simulation-enabled evaluation of the integrated product, forming process, and system design thus becomes a practical and feasible metal-forming design and development paradigm for solution generation in metal-forming industries.

3.2.2 Design of Deformed Parts

Most parts fabricated by metal-forming processes are usually semi-finished products despite of the fact that the deformed parts are generally net or near net shape in nature. In design of the metal-deformed part, it involves a mapping which converts the design of final product into the design of metal-deformed part. Since the metal-deformed part may not be the final product and the subsequent manufacturing operations are needed to further work on the semi-finished deformed part such that the geometry and shape requirements of the final product can be met. In this design process, the geometry and shape requirements, which could include dimensional accuracy, tolerance requirement, and surface condition of the final product, are an important factor to be considered. In addition, the workability of the deformed part and the formability of the materials used are also needed to be considered. However, to consider these two factors in design process, there are guidelines summarized in different formats of knowledge or empirical data in engineering handbooks. On the other hand, know-how and experience also play an important role in decision-making in design. Taking the case as shown in Fig. 3.3 as an instance, Fig. 3.3a shows a product with a through hole at its central axis and the material is aluminum alloy 6061. By using metal-forming process to fabricate this part, the designed metal-deformed part is shown in Fig. 3.3b, where there is a patch-up to convert the through hole into two blind holes with the same diameter by a patch-up with the thickness designated as H and the location, which is the height away from the bottom surface, is denoted as L, and in such a way, the designed metal-deformed part can be fabricated via plastic deformation of the cylindrical billet in one stroke. In addition, the two blind holes can be formed based on the geometries of punch and the convex feature in the cavity of die. Therefore, two geometrical related design parameters are raised up. One is the location of patch-up L, and the other is its thickness H. Different locations and thicknesses of the patch-up would form different design scenarios of the metal-deformed part.

To configure the location of patch-up and its thickness, as shown in Fig. 3.3b, L can have any value in the range of 0 to (25-H), and on the other hand, H can have a certain value of the thickness such as 3 mm to any value less than 25 mm as the total height of the part is 25 mm. For these two design parameters, the design configuration can be infinite as they can have any value in their value ranges. Furthermore, for different design configurations of the two parameters, the workability of the design is different. The deformation behavior of the workpiece in different designs is different too. So are the physical behavior and performance of the corresponding forming system. In detail, the deformation load, material flow,

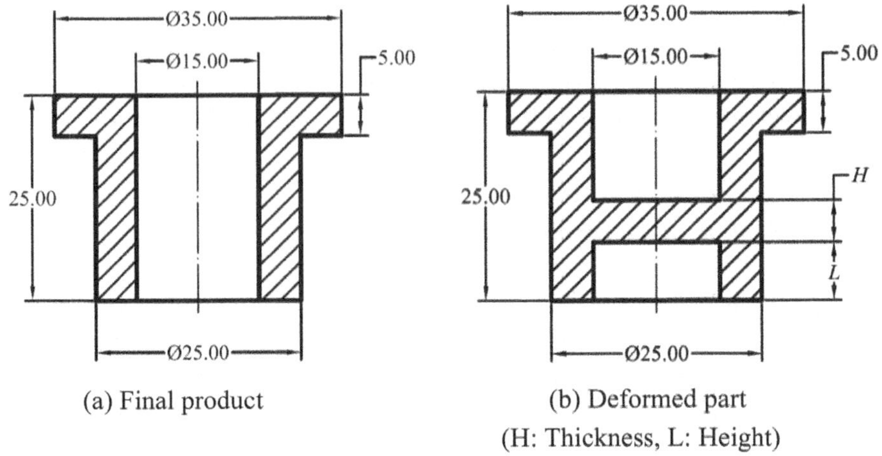

Fig. 3.3 Product and its deformed part design

working stress magnitude, and distribution in die structure are all different. These different behaviors and performances would lead to different outcomes of the design in terms of product quality, material utilization, energy consumption, tooling service life, and productivity and production cost. Therefore, it is difficult but a must to assess and evaluate the designs in up-front design process. In such a way, the good, better, and best designs can be identified and the best one can be implemented thereafter.

For the two design parameters, different design alternatives can be determined based on their different configurations, as mentioned above. For instance, *H* can have three values, i.e., 3, 5, and 7 mm, and *L* has four positions with the value of 0, 4.25, 8.5 and 17 mm, where the latter value doubles its previous value except for the second one. There will be twelve design parameter configurations and design scenarios, as shown in Tables 3.1 and 3.2. For the different design scenarios, the geometries and dimensions of billet and die set are different. The deformation behaviors including deformation load, material flow mode and pattern, and the stress and strain generated in punch and die cavity could be totally different. Therefore, how to accurately determine these parameters, evaluate, and identify the good, better, and best design scenarios is quite difficult, if not impossible, by the traditional approaches.

Table 3.1 The dimension configuration for the deformed part design

H	L1	L2	L3	L4
3	0	4.25	8.5	17
5	0	3.75	7.5	15
7	0	3.25	6.5	13

Table 3.2 Twelve design scenarios with different patch-up thicknesses and height levels

Design scenarios	L1	L2	L3	L4
$H = 3$	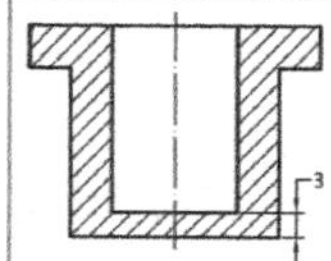	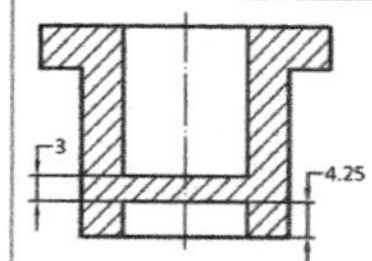	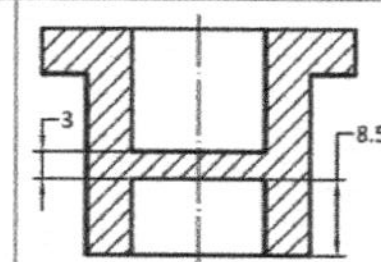	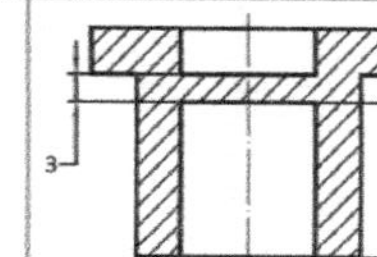
$H = 5$	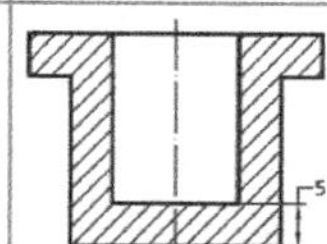	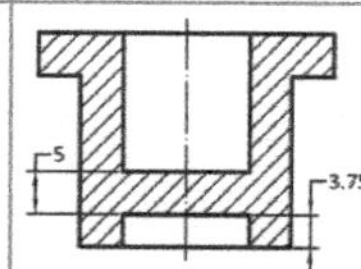	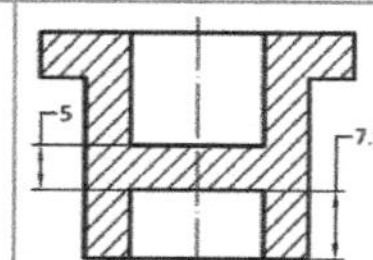	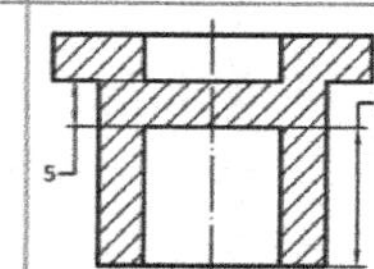
$H = 7$	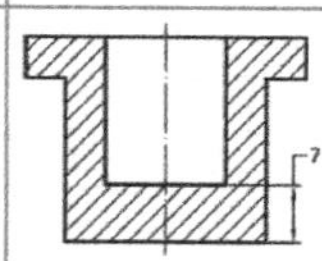	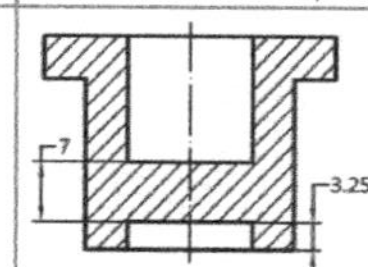	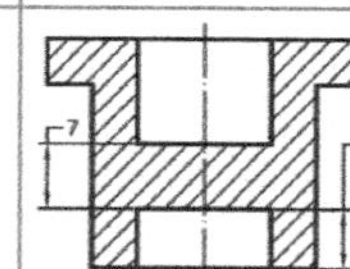	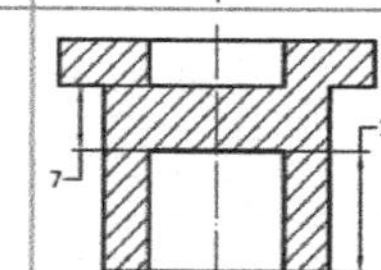

3.2.2.1 Experimental Realization of the Twelve Design Scenarios

To figure out the real differences among these design scenarios in terms of deformation behavior, flow mode, and the quality of the deformed parts, physical experiments were designed and implemented. Figure 3.4 shows a general-use die set designed for forming of the twelve design scenarios via change of the forming punch. Figure 3.5 presents different fabricated punches for forming of the different designed parts listed in Table 3.2. Figure 3.6 gives the deformation load–stroke curves of the twelve design scenarios and shows the different deformation loadings and the load variation in deformation process when the forming material is Al 6061. In the case of $H = 3$ mm and $L = 17$ mm, the deformation load for the scenario is much greater than that in other three cases. For the cases with $H = 5$ and 7 mm, and L has different values, the difference in deformation load is not much different and the variation trend is almost the same. In addition, the material flows are also different. Figure 3.7 shows the parts formed in the twelve design scenarios and their

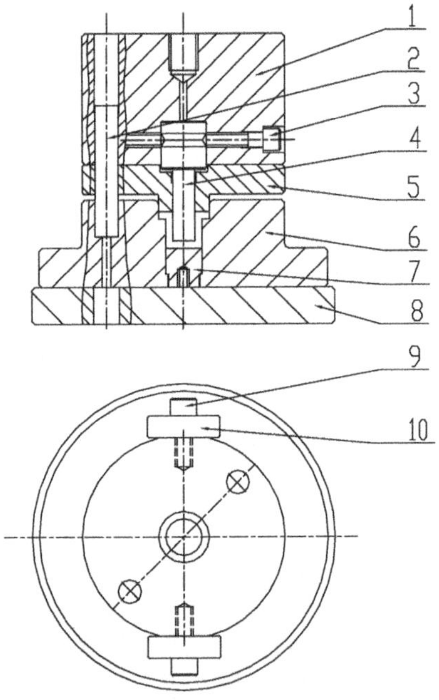

1—Top mounting bushing; 2—Guide pin; 3, 9—Screw; 4—Punch; 5—Punch stripper plate; 6—Die; 7—Die insert; 8—Bottom plate; 10—Connecting plate.

Fig. 3.4 A general-use die set for forming the twelve design scenarios by the change of different punches

Fig. 3.5 Different punches and die inserts fabricated for forming different designed parts

material flows in each scenario. It clearly shows that the defects exist in the internal structure despite of the fact that the external appearance of the deformed parts is quite satisfactory.

3.2.2.2 Simulation and Analysis of the Twelve Design Scenarios

To efficiently and economically reveal the detailed deformation behaviors of different design scenarios, FE simulation was conducted. The testing material is Al 6061, and the stress–strain curve is shown in Fig. 3.8. The FE simulation system is DEFORM 3D [18]. The Young's modulus of the tooling material is 250,000 MPa, and the Poisson's ratio is 0.33. The billet is meshed with about 30,000 tetrahedral elements. The punch speed is set to 1.0 mm/s, and the shear friction in between the die and workpiece is specified to be 0.1.

A. Deformation Load

The influence of patch-up thickness on the deformation load of the designed metal-deformed parts is studied. Figure 3.9 shows the load–stroke curves in the design scenario L2 with different patch-up thicknesses (referring to Table 3.2). The thicker the patch-up, the shorter the punch head is, which would lead to the earlier stage of the material to reach the punch shoulder. In addition, for the design with larger patch-up thickness, a sudden change of deformation load occurs earlier, as shown in Fig. 3.9. Based on the simulated forming status and the shape of deformation load, it can be found that the forming of the flanged feature significantly affects the deformation load. Figure 3.10 shows the maximum forming loads in

Fig. 3.6 Deformation loads in different design scenarios measured in experiment

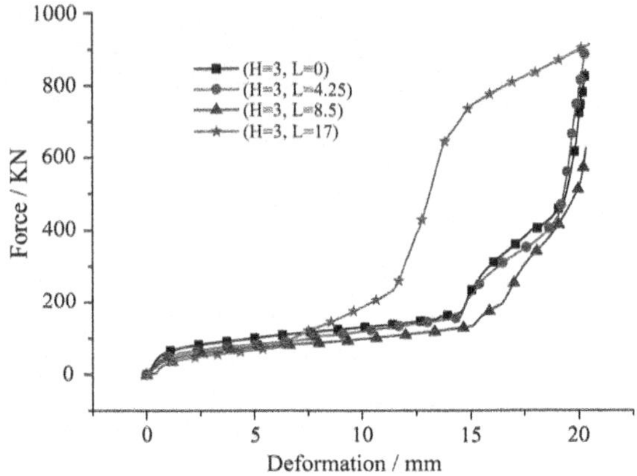

(a) The cases with H=3 mm

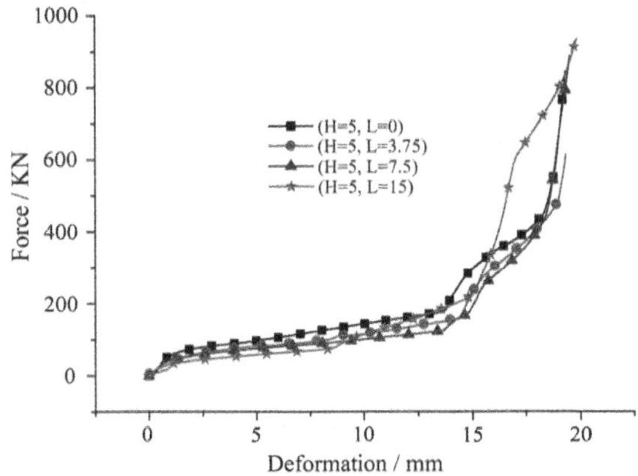

(b) The cases with H=5 mm

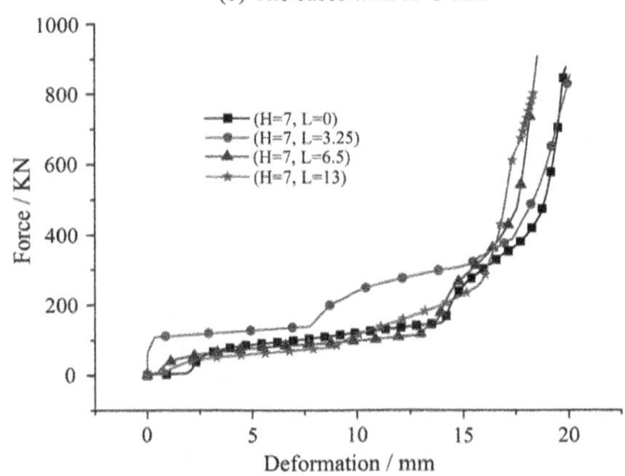

(c) The cases with H=7 mm

(a) H=3, L=0, 4.25, 8.5, 17 (mm)

(b) H=5, L=0, 3.75, 7.5, 15 (mm)

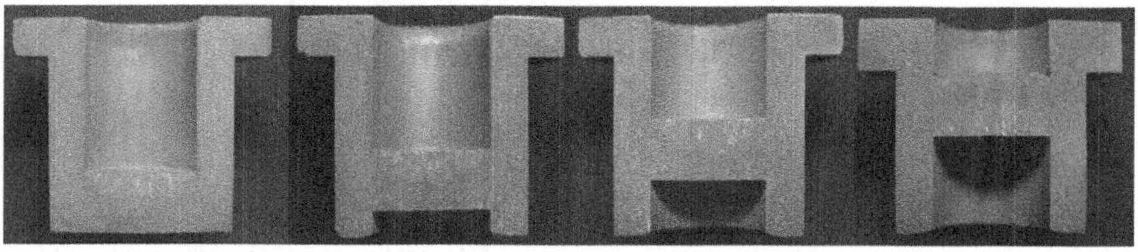

(c) H=7, L=0, 3.25, 6.5, 13 (mm)

Fig. 3.7 The experimental material flow modes and patterns in the twelve design scenarios

different patch-up design scenarios. It can be seen that the deformation load decreases with the increase of patch-up thickness, except for the case of L1 in which the thickness of 5 mm has the maximum deformation load. When the patch-up thickness is constant, the scenarios with the patch-up height level of L3 require the largest deformation load.

B. Energy Consumption

The energy consumption shown in Fig. 3.11 shows the change of energy consumption with patch-up thickness and height level. In the scenarios with the same patch-up thickness, the patch-up height level of L3 requires the lowest energy to form the part as shown in Fig. 3.11a. This could be caused by its unique flow pattern in which the patch-up divides the material more equally into two streams to flow up and down with the least travel distance such that the energy needed is least. In addition, Fig. 3.11b shows the energy consumption decreases with the increase of patch-up thickness. With the increase of patch-up, the amount of material to flow up and down is thus becoming less, such that the energy needed to deform the part is getting less.

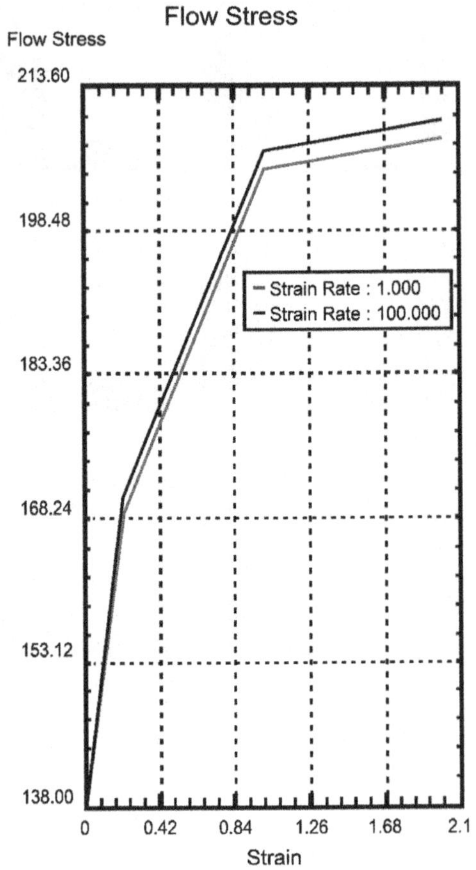

Fig. 3.8 Stress–strain curve of the testing material Al 6061

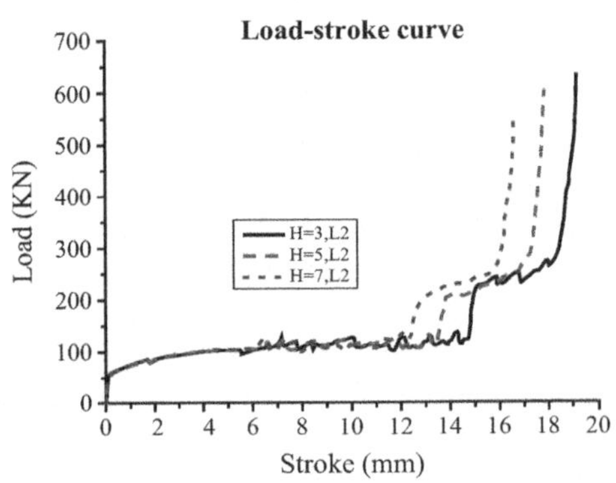

Fig. 3.9 Influence of patch-up thickness on the deformation load

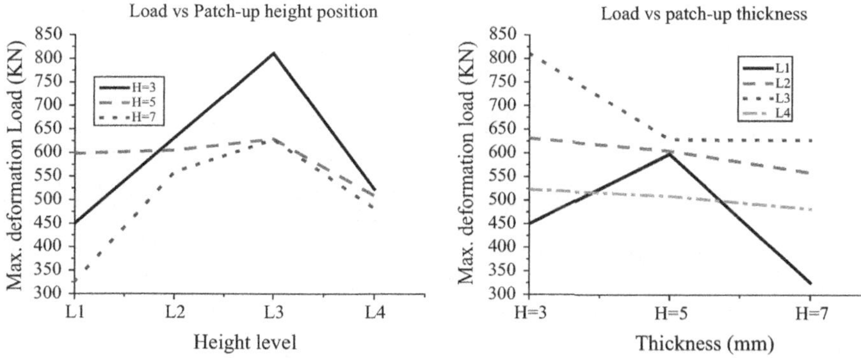

Fig. 3.10 Influence of patch-up thickness and height level on the maximum deformation load

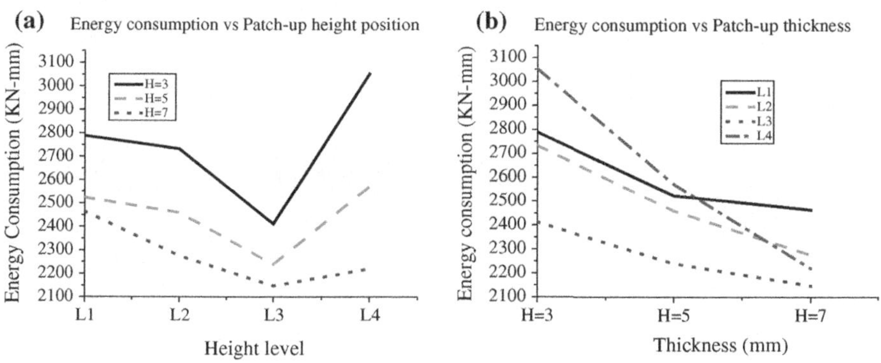

Fig. 3.11 Change of energy consumption with patch-up

C. Effective Stress Distribution

Figure 3.12 shows the change of stress and its distribution in forming process. It can be seen that the high stress is mainly concentrated at the upper inner surface of the deforming part. The lower patch-up location, the more materials with high stress it has. Figure 3.13 shows the high stress distribution in the scenarios with different patch-up thicknesses. It can be found that the patch-up thickness does not have a significant effect on stress magnitude and distribution.

D. Effective Strain Distribution

Figure 3.14 shows the effective strain distribution of the deformed part in the scenarios with different patch-up locations. From the figure, it can be found that good deformation uniformity can be achieved in the scenario of L4. In other scenarios of L1 to 3, there is a high strain at the inner hole surface and flange edge. Figure 3.15 shows the effective strain distribution in the scenarios with different patch-up thicknesses. The patch-up thickness does not have a significant influence

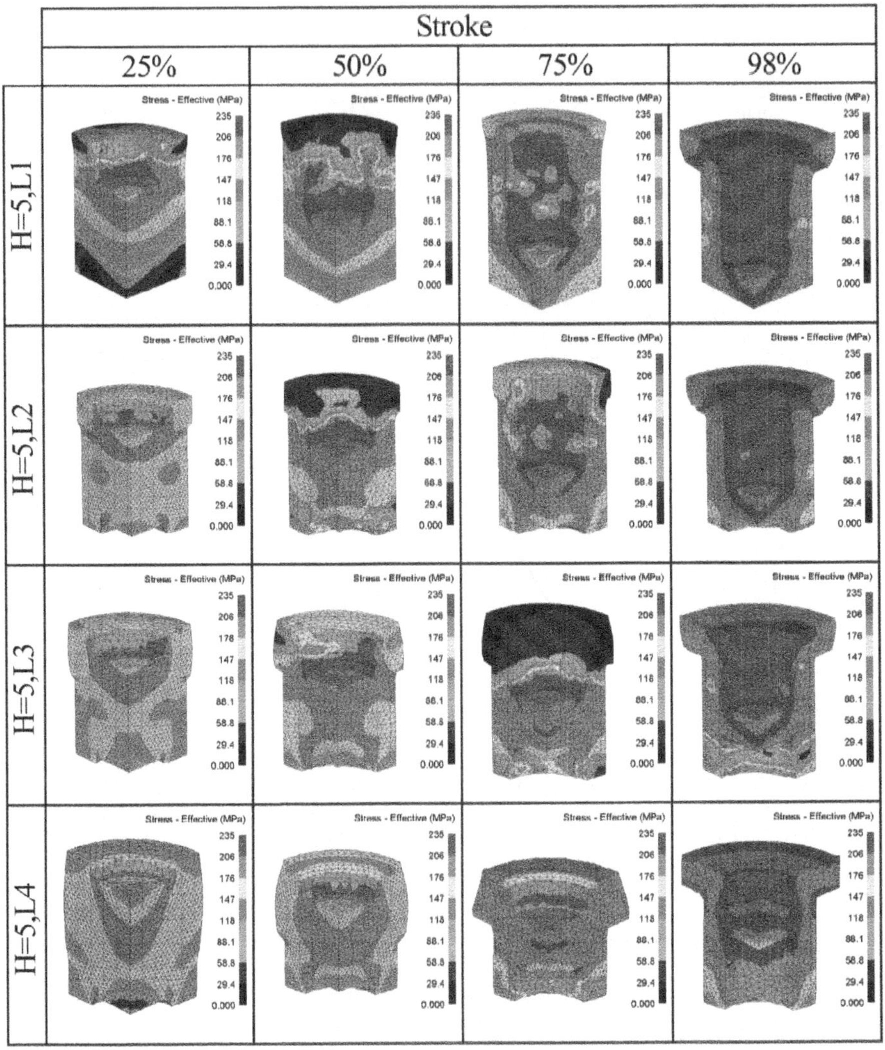

Fig. 3.12 Stress level and its distribution in the design scenarios with different locations of patch-up

on strain distribution. In addition, based on the graphs presented in Fig. 3.16, there is no obvious trend found regarding the effect of patch-up thickness and location on the maximum effective strain. Among those different design scenarios, the scenarios with the patch-up thickness of 7 mm and the height level of L4 give the smallest effective strain.

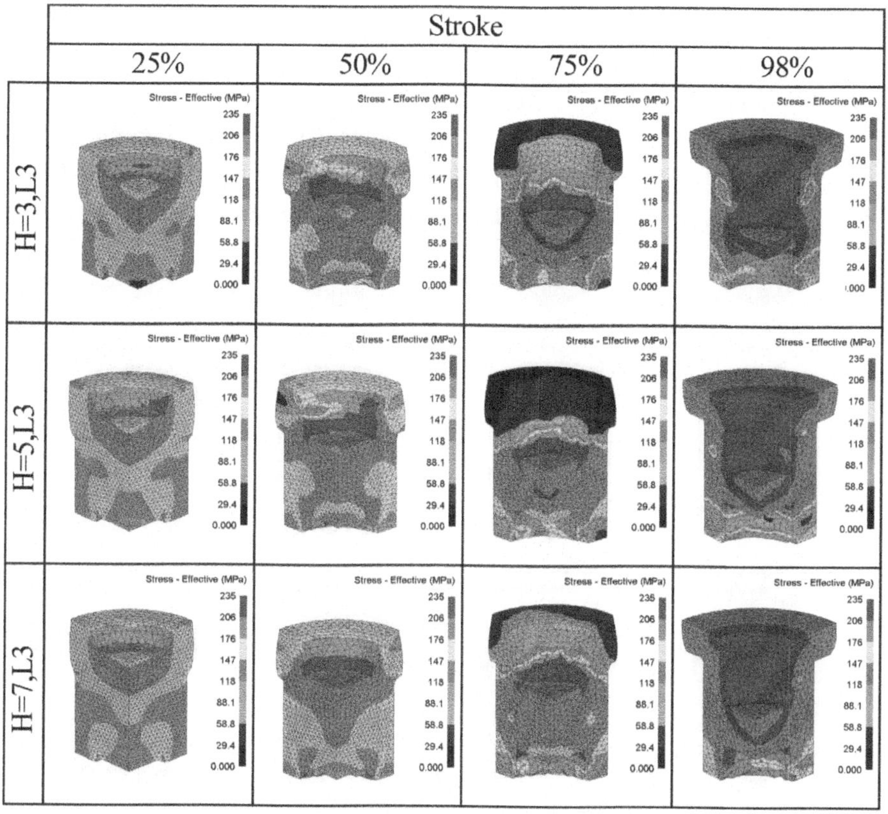

Fig. 3.13 Stress distribution in the scenarios with different patch-up thicknesses

E. Damage Factor

In metal-forming process, if the deformation exceeds the deformation capacity of materials, ductile fracture would happen, which is usually generated via the formation and coalescence of voids to form microcracks, and their further growth and propagation would lead to the occurrence of fracture and failure. In plastic deformation of materials, damage factor is defined to predict fracture occurrence in deformation process; meanwhile, it increases with deformation. When the damage factor reaches its critical value, fracture occurs. The critical value of damage factor can be determined through physical experimentation. In addition, damage factor, designated as D_f, is defined based on the Cockcroft & Latham (C&L) criterion in the following [18]

$$D_f = \int \sigma^* / \sigma \, de \tag{3.1}$$

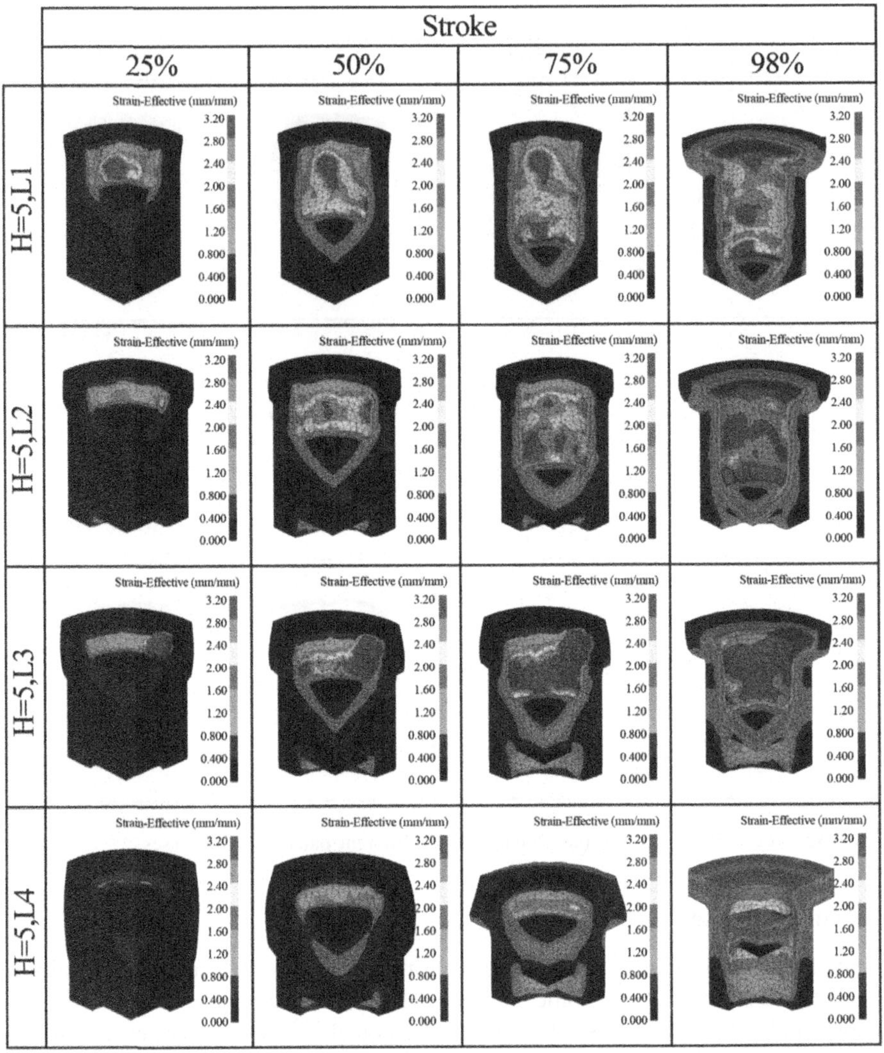

Fig. 3.14 The effective strain distribution in the scenarios with different patch-up locations

where is σ^* the tensile maximum principal stress, σ is the effective stress, and *de* is the effective strain increment. The larger the damage factor, the more potential the ductile fracture happens. Through comparison of its value, the design can be evaluated and assessed.

As shown in Fig. 3.17, high damage factor is located at the upper inner cup surface and the inner surface of the flange edge. The change of patch-up thickness does not have a significant effect on the distribution of damage factor of the formed part, as shown in Fig. 3.18. The graphs in Fig. 3.19 show the influence of the

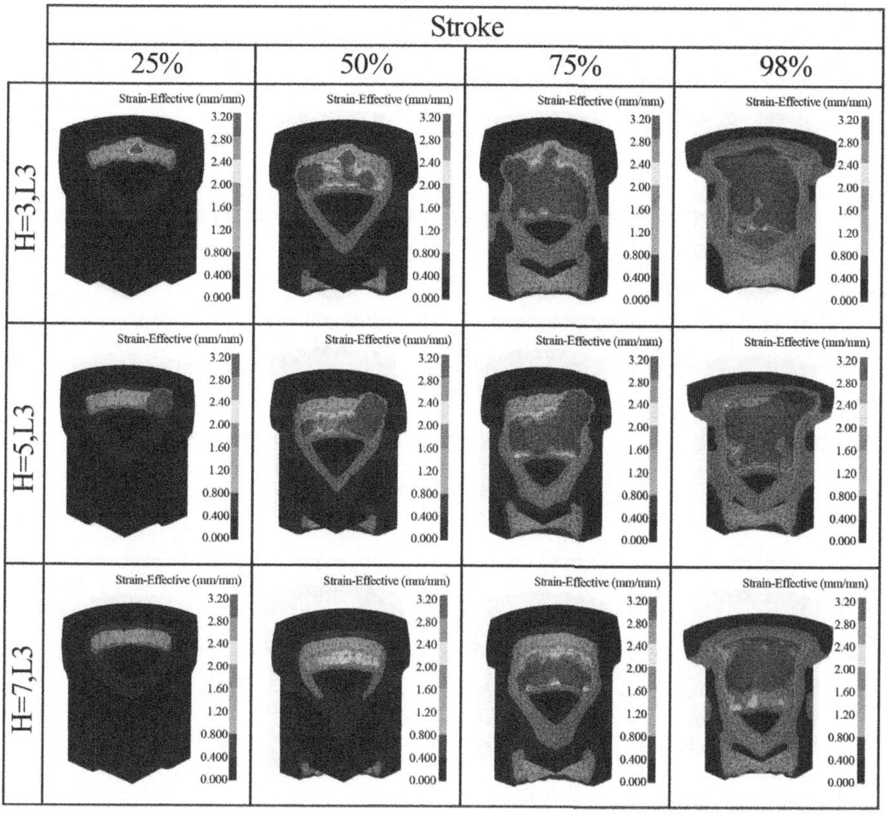

Fig. 3.15 The effective strain distribution in the scenarios with different patch-up thicknesses

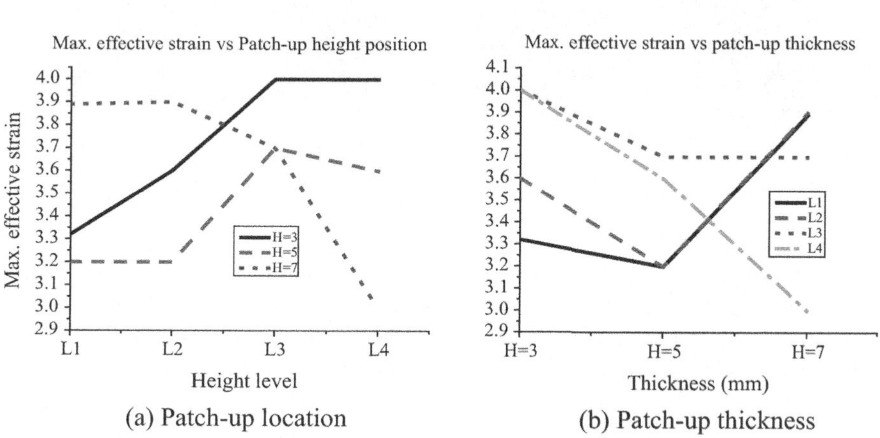

Fig. 3.16 The maximum effective strain in different design scenarios

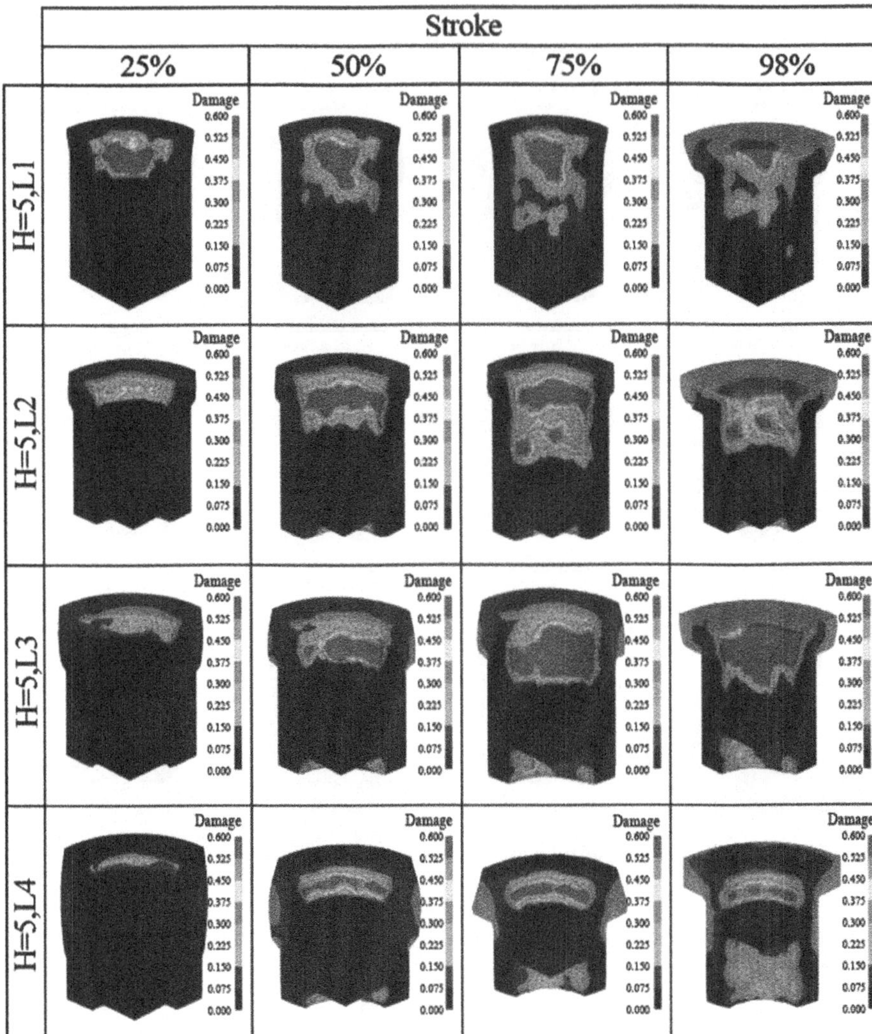

Fig. 3.17 The damage factor distribution in the scenarios with different patch-up locations

patch-up thickness and height level on the maximum damage factor. There is no an obvious tendency found in terms of the influence of patch-up thickness and height level on the maximum damage factor. Among different simulated design scenarios, the case with the patch-up thickness of 7 mm and the height level of L4 gives the lowest damage factor value.

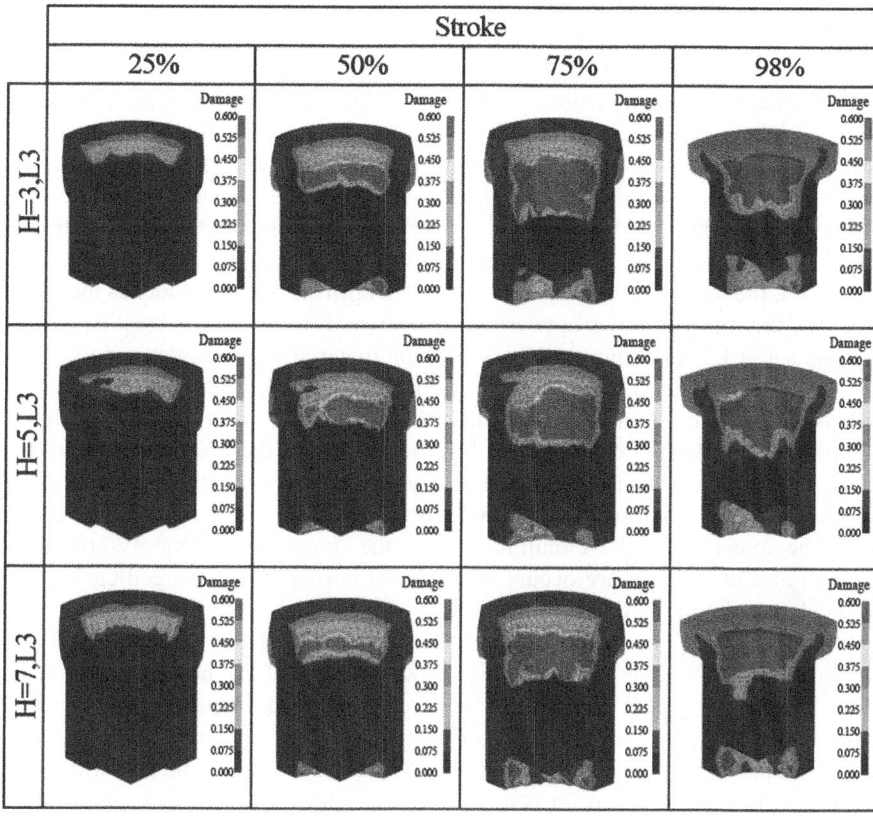

Fig. 3.18 The damage factor distribution in the scenarios with different patch-up thickness designs

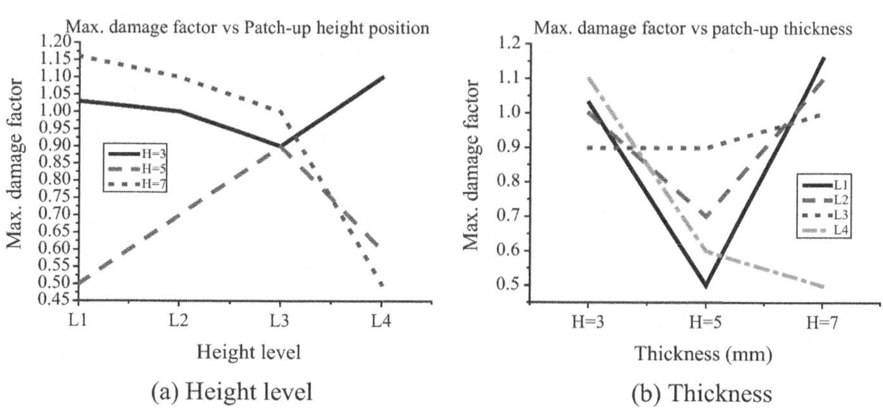

Fig. 3.19 Influence of the patch-up location and height on the maximum damage factor

F. Material Flow

The material flows in different design scenarios are also different. How the patch-up location affects the material flow is studied via FE simulation. Figure 3.20 shows the material flows in the cases with different patch-up locations. It is found the material flow pattern does not change much in different designs, but a slight difference in material flow could generate different flowline and deformation texture, which could further affect the quality of the deformed parts, or even formation of defect. For the case with the patch-up located at bottom level (L1), as shown in Fig. 3.20a, the material flows upward. Once the up-front material touches the punch shoulder, the material flows outward and downward. In the case with the patch-up location set at L2, as shown in Fig. 3.20b, the material flows downward to fill the bottom cavity first and then flows upward and outward to the cavity under the shoulder. Once the bottom cavity is fully filled up, all the materials flow upward to fill up the upper flanged cavity. In the last forming stage, the materials at the corner of the bottom surface in the flanged cavity experience the strong friction at the bottom surface of the upper flanged cavity between the material and tooling surface, while the materials at the circumference in the upper flanged cavity still move outward and thus a fracture occurs at the lower corner of the flanged cavity. The physical experiment shows this in the case of [H5, L2 ($L = 3.75$)] in Fig. 3.7.

For the case with the patch-up height of $L = 7.5$, viz. [$H = 5$, $L = 7.5$ (L3)], as shown in Fig. 3.20c, the material flows downward to fill the bottom cavity at the filling percentage of 25, 59, and 75 %. Even though the materials flow into the upper flanged cavity, the materials still flow downward once the bottom cavity is not filled up. In the last stage, the material flows quite smoothly and uniformly such that defect-free part can be formed. In the case with the patch-up location of L4 ($L = 15$ mm), as shown in Fig. 3.20d, in the early stage such as the filling percentage of 25 and 50, all the materials flow downward and outward to fill the bottom cavity and the upper flanged cavity, respectively. At the filling percentage of 75, a curved surface is formed at the inner surface of the flange of the workpiece. When the material touches the punch shoulder before the bottom cavity is fully filled, the materials under the punch flanged shoulder flow downward and outward, while the materials under the punch head also flow downward and outward. However, the unbalance of the downward flows and curved inner surface lead to a flow-induced defect near the corner of punch. This is collaborated by the photograph shown in Fig. 3.7.

Based on the presented case study, it is found that the design of the deformed part is the first and most critical step in the whole design cycle. A subtle change of metal-deformed part design would lead to totally different material flow behaviors and deformation characteristics. For the given case, there are only two design parameters with the levels of 3 and 4 to be considered. The process performance and the quality of the deformed parts, however, are quite different. For the metal-deformed part design with more parameters and more levels involved, the design space will be larger, or even infinite, and the evaluation of the designs by using the traditional approaches such as analytical method or know-how based

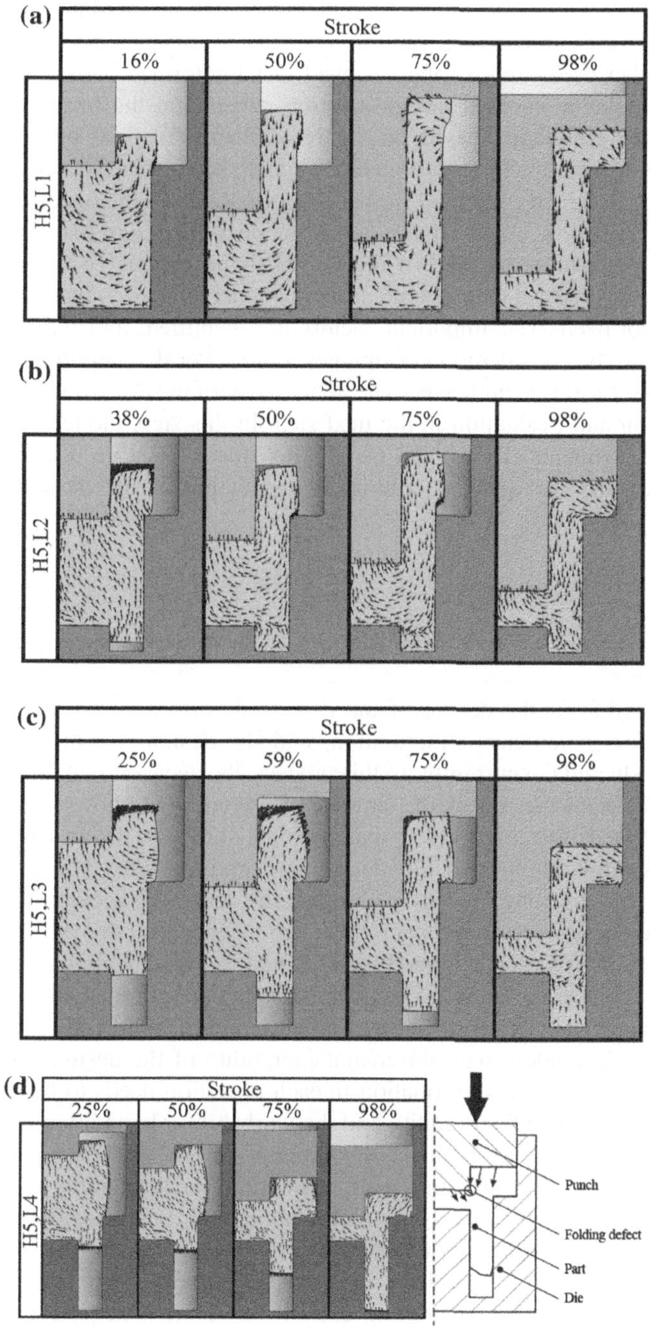

Fig. 3.20 The material flows in different design scenarios

empirical approach would be quite difficult, if not impossible. FE simulation, however, provides an efficient solution to reveal the deformation and flow behaviors and further the quality of the designed metal-deformed parts.

In FE simulation-enabled design solution generation, the metal-deformed part design can be conducted based on the requirements of final products from the perspectives of manufacturability, geometry and shape, dimensional accuracy, the quality of the designed parts, and further the formability of the materials to be used. In this design process, how to figure out all the possible design parameters, as those used in the case study part shown in Fig. 3.3, is important. To reduce the design space to be explored, all the possible design parameters need to be evaluated, sorted out, and prioritized. The important factors are identified, and the less important factors will not be considered in design generation. For the identified design space, the preliminary assessment needs to be done to determine the promising designs for FE simulation and evaluation in the next step. In this step, the techniques such as design of experiments (DoEs) can be used to efficiently scale down the potential design space and the detailed criteria for evaluation of the designs need to be developed.

3.2.3 Process and Die Design

Upon completion of the design of the deformed parts, forming process determination, process parameter configuration, and die design can then be conducted simultaneously and concurrently. In this process, the decision-making in design can also be made with the aid of FE simulation. Figure 3.21 shows the workflow in simulation-based metal-deformed part design, process determination, process parameter configuration, and die design and optimization. The process design determines what forming process is used and the detailed process route. In detail, it specifies how many forming operations are needed. In this design stage, the manufacturability of each designed preform and the workability of the material to be used in each operation need to be quantitatively determined such as the operation in this process route can be efficiently evaluated and the defect-free preform or part can be made. In tandem with the manufacturability of the design and the workability of the material, the deformation in each operation needs to be maximized in terms of both the manufacturability and the workability. In addition, the predefined criteria are employed to evaluate the process design from the perspectives of manufacturability of the designed part, workability or formability of the material, material flow behaviors, defect-free deformation, and dimensional accuracy of the deformed part, etc. On the other hand, it must be highlighted that FE simulation cannot generate design solutions in terms of process determination, process parameter configuration, and tooling design; however, it helps evaluate the performance of the designs against the predefined criteria. Through evaluation and assessment of the designs in different aspects as mentioned above and the entire

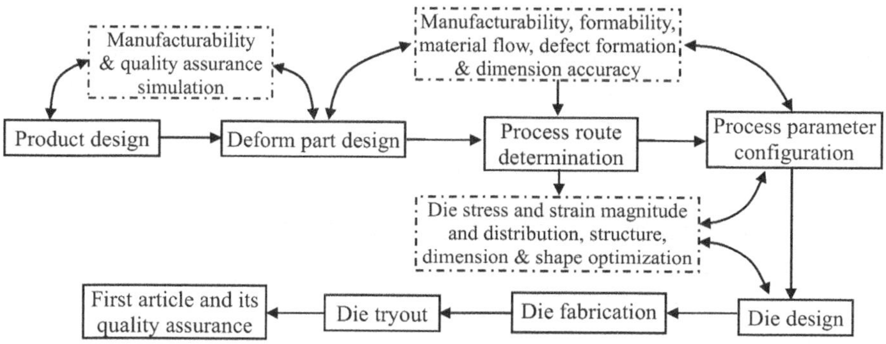

Fig. 3.21 Workflow in simulation-based process and die design

forming system, the design of the deformed parts may be necessary to be revised and the revision of the original product design needs to be done if necessary.

Regarding the process parameter determination, it details the process working parameters such as working temperature, deformation velocity, friction condition, and the deformation amount in each operation. These parameters are considered as the external process parameters since they are preset as the input to the forming system. In addition, these process parameters determine the internal process state variables, which are the physical variables of the deformation body such as strain, strain rate, stress, temperature, and microstructure, in the deformation process. Process state variables affect the process behavior and performance and further the dimensional accuracy and product quality of the deformed parts. How the external process working parameters affect the deformation behaviors of materials and the performance of a forming system can be represented by the internal process state variables.

For determination of process working parameters, different specified values of the process parameters constitute different configurations and the process performances of different configurations are different. FE simulation can evaluate those different process parameter configurations and identifies the good, better, and best performance of the process parameter configurations in terms of material flow, defect-free forming, material filling, and dimensional accuracy of the deformed parts. Eventually, the goal of this simulation-aided design process is to determine the best process parameter configuration or the reasonable, feasible, or even optimal process parameter configuration window for fabrication of the similar parts with the same or similar material and geometry for practical applications.

For FE simulation-aided design of die structure and components, the deformed part design, process route, and process parameter configuration are considered to be the input to this design process. FE simulation reveals the physical state variables of the die and its components to be simulated in the course of forming process and determines whether the geometry and shape of die components and the assembly and structure relationship are reasonable from the perspectives of deformation load, die stress and strain distribution, and the dimensional accuracy and defect-free of the deformed parts. The detailed FE simulation-aided design of die will be discussed in Chap. 4.

3.2.4 Simulation-Aided Evaluation of Metal-Forming Systems

Figure 3.22 shows the FE simulation-aided design solution generation in integrated product and process development. In the figure, the product design, process determination, and tooling and forming system design constitute the whole forming system design. The design of metal-formed products needs to consider the voice of the customer (VoC) and the detailed functional requirements and design specifications of the products. The product geometry conceptualization needs to consider the manufacturability of the designed part, the efficient forming process by which the design can be fabricated, and the formability of the materials to be used for making the designed part. Therefore, it is a process and know-how based conceptualization. After that, the process and tooling design needs to be conducted to realize the preliminary design of the forming system, as shown in Fig. 3.22. In this stage, the CAD representation of the whole designed system from geometry perspective is needed. The entire system performance and the final outcome of the system can then be evaluated and verified through CAE simulation.

In this simulation process, the whole system is first modeled through establishing the physical, mathematical, and computational models of the forming system and these models are then considered as an input to the simulation conducted by a FE simulation system. Upon completion of the simulation, the geometric-, physical-, deformation-, microstructure- and property-related data and information are extracted out and available for design verification and evaluation. How to use these data and information to evaluate the designs, however, needs systematic evaluation criteria. The quantitative evaluation criteria are thus needed to be developed based on the domain-specific know-how and engineering knowledge of metal-forming processes and forming system design. The system performances from the perspectives of product manufacturability and quality, process efficiency, tooling service life and cost, etc., are assessed based on the evaluation criteria, and the optimal system design can eventually be identified from all the potential design alternatives.

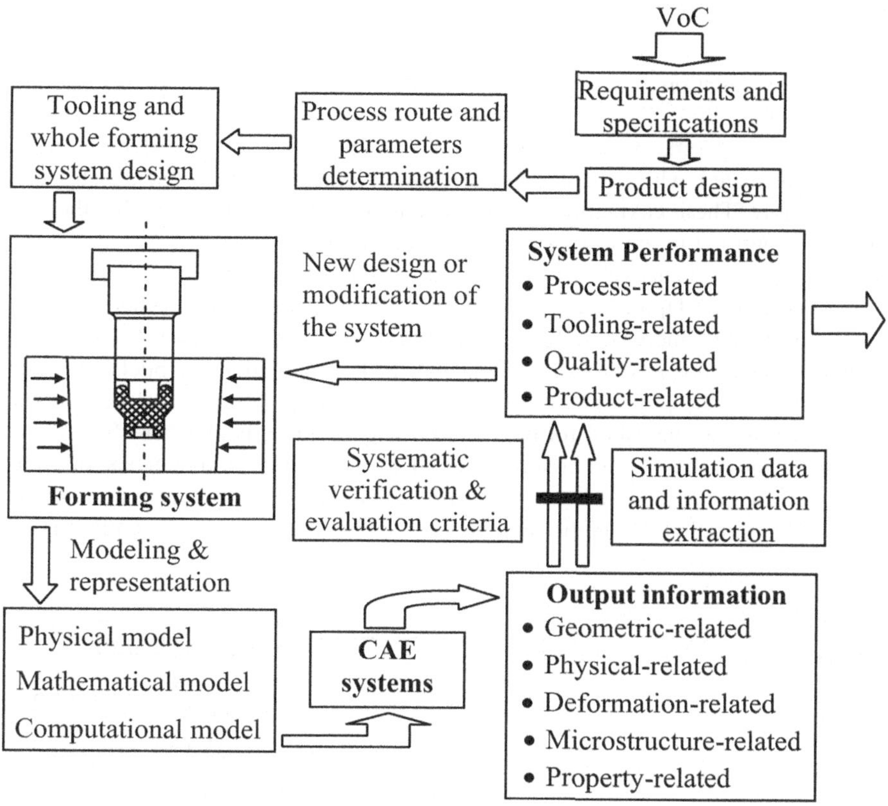

Fig. 3.22 FE simulation-enabled integrated design and development of product, process, tooling, and forming system [7]

3.3 Realization of CAE Simulation

Under the above-proposed integrated product, process, tooling and forming system design, and development paradigm, CAE or more detailed FE simulation is an enabling technology to reveal the physical variables and system performances of the integrated forming system. The deformed part design, process determination and process parameter configuration, tooling structure, and the entire forming system can be simulated and evaluated through the integrated simulation. The good, better, and best solutions can then be systemically evaluated and identified against the predefined criteria.

3.3.1 Simulation Procedure

The CAE simulation procedure for this simulation-enabled design paradigm is shown in Fig. 3.23. First of all, the metal-forming system to be simulated is modeled from geometry perspective. In detail, the CAD models of the forming system are created, which include deformation billet, tooling components, structure, and assembly. These CAD models should be in a suitable format such that they can easily be imported into a CAE system for simulation. In general, CAD models are converted into a data exchange format such as IGES, STEP, or STL for model export and import. Upon completion of the CAD model generation, there are three simulation-related models to be generated for the given forming system to be evaluated. They are physical, mathematical, and computational models. As mentioned in Chap. 2, physical model idealizes and simplifies the real engineering problems such as the metal-forming systems to be investigated here based on a certain physical theory and assumptions. The mathematical model specifies the mathematical equations such as governing equations, the physical model should follow. This model also specifies and details the boundary and initial conditions and

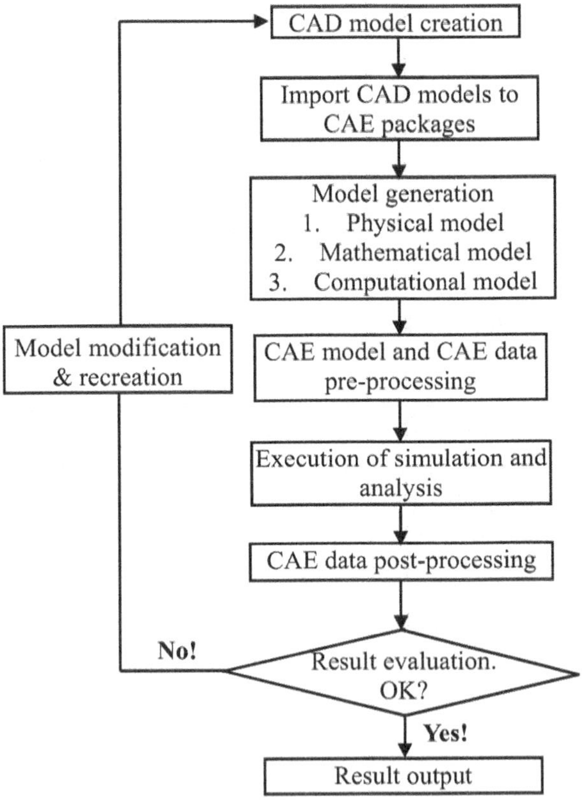

Fig. 3.23 CAE simulation procedure [4]

constraints of the systems to be studied. The computational model, on the other hand, describes the element types, mesh density and solution parameters, and the data related to implementation and execution of simulation. The solution parameters further provide the detailed calculation tolerance, error bound, iteration specification, and convergence criterion. Generally, most FE simulation systems have part of built-in content of these models, but users still need to prepare and input most of the model information into the FE simulation systems to conduct simulation.

In CAE simulation process, there are four main steps:

(1) Preprocessing.
(2) Simulation execution.
(3) Post-processing.
(4) Results analysis and evaluation.

Upon completion of the simulation, the calculated results are analyzed and evaluated. If the results and solutions are not satisfactory, the changes and modifications are generally suggested for the revision of the metal-forming system design in terms of metal-deformed part design, tooling design, process configuration, and material selection, and the next round simulation can be conducted. This design process is iterative until all the system design requirements and specifications are met and the optimal design is generated.

3.3.2 *Integrated Simulation Framework*

The integrated numerical simulation of a metal-forming system, mainly including metal-forming process and tooling deformation, provides a systematic approach for simultaneously modeling mechanical deformation behaviors, metallurgical phenomena, and thermal phenomena in the forming process of the system. Therefore, the interaction and interplay of these physical behaviors and phenomena are taken into consideration. Simulation and analysis of tooling deformation and stress and strain distribution during the forming process can be conducted simultaneously with the simulation of forming process. From the perspective of metal-forming process, the physical phenomena and mechanical flow behaviors of the deformation workpiece affect die deformation and the stress and strain distribution in the die. Die deformation, on the other hand, also affects the plastic deformation and flow of the workpiece material. The systematic design of metal-forming systems thus needs an integrated CAE simulation approach to simulating the forming of the deformed part and tooling deformation simultaneously.

Figure 3.24 schematically shows an integrated simulation and analysis framework for simulation of forming process of workpiece and deformation and stress analysis of die components and its structure in forming process. In this simulation process, there are three main issues to be addressed, viz. metallurgical phenomena

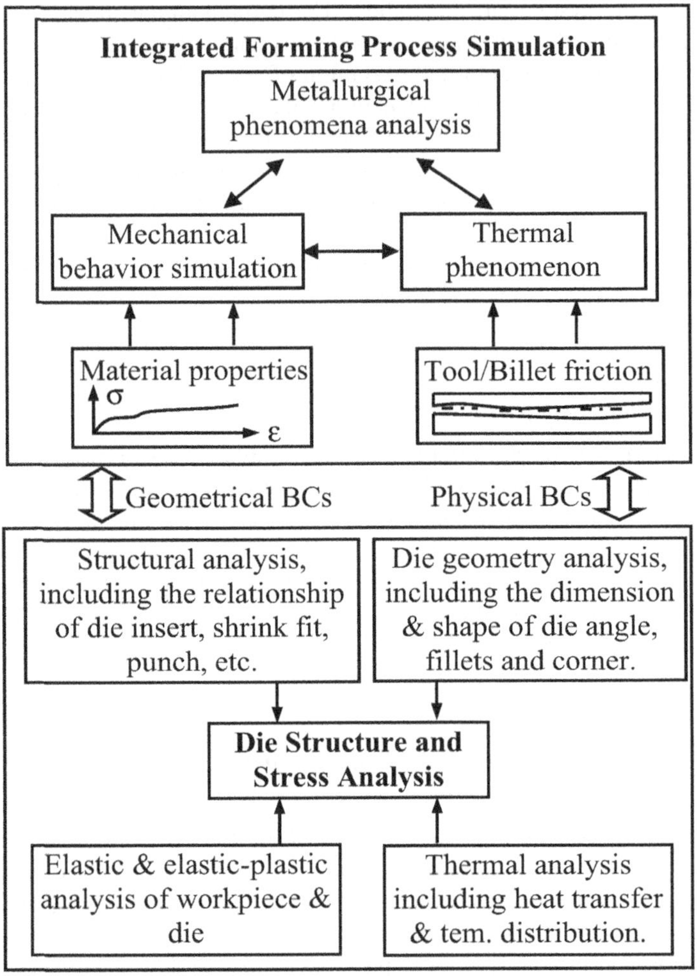

Fig. 3.24 Integrated simulation framework [4]

modeling, mechanical behavior analysis, and thermal phenomena simulation. The metallurgical phenomena modeling is focused on microstructure prediction and evolution analysis of the deforming body in the forming process. The mechanical behavior analysis reveals the plastic flow behavior of the workpiece. The thermal phenomena analysis, on the other hand, investigates the heat convection, conduction and radiation, and the temperature distribution in workpiece and tooling.

The interaction and dependency of the above-mentioned phenomena and behaviors are also shown in the framework. The plastic work generated in the forming process can be converted into thermal energy, which could result in the change of thermal behaviors and phenomena, including the heat transfer and temperature distribution in the deforming body. The change of thermal behavior could further lead to the evolution of microstructure and metallurgical characteristics of the deforming body in forming process. The evolution of microstructure and the change of properties of the deforming body would in turn affect its mechanical and plastic flow behaviors of the workpiece due to the variation of the material properties and the tribological characteristics in-between the tooling and the deforming body. These mutual interaction and dependency of the physical phenomena and behaviors need to be considered in simulation process to ensure the high simulation fidelity.

From die structure and stress analysis perspective, there are a few issues to be addressed. Die structural analysis studies the structure relationship of die insert, casing, punch, and other working components and their interference such as shrink fit in between die insert and casing. Die geometry analysis determines the shape and geometry of die components such as the dimension and shape of die angle, fillet, and corner. Stress and strain analysis reveals the magnitude and distribution of stress and strain in die set. Thermal analysis investigates the heat transfer and temperature distribution inside and in between the workpiece and die. From optimal design point of view, these analyses help generate the best solution in metal-forming system design.

In the above-stated simulations and analyses, the dynamic physical and geometrical boundary conditions in-between the simulations of workpiece deformation and die structure and stress analysis need to be established in each simulation step. The physical boundary conditions include mechanical boundary conditions (BCs) and thermal BCs. To determine the BCs in this integrated simulation and analysis scenario, the interpolation of the state variables such as stress, strain, pressure, or temperature from one simulation kernel to the other needs to be conducted. In detail, the determination of boundary conditions for die structure and stress analysis is done based on the extracted BC's information from the deformation process simulation of the workpiece. In the deformation process simulation of the workpiece, on the other hand, the BC's information from die structure and deformation analysis is interpolated for the establishment of the BCs in simulation of the deformation of workpiece.

3.4 Evaluation Methodology

To systematically evaluate the integrated conceptual design of metal-formed part, forming process, tooling design, and the entire forming system, an efficient evaluation methodology is needed to be developed. The methodology presented in the following considers different factors from different perspectives to evaluate the general performance of the designed forming system. In detail, a few main evaluation parameters are listed in the following.

3.4.1 Deformation Load

Deformation load is one of the key parameters which affects die design and forming equipment selection. It should be as small as possible since the smaller deformation load requires a smaller capacity of forming equipment and generates lower die stress and strain. The deformation load would further affect production cost and energy consumption. It is the combined performance representation of billet design, process route determination, process parameter configuration, the deformed part geometry and shape design, die structure and component assembly relationship, and geometry configuration. It also affects the selection of die and billet materials. The deformation load is thus selected as an evaluation parameter in the used evaluation methodology in this book. The deformation load can be determined based on the following equation [17]:

$$F = \sum_{e} \int_{Ve} (2/3\bar{\sigma}/\bar{\varepsilon}B^T\mathrm{DB})\mathrm{d}V + \sum_{e} \int_{Ve} (\mathrm{GB}^T\mathrm{CC}^T)\mathrm{d}V \tag{3.2}$$

where $\bar{\sigma}$ is the effective stress and equals to the flow stress of material in forming process. $\bar{\varepsilon}$ is the effective strain. B is the strain rate–nodal velocity relationship matrix in viscoplastic FEM or nodal displacement in rigid-viscoplastic FEM. G is a penalty constant or Lagrange multiplier to ensure the volume constancy to be met in forming process. D and C are a matrix and a vector of constant components, respectively. The maximum deformation load $F_{\max}$ in the entire forming process is considered as a critical evaluation parameter.

Equation (3.2) is used to calculate the deformation load by numerical discretization via FEM as it is a nonlinear equation from the perspectives of nonlinearity of both geometry and material and can only be solved by numerical method. FEM is an efficient numerical method. Currently, the determination of deformation load is available in commercial metal-forming FE simulation systems, and the results can be directly provided through simulation.

To verify the accuracy of the deformation load predicted by simulation, Fig. 3.25a shows a combined extruded part of aluminum alloy 6061 and its deformation load is predicted by simulation and experiment. Two FE simulation systems, viz. LS-DYNA and DEFORM, are used. Figure 3.25b shows the deformation loads. The figure illustrates that the experimental and simulation results have a good agreement and the maximum difference is about 8 %, which is an acceptable accuracy for deformation load determination in metal-forming arena.

3.4.2 Effective Strain

The effective strain represents an accumulated deformation quantity in plastic deformation body in deformation process. The energy needed to deform the billet to

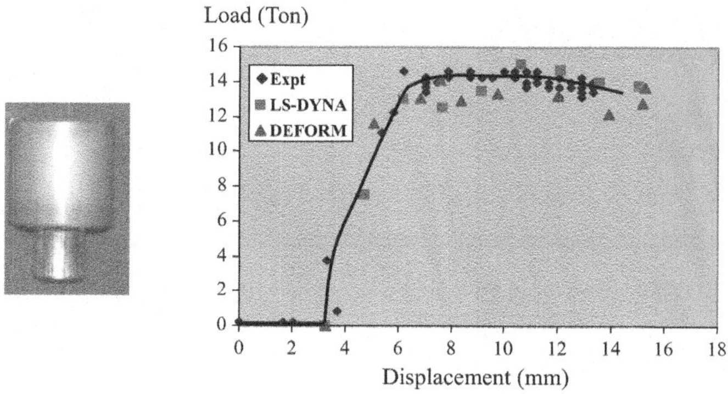

(a) Deformed part (b) Deformation load

Fig. 3.25 Comparison of deformation loads predicted by simulation and experiment [7]

a specific geometry can also be represented by the effective strain defined in the following:

$$\bar{\varepsilon} = (2/3)^{1/2}(\varepsilon_{ij}\varepsilon_{ij})^{1/2} \tag{3.3}$$

where ε_{ij} is the detailed strain component. The effective strain should be as small as possible. The maximum effective strain $\bar{\varepsilon}_{max}$ is thus considered as an evaluation parameter from the deformation point of view.

Figure 3.26a shows a cross section of the forward–backward extruded part, as shown in Fig. 3.25a, where two shear bands are clearly illustrated. The shear band streams the material flow into two. One is the material flow into the upper cup of the part, while the other is the material flow into the lower bar of the part. The shear band is actually the severe deformation zone in the workpiece, which can be represented by the effective strain. The figure illustrates a clear shear deformation zone in-between the punch and extrusion die. The simulation result of the effective strain and its distribution is shown in Fig. 3.26b, which reveals this shear deformation zone represented by the effective strain and its distribution. It can be found a good agreement in-between the experimental and simulation results.

3.4.3 Damage Factor

In metal-forming process, ductile fracture is a common defect. When the deformation of material exceeds its ductility limit, ductile fracture may occur, which starts from microvoid to microfracture. Its further growth and coalescence would lead to the formation of macrofracture. Any microcrack or fracture in the deformed parts would cause the loss of load-carrying capacity of the parts. Therefore, product

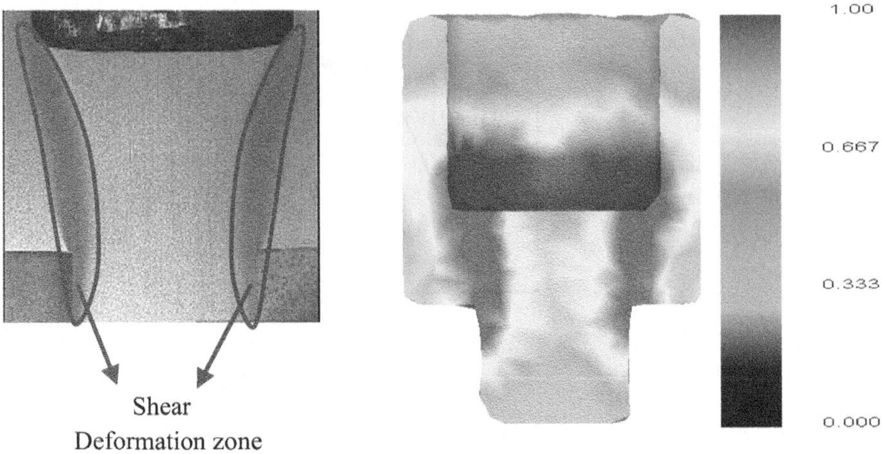

(a) Experiment (b) Effective strain distribution predicted by FEM

Fig. 3.26 Effective strain prediction and experimental analysis [7]

quality assurance requires that there is no any microfracture in the entire deformed part. In forming process, microfracture generally appears in the format of ductile fracture, whose appearance is generally decided by the manufacturability of the designed metal-deformed part and the formability of material in a specific forming process. To model the ductile fracture in forming process, a lot of efforts have been provided to establish different criteria for determination of the occurrence of ductile fracture in plastic deformation process from macroscale to microscale [19–28]. From application perspective, simple and accurate criteria could be easy to use. In this book, Cockcroft & Latham ductile fracture criterion is used by the introduction of the damage factor defined in Eq. (3.1). Through simulation, the distribution of damage factor is revealed. For a given material, its detailed damage factor value at which ductile fracture occurs is determined based on experiment. The smaller the damage factor value, the less potential the ductile fracture to occur. The simulated distribution of damage factor, however, provides a good quantitative parameter for comparison and evaluation of the designs.

3.4.4 Maximum Effective Stress

The effective stress is a combined representation of stress status in deformation process. In a deformation body, the stress tensor is designated as σ_{ij}, which represents the six independent stress variables. The effective stress formulated in the following systematically and uniquely represents the combined stress level and determines whether the yield deformation happens.

$$\bar{\sigma} = (3/2)^{1/2}(\sigma'_{ij}\sigma'_{ij})^{1/2} \tag{3.4}$$

where $\sigma'_{ij} = (\sigma_{ij} - \delta_{ij}\sigma_m)$. σ'_{ij} is the deviatoric stress tensor, σ_{ij} is the stress tensor, δ_{ij} (=1 for $i = j$ and = 0 for $i \neq j$) is the Kronecker delta, and σ_m is the hydrostatic component of the stress σ_{ij}.

In tooling service life design and analysis, tooling service life is determined by its cyclic stress for a given tooling material. The effective stress of tooling should be as low as possible such that the maximum stress does not exceed the limit of yield stress of tooling material and ensures the working stress represented by effective stress in all the tooling service cycles does not exceed its yield stress and the working condition is in elastic range. In addition, the working stress of tooling has a close relationship with its service life. It is known that 10 % decrease of working stress would lead to an increase of tooling service life for a few folds. Therefore, the effective stress $\bar{\sigma}_{\max}$ is considered as an evaluation parameter from the perspectives of judging of yielding of material and tooling service life.

3.4.5 Deformation Homogeneity

Deformation uniformity reveals the distribution of deformation in a deforming body. The design of forming process and process route should ensure the deformation as uniform as possible. The non-uniform deformation would lead to deformation localization and lead to local deformation, which could affect the microstructure and mechanical properties of the deformed part. Deformation uniformity is thus considered as an evaluation parameter in this book for the evaluation of the forming system design. The deformation homogeneity (DH) is thus proposed to represent the uniformity and designated as:

$$\mathrm{DH} = \left(\frac{1}{n}\sum_{i=1}^{n}\bar{\varepsilon}_i^2\right)^{1/2} \tag{3.5}$$

where n is the number of the sampling places defined in the deformation zone to take the effective strain. $\bar{\varepsilon}_i$ is the effective strain extracted out at the sampling location. i is the number of sampling. In real implementation of DH, the maximum and minimum effective strains in the deformation body need to be identified and determined for the determination of deformation homogeneity. For simplification, only the maximum and minimum effective strains are used for the calculation.

The above evaluation parameters represent the behaviors and performances of a forming system from different perspectives. Their maximum values are used for the evaluation of different design alternatives. To quantitatively articulate the performance of the design, a detailed evaluation criterion is introduced in the following section.

3.4.6 Evaluation Criterion

To quantitatively represent the above evaluation parameters, the following criterion is defined to evaluate the design alternatives based on the above-defined evaluation parameters. The criterion is called design index and designated as DI [4]:

$$\mathrm{DI} = k(P_i - (P_{\max} + P_{\min})/2)/(P_{\max} - P_{\min})/2 \tag{3.6}$$

where DI is the design index. P_i is the output value of a specific evaluation parameter for design alternative i. The specific evaluation parameters are defined previously. $P_{\max}$ is the maximum output value of the evaluation parameter in all the design alternatives, and $P_{\min}$ is the minimum output value of the design parameter. k is a characteristic factor, which is assigned by + (plus) when the bigger the output value, the better the designed system is; and it is − (minus) if the smaller the output value, the better the design system is. Therefore, the characteristic factor is "−" for the deformation load output. In addition, DI varies between +1 and −1. For a given evaluation parameter, it is the best design scenario when it has a value of +1 and the worst case when its value is −1.

To systematically evaluate the whole forming system based on the above-defined criteria for different evaluation parameters, the following system evaluation index is defined and designated as SI.

$$\left.\begin{aligned} \mathrm{SI}_{\mathrm{system}} &= \sum_{j}^{N} K_j \mathrm{DI}_{p_j} \\ \sum_{j}^{N} K_j &= 1 \end{aligned}\right\} \tag{3.7}$$

where $\mathrm{DI}_{\mathrm{system}}$ is the system evaluation index and DI_{p_j} is the design index for the evaluation parameter P_j. N is the number of evaluation parameters. In this evaluation approach, DI_{p_j} is $\mathrm{DI}_{F_{\max}}$, $\mathrm{DI}_{\bar{\varepsilon}\max}$, $\mathrm{DI}_{\mathrm{DF}}$, $\mathrm{DI}_{\bar{\sigma}\max}$, and $\mathrm{DI}_{\mathrm{DH}}$ for the evaluation parameters defined from Sects. 3.4.1–3.4.4. In addition, K_j is the weight number, which specifies the important level of the corresponding evaluation parameter and represents how much the evaluation parameter contributes to the whole system quantitatively. How to determine the value of weight number needs to consider the detailed priority of the individual parameter. For simplicity, the equal weight number value is chosen as 0.2 for each of the five evaluation parameters. According to the above definition, the system design index has the value between +1 and −1. When the value of SI_{system} is +1, the designed system is the best solution among all the design scenarios and it is the worst case when its value is −1.

3.5 Case Studies

To demonstrate the application of the above-presented criterion, a case study shown in Fig. 3.27 is used to illustrate how to use the above-described methodology. The case is a simple metal-deformed part, but different part designs can be generated and lead to different forming processes and forming systems based on the location of patch-up in the deformed part. For simplification, three design alternatives are generated and shown in Fig. 3.27b–d. For these three designs, each design has its own forming system and process configuration. In these three designs, the patch-up location is different. In design I, the patch-up is located at the bottom of the part. In design II, it is located at the middle of the inner small hole feature, and in the last case, the patch-up is designed at the top of the inner small hole feature. Figure 3.28 shows the forming system for the given metal-deformed part design.

In this case study, FE simulation is employed to evaluate the design performance by using DEFORM system as the simulation platform for the integrated simulation

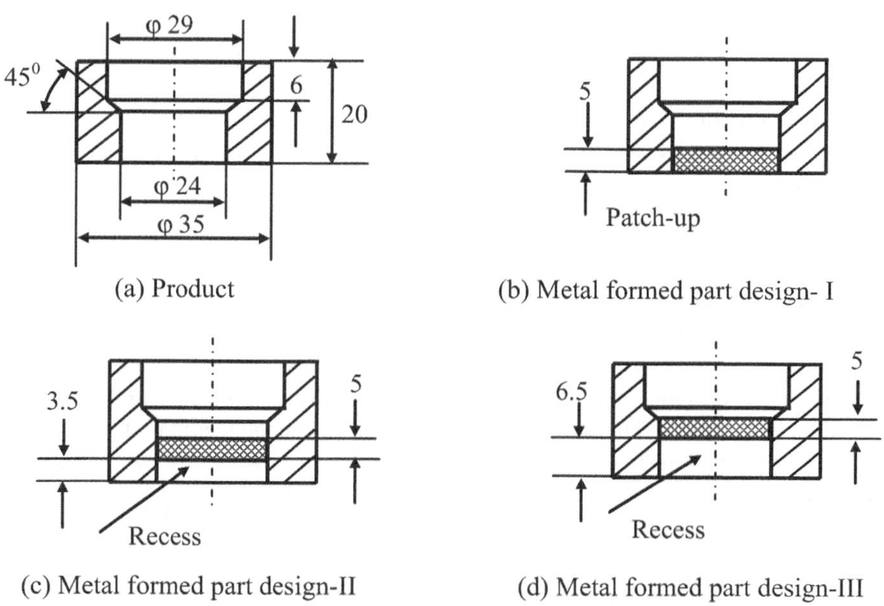

Fig. 3.27 A product and the related metal-formed part designs [7]

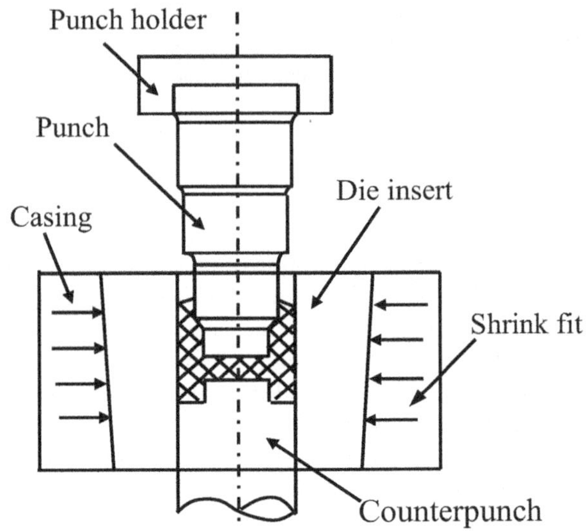

Fig. 3.28 A forming system based on the part design and process configuration [7]

of tooling and workpiece deformation. The tooling material is *M2,* and the billet material is a structural steel. For the tooling, it is considered as an elastic body and its Young's modulus is 250,000 (MPa) and Poisson's ratio γ is 0.3. For the billet material, it is considered as a plastic deformation body, and the stress–strain relationship is as follows:

$$\sigma = 150\varepsilon^{0.1} + 547 \tag{3.8}$$

where σ is the effective stress and ε is the effective strain.

For the interfacial relationship between tooling and billet, a constant friction condition is employed and represented in the following:

$$\tau = mK \tag{3.9}$$

where m is the friction factor, and K is the shear strength of the billet material. m is 0.1 in this case of the cold-forming process.

Upon completion of the FE simulation, the values of the evaluation parameters defined in the previous section can be extracted and the related *DI* and *SI* can be determined for the evaluation and validation of those conceptual designs. The detailed simulation results for the above evaluation parameters are summarized in the following.

(a) Deformation load: The deformation load in the forming process varies and thus the maximum load should be selected for the evaluation and validation of the design concept. Generally, the deformation load has the maximum when the final

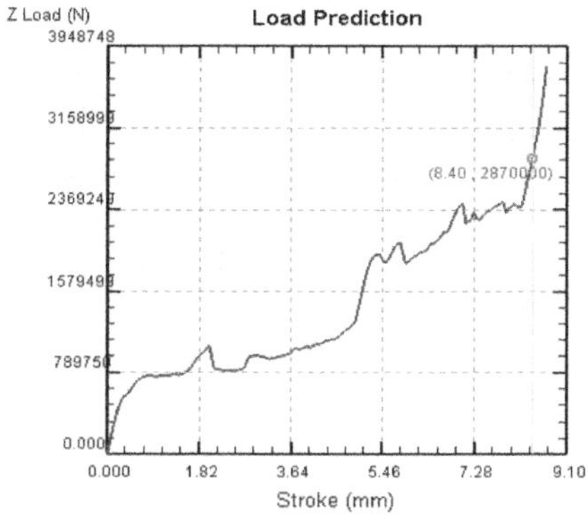

(a) Deformation load for Design I

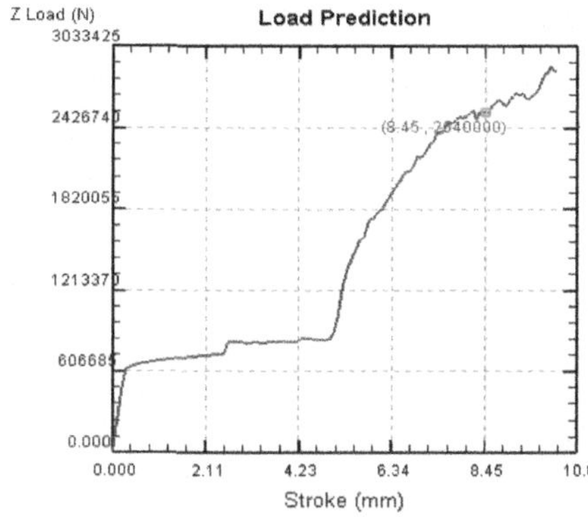

(b) Deformation load for Design II

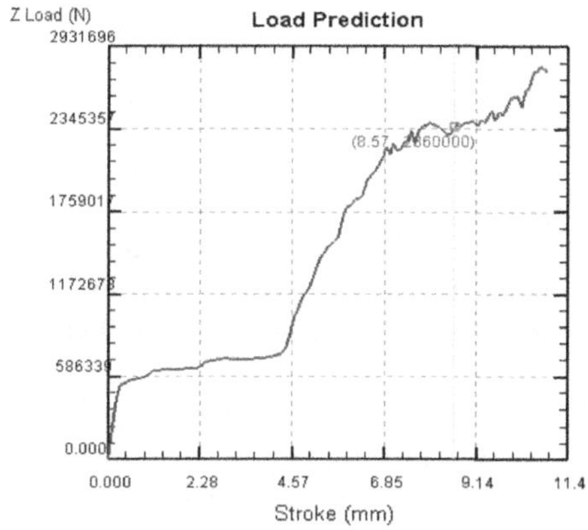

(c) Deformation load for Design III

Fig. 3.29 Deformation loads of designs I, II, and III [7]

design dimensions are achieved. Therefore, the deformation loads shown in Fig. 3.29 are extracted from designs I, II, and III when the final patch-up dimensions are achieved.

(b) Effective strain: The maximum effective strain needs to be extracted for the evaluation of the design concept. Since it is an accumulated field variable, the maximum value should appear at the last stage of the deformation when the final part dimensions are obtained. The effective strain defined here is the deformation strain of the workpiece. The detailed values of the maximum effective strain for the above three design scenarios are summarized in Table 3.3.

(c) Damage factor: The damage factor varies during the forming process. Its maximum value represents the greatest potential of the occurrence of ductile

Table 3.3 Evaluation parameters [7]

Design alternatives	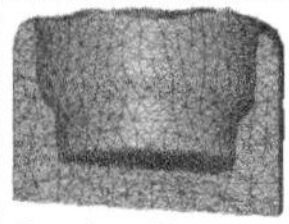 Design I	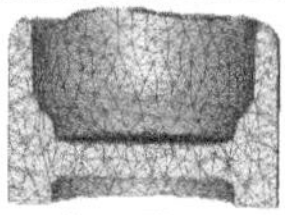 Design II	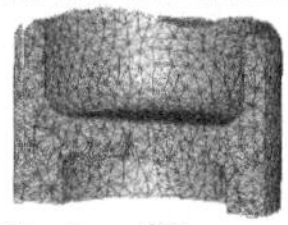 Design III
Deformation (Ton)	287	236	254
$DI_{F_{max}}$	−1.000	1.000	−0.294
Max $\bar{\varepsilon}$	2.56	2.41	2.88
$DI_{\bar{\varepsilon}\,max}$	0.181	1.000	−1.000
DF	1.52	0.90	1.72
DI_{DF}	−0.5	1.000	−1.000
Max $\bar{\sigma}$	3430	3270	3590
$DI_{\bar{\sigma}\,max}$	0	1.000	−1.000
DH	(2.56 − 0.979/2 = 0.791	(2.41 − 1.03)/2 = 0.690	(2.88 − 1.01)/2 = 0.935
DI_{DH}	0.18	1.000	−1.000
System design index	−0.228	1	−0.859

fracture in the deforming body in the specific deformation stage. The damage factor here refers to the deformation body. The maximum values of the damage factor are extracted for the three design scenarios, and the detailed values are also listed in Table 3.3.

(d) Effective stress: The effective stress defined represents the stress status of tooling in the forming system. Since tooling service life is determined by its stress. The effective stress of tooling should be as small as possible. In this case study, the maximum effective stress of the punch is selected as the representative stress of the tooling as the punch is the most vulnerable part in the entire forming system. Figure 3.30a shows the location of the maximum effective stress for the punch in design III, and Fig. 3.30b presents the variation of the maximum effective stress during the forming process. For the other two design scenarios, the maximum effective stresses are summarized in Table 3.3.

(e) Deformation homogeneity: The value of the deformation homogeneity is extracted and presented in Table 3.1 for the three design scenarios. Since the DH represents the difference level of the maximum and minimum strains, these values need to be extracted from the simulation results, and thus, the DH value can be calculated.

For the above extracted five evaluation parameters, their corresponding design indexes are calculated and listed in the table. The design indexes in the system level are then determined and listed in Table 3.1. Based on the definition of the system design index, the detailed design index values in the system level are −0.228, 1, and −0.859. Therefore, the design II is the best case, and the worst design alternative is design III.

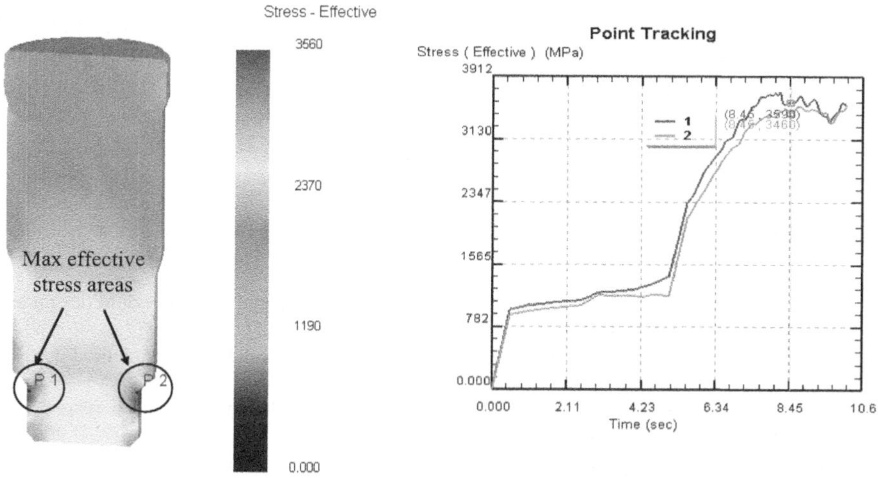

(a) σ location and the tracking points defined

(b) σ distribution at tracking points

Fig. 3.30 The maximum effective stress in the punch for design III [7]

3.6 Summary

The traditional design and development of metal-forming systems including metal-deformed part design, process determination, and tooling design, is generally a trial-and-error process based on heuristic know-hows, experiences of designers, and rule of thumb adopted by practitioners. The design generated in such a way is lack of scientific calculation and analysis and needs more times of design change and physical tryout in workshop. There are three main reasons. The first one is the potential design space may not be fully identified and explored; the second one is all the promising design solutions in the determined design space may not be figured out as potential design alternatives. The last one should be the evaluation methodology, which may not be able to thoroughly evaluate and identify the good, better, and best design solutions in up-front design stage by using the traditional approaches and methodologies.

By using the simulation-enabled design solution generation approach in metal-forming arena, the simulation-provided information can help determine design space and figure out potential designs, despite of the fact that it cannot directly address these two issues. The evaluation and validation of the designs in up-front design process, one the other hand, can be efficiently conducted with the aid by finite element simulation. The simulation-enabled forming system design and evaluation is thus able to reduce experimental work, shorten time-to-market, and cut development cost. The chapter addresses the issues related to metal-forming system design, the methodology for evaluation of the system design via integrating metal-deformed part design, forming process determination, and forming system design as a whole design, and the evaluation process. Through case studies, how to implement the simulation-enabled design evaluation is illustrated.

References

1. Fuh JYH, Zhang YF, Nee AYC, Fu MW (2004) Computer-aided injection mould design and manufacture. Marcel Dekker Inc, New York
2. Kobayashi S, Oh SI, Altan T (1989) Metal forming and the finite-element method. Oxford University Press, Oxford
3. Bariani PF, Negro TD, Bruschi S (2004) Testing and modeling of material response to deformation in bulk metal forming. Ann CIRP 53(2)
4. Fu MW, Yong MS, Tong KK, Muramatsu T (2006) A methodology for evaluation of metal forming system design and performance via CAE simulation. Int J Prod Res 44:1075–1092
5. Tong KK, Yong MS, Fu MW, Muramatsu T, Goh CS, Zhang SX (2005) A CAE enabled methodology for die fatigue life analysis and improvement. Int J Prod Res 43:131–146
6. Fu MW, Yong MS, Muramatsu T (2008) Die fatigue life design and assessment via CAE simulation. Int J Adv Manuf Technol 35:843–851
7. Fu MW, Yong MS, Tong KK, Danno A (2008) Design solution evaluation for metal forming product development. Int J Adv Manuf Technol 38(3–4):249–257
8. Lange K, Cser L, Geiger M, Kals JAG (1992) Tool life and tool quality in bulk metal forming. Ann CIRP 41(2):667–675
9. Jirathearanat S, Hartl C, Altan T (2004) Hydroforming of Y-shapes—product and process design using FEA simulation and experiments. J Mat Proc Tech 146(1):124–129
10. Geiger M, Merklein M, Kerausch M (2004) Finite element simulation of deep drawing of tailored heat treated blanks. CIRP Ann 53(2):223–226
11. Petty DM (1996) Application of process modeling—an industrial view. J Mat Proc Tech 60:421–426
12. Bariani PF, Bruschi S, Dal NT (2004) Integrating physical and numerical simulation techniques to design the hot forging process of stainless steel turbine blades. Int J Mach Tools Manuf 44(9):945–951
13. Falk B, Engel U, Geiger M (1998) Estimation of tool life in bulk metal forming based on different failure concepts. J Mat Proc Tech 80–81:602–607
14. Mungi MP, Rasane SD, Dixit PM (2003) Residual stresses in cold axisymmetric forging. J Mat Proc Tech 142:256–266
15. MacCormack C, Monagham J (2001) Failure analysis of cold forging dies using FEA. J Mat Proc Tech 117:209–215
16. Fu MW, Luo ZJ (1992) The prediction of macro-defects during the isothermal forging process by the rigid-viscoplastic finite-element method. J Mat Proc Tech 32:99–608
17. Fu MW, Luo ZJ (1995) The simulation of the visco-plastic forming process by the finite-element method. J Mat Proc Tech 55:442–447
18. Scientific forming Technologies Corporation (2014) DEFORM™ 3D manual, Version 8.11
19. Cho H, Ngaile G, Altan T (2003) Simultaneous determination of flow stress and interface friction by finite element based inverse analysis technique. Ann CIRP 52:221–224
20. Mamalis AG, Johnson W (1987) Defects in the processing of metals and composites. In: Predeleanu M (ed) Computational methods for predicting material processing defects. Elsevier, Amsterdam
21. Gelin JC, Oudin J, Ravalard Y (1985) An improved finite element method for the analysis of damage and ductile fracture in cold forming processes. Ann CIRP 34:209–212

22. Reddy NV, Dixit PM, Lal GK (1996) Central bursting and optimal die profile for axisymmetric extrusion. ASME J Manuf Sci Eng 118:579–584
23. Clift SE, Hartley P, Sturgess CEN, Rowe GW (1990) Fracture prediction in plastic deformation processes. Int J Mech Sci 32:1–17
24. Reddy NV, Dixit PM, Lal GK (2000) Ductile fracture criteria and its prediction in axisymmetric drawing. Int J Mach Tools Manuf 40:95–111
25. Gouveia BPPA, Rodrigues JMC, Martins PAF (1996) Fracture predicting in bulk metal forming. Int J Mech Sci 38:361–372
26. Gupta S, Reddy NV, Dixit PM (2003) Ductile fracture prediction in axisymmetric upsetting using continuum damage mechanics. J Mater Process Technol 141:256–265
27. Ran JQ, Fu MW (2014) A hybrid model for analysis of ductile fracture in micro-scaled plastic deformation of multiphase alloys. Int J Plast 61:1–16
28. Li H, Fu MW, Lu J, Yang H (2011) Ductile fracture: experiments and computations. Int J Plast 27:147–180

Chapter 4
Die Design and Service Life Analysis

4.1 Introduction

In metal-forming industries, die performance and service life commit to product quality, time-to-market, and production cost. Die design and manufacturing is thus an important issue in metal-forming product development. Without suitable and efficient die, metal-forming processes would be crippled or rendered totally inefficient. To have a high-quality die with long service life, the die design with low working stress, long service life, and being able to produce high-quality metal-deformed parts is crucial. Therefore, die design and die service life assessment in up-front design stage need to systematically consider different factors, which could interactively affect the performance of die and its service life, and further the quality of the metal-deformed parts using the designed die. In die design and development cycle, the eventual goal of all the design activities is to generate the optimal design with the best parameter configuration. Therefore, the simulation-enabled die design and development paradigm needs to overcome the disadvantages of the traditional die design practice and approaches as they cannot well and accurately predict and evaluate the performance and service life of die in design stage due to the difficulty to fully explore and determine die working stress and its distribution in working condition and lack of efficient methods for die life assessment.

To implement the simulation-enabled die design paradigm aiming at long service life, there are two issues to be addressed. One is to develop an efficient approach for die life assessment and the other is to implement the developed approach. To address the first issue, it is necessary to represent and formulate three physical variables, viz. die working stress and deformation strain in die working process, and the preprocessed and configured die material properties. In addition, the correlation between die service life with these factors needs to be established and realized in the developed

M.W. Fu, *Design and Development of Metal-Forming Processes and Products Aided by Finite Element Simulation*, Engineering Materials and Processes,
DOI 10.1007/978-3-319-46464-0_4

approach. Regarding the implementation of the developed approach, it needs to dynamically determine the instantaneous values of these variables and their cyclic variation in die service cycle in such a way that the die service life can be determined.

To achieve this goal, many researchers in metal-forming arena have conducted extensive exploration in this area. To name a few, Falk et al. used different concepts including local stress, local strain, and local energy approaches to estimating die service life [1]. As the approaches need to determine the local stress and strain of die in forming process, FE simulation technology is employed for revealing the dynamic stress and strain in their research. In addition, Knoerr et al. employed fatigue techniques to identifying the root causes of fatigue failure and developed a strain-based approach for estimation of die fatigue life [2]. The approach uses an elastic–plastic analysis model to determine the maximum strain amplitude and further analyzes the fatigue life based on the damage analysis concept. Fu and his colleagues employed the stress-based approach, together with off-line FE simulation, for identifying the potential fatigue location of die according to the distribution of the maximum principal stress [3]. The die service life is then assessed based on the amplitude and mean stresses at the potential fatigue area of die. All of these efforts are good exploration on die service life assessment.

From die service life improvement perspective, on the other hand, a lot of efforts have been provided for development of die life improvement methods [4–13]. Despite of the fact that there is no a general way to increase die service life, there are some tactics, which do help increase die service life. Therefore, identification and exploration of these "tactics" needs to systematically correlate the die service life with various affecting factors and mathematically represent the correlation.

In this chapter, die performance and service life are firstly summarized. A stress-based die design is presented. A general simulation-based die life assessment approach is described. The approaches employing both the stress-based and strain-based fatigue analysis techniques [14–17] to develop the relationship of die life with its various affecting factors are articulated. How the approaches are correlated with die life design is formulated and presented. Furthermore, an implementation framework of the approaches is orchestrated. From simulation point of view, the simulation process and the iteration procedure for die life estimation is described. In the end of this chapter, case studies are used to illustrate how to use the simulation-enabled die design approaches aiming at enhancing the die service life.

4.2 Die Performance and Service Life

Die performance and service life is closely linked to die design, manufacturing, die and workpiece materials, and many other factors. Figure 4.1 shows these factors from different perspectives including die design, die geometry and material selection, heat treatment of die components and die material properties configuration, workpiece and its material properties, and further to manufacturing and the tribological condition between die and workpiece [5]. Therefore, how to systematically

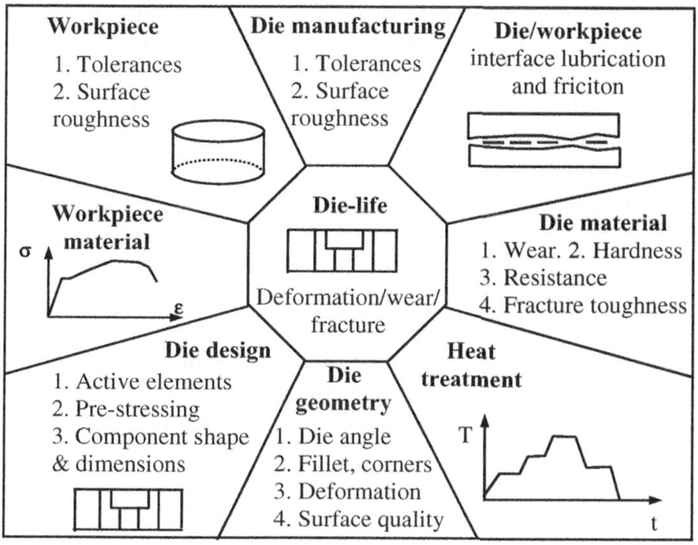

Fig. 4.1 Factors affecting die service life [5]

and integrally design die together with metal-deformed product, process sequence, and the detailed forming process conditions is a non-trivial issue.

Die service life can be defined as the number of parts produced by a die. In bulk metal-forming processes, die service life is mainly determined by three failure causes, viz. fracture, plastic deformation, and wear. In different forming processes, die failure mode and root cause could be different. In cold forging processes, the common failure modes shown in Fig. 4.2a, which include fracture, surface welding, connected defect, wear, plastic deformation, etc., are the main root causes. Among

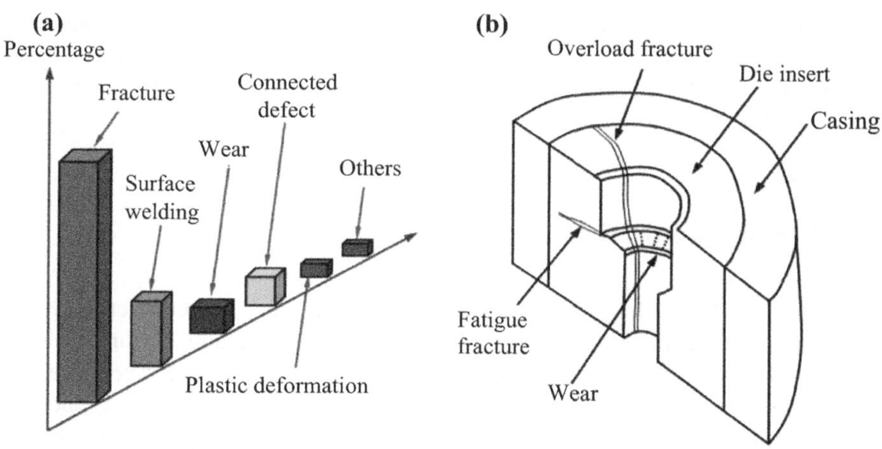

Fig. 4.2 Die failure modes in cold forging [5]

these failure modes, fracture is a dominant mode for die failure in cold forming processes, while wear is the most frequent format for die failure in warm and hot forging processes [5].

In die fracture failure, there are two common modes. One is overload fracture and the other is fatigue fracture, as shown in Fig. 4.2b. Overload fracture can be caused by a large deformation load, which exceeds the strength limit of die material in forming process. In addition, catastrophic failure via brittle fracture also happens frequently in forming processes. From fatigue failure perspective, low-cycle fatigue (LCF) failure caused by cyclically thermal and mechanical loading often occurs even the die working stress is below the yield stress of die material. But over many cycles, cyclic tensile stresses can lead to die failure. The approach to avoiding the occurrence of fatigue failure is to reduce or eliminate the tensile stress in the die. This can be done via careful design of die, control of working stress distribution and magnitude, and appropriate configuration of forming process parameters and material properties of die [6].

For die fatigue fracture, on the other hand, it is generally caused by the fact that die works under severe loading condition, which helps microcrack initiation and growth in forming process. Fatigue fractures generally undergo four stages, i.e., crack initiation, slow crack growth, accelerated crack growth, and finally rapid fracture. To avoid this phenomenon, careful design of die stress is critical as a subtle design change and a slight reduction of die working stress can significantly increase die fatigue life. Sometimes, it could be more than a few folds with just 10–15 % decrease of the working stress.

Furthermore, the detailed stress component analysis is also important in die stress design and analysis. In general, the effective stress is considered as an accepted parameter to determine the initial yielding of die if it exceeds the yield strength of die material. In addition, the maximum principal stress is also needed to be considered in some cases in which tensile stress plays a big role in die failure, especially for brittle materials. Taking carbide die material as an instance, the tensile stress of die could result in a premature LCF failure and thus the first principal stress of tensile stress should be reduced as smaller as possible. Therefore, stress component analysis is critical in troubleshooting die performance problems, determining the root causes of die failure, and studying different design alternatives for die service life improvement.

4.3 Stress-Based Die Design

In die design, the working stress of die is one of the most critical factors, which affects die service life and performance. Die service life and performance are generally determined by working stress distribution and magnitude. Therefore, one of the main design goals is to reduce die working stress in such a way to increase die service life and performance. How to design the working stress of die thus then becomes a critical and non-trivial issue. Since the working stress distribution and magnitude of die are two important design variables, the focus of die stress design is to determine the reasonable stress distribution and the small stress magnitude. For the former, FE simulation helps reveal the stress distribution of die in the entire working cycle and determines the most critical and severe stress location. For the latter, the optimal die structure design is a perfect solution. The multilayer combined die structure, such as two-layer and three-layer die structure, is widely used in industries, which can reduce the stress distribution and magnitude of die. In multilayer die design, prestress plays a crucial role in reducing the magnitude of die working stress. Therefore, the determination and design of prestress in design of multilayer die structure are discussed first in the following.

4.3.1 Prestress in Design

Die prestress is generated by shrink fitting between two interfacing components in a multilayer die structure via interference fit. The two interfacing components are generally the shrink ring and die insert. The interference fit is usually realized by the relative dimension change after the assembly between die components. By employing the phenomenon of thermal expansion, shrink ring is usually heated up and die insert is then put into the hole of shrink ring by an external force. After cooling down of the shrink ring, the die insert is strained, while the shrink ring is expanded compared to its original dimensions before fitting. A compressive prestress in die insert and a tensile stress in shrink ring are thus generated. Due to the existence of prestress in die components, the actually working stress would thus be reduced. For the die insert, the actual working stress is eventually the difference between the stress generated in forming of workpiece inside the die cavity and the compressive prestress generated by shrink fitting. The working stress of die insert generated in forming process is generally tension stress, while its prestress is compression stress. In this way, part of the working stress (tension stress) of die insert is canceled out by the prestress and the actually die working stress is thus reduced.

Figure 4.3 shows the shrink fitting between die insert and shrink ring via interference fitting. Based on the elastic deformation analysis of shrink ring and die insert, the radial and tangential stresses represented by polar coordinates system (r, θ) are $\sigma_r(r)$ and $\sigma_\theta(r)$, respectively. The following equations establish the relationship among the pressure or the so-called stress, the magnitude of interference and the geometry dimensions of die components [18]:

$$\sigma_r = \frac{b^2a^2(P_2 - P_1)}{(b^2 - a^2)r^2} + \frac{a^2P_1 - b^2P_2}{b^2 - a^2} \tag{4.1a}$$

$$\sigma_\theta = -\frac{b^2a^2(P_2 - P_1)}{(b^2 - a^2)r^2} + \frac{a^2P_1 - b^2P_2}{b^2 - a^2} \tag{4.1b}$$

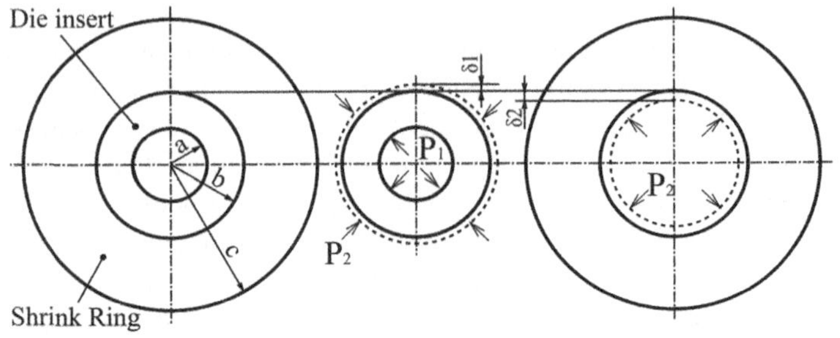

Fig. 4.3 Shrink fitting between die insert and shrink fit

where σ_r is radial stress. σ_θ is tangential stress. a is the inner radius of die insert; and b is the outer radius of die insert and also the inner radius of shrink ring after fitting. p_1 is the inner pressure at the inner hole surface while p_2 is the contact pressure at the interface in between the die insert and shrink ring. The details are shown in Fig. 4.3.

If the die insert is only subject to the outer pressure and without inner pressure, viz. $p_1 = 0$, which is the force boundary condition in shrink fitting scenario, the following equations are then obtained:

$$\sigma_r = \frac{b^2 P_2}{b^2 - a^2}\left(\frac{a^2}{r^2} - 1\right) \tag{4.2a}$$

$$\sigma_\theta = -\frac{b^2 P_2}{b^2 - a^2}\left(\frac{a^2}{r^2} + 1\right) \tag{4.2b}$$

Assuming that the magnitude of interference of single side is δ ($\delta = \delta_1 + \delta_2$, as shown in Fig. 4.3) and the contact pressure p_2 in between the shrink ring and die insert is p, it can thus be determined in the following:

$$P = \frac{E\delta}{b}\left[\frac{(b^2 - a^2)(c^2 - b^2)}{2b^2(c^2 - a^2)}\right] \tag{4.3}$$

where E is Young modulus and c is the outer radius of shrink ring.

The displacement u in radial direction of the shrink ring under elastic deformation is

$$u = -\frac{Pb}{E}\left(\frac{b^2 + a^2}{b^2 - a^2} - v\right) \tag{4.4}$$

where v is the Poisson's ratio.

For the die insert, the outer displacement at $r = b$ is designated as $u|_{r=b}$ and determined by:

$$u|_{r=b} = \delta_1 = -\frac{Pb}{E}\left(\frac{b^2 + a^2}{b^2 - a^2} - v\right) \quad (4.5)$$

For the shrink ring, the inner displacement is

$$u|_{r=b} = \delta_2 = \frac{Pb}{E}\left(\frac{c^2 + b^2}{c^2 - b^2} + v\right) \quad (4.6)$$

To compare the analytical solutions using the above equations and the outcome of FE simulation, a case study shown in Fig. 4.4 is employed. In this case study, different magnitudes of interference fit are used in analytical analysis and FE simulation. By comparing the results of both of them, the FE simulation is validated and the robustness of the simulation results is verified. This would show that the FE simulation could be directly and efficiently used in die shrink fitting design.

In addition, the dimensions and material properties of the shrink fitting case study are listed in Table 4.1. Two FE simulation systems, viz. DEFORM and ABAQUS, are used. The analytical and simulation results are presented in Figs. 4.5, 4.6, 4.7, and 4.8.

Figure 4.5 shows the comparison of the radial stress σ_r determined by analytical approach and FE simulation. The stress σ_r located at the plane passing through the cylindrical axis of the die insert is calculated. When r is 10 mm, it is located at the inner surface of the die insert, and when r is equal to 20 mm, it is the outer surface of the die insert, which is also the interfacing surface between shrink ring and die insert. The agreement between the analytical solution and simulation result is quite satisfactory, and the maximum deviation is less than 1 %. With the increase of the magnitude of interference, the absolute value of stress gradually increases as σ_r is a

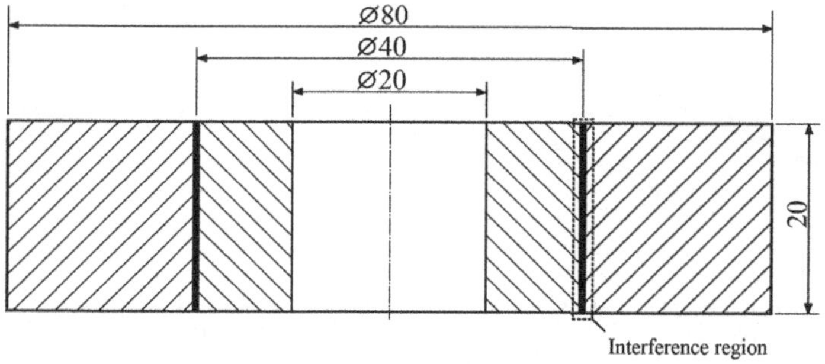

Fig. 4.4 A case study of shrink fitting

Table 4.1 Dimensions and material properties of the case study

a (mm)	b (mm)	c (mm)	E (MPa)	v
10	20	40	210,000	0.3

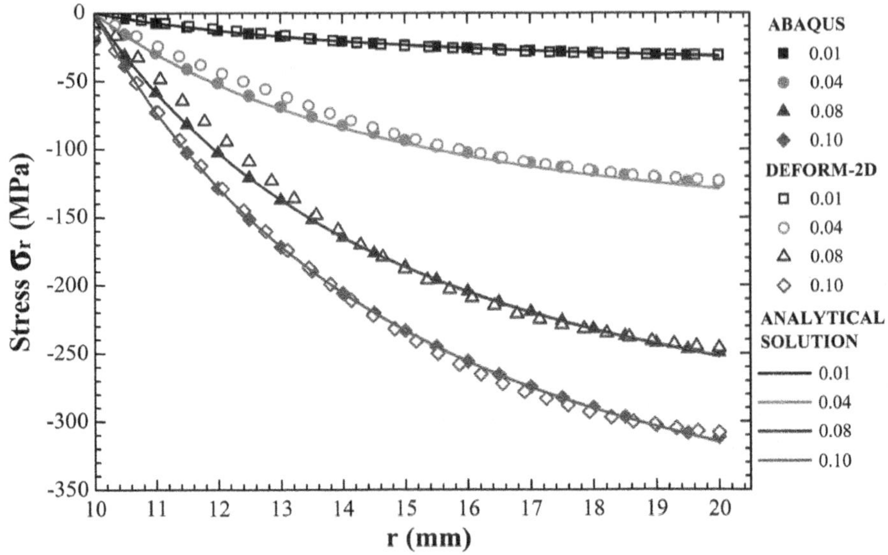

Fig. 4.5 Comparison of σ_r and its distribution in die insert between analytical solution and FEM calculation with different magnitudes of interference

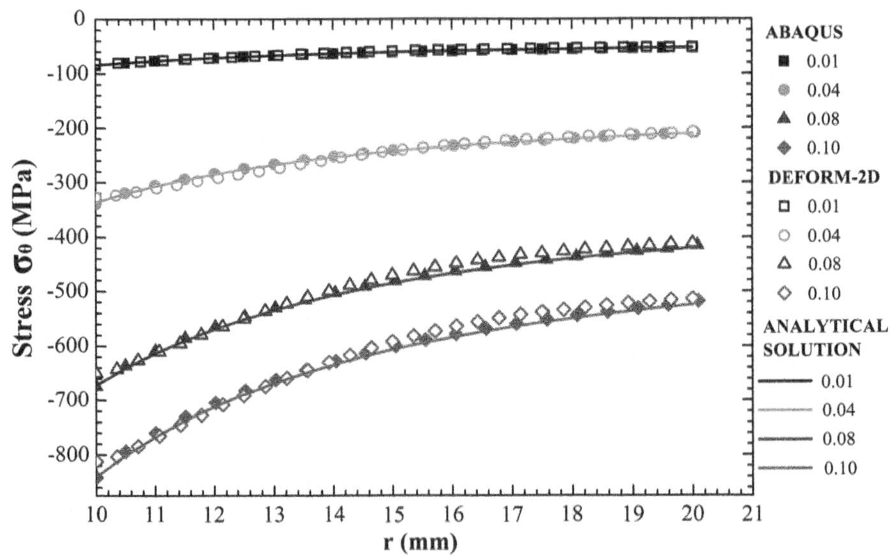

Fig. 4.6 Comparison of σ_θ and its distribution in die insert between analytical solution and FEM calculation via different magnitudes of interference

compressive stress. In the inner surface of die insert, the stress is zero, but it reaches the maximum in the outer surface of the die insert.

Figure 4.6 presents the comparison of σ_θ and its distribution in the die insert between analytical solution and FEM calculation via different magnitudes of interference at the same plane of the die insert as mentioned previously in Fig. 4.5. From the figure, it is found that σ_θ of the die insert is also a compressive stress and decreases with radius. In the inner surface of the die insert, the absolute value of σ_θ is maximum, which can effectively cancel out the tension stress generated in the die insert in forming process to a certain extent. Comparing the analytical solution with the simulation results, the maximum deviation is found less than 1.2 %.

As shown in Figs. 4.5 and 4.6, it can be seen that the magnitude of interference has a great influence on the distributions of σ_r and σ_θ and their detailed magnitudes. A suitable magnitude of interference should be determined based on the principle that the stress condition of the shrink ring is not too severe in terms of the yielding condition of its material.

Furthermore, Fig. 4.7 gives the comparison of the contact pressure in the interference surface between shrink ring and die insert calculated by analytical approach and FE simulation with different magnitudes of interference. The simulation results have a perfect agreement with analytical calculation. In addition, the contact pressure is linearly increased with the magnitude of interference. Figure 4.8 presents the comparison of the displacement determined by simulation and the analytical calculation with the magnitude of interference. The maximum deviation in between is about 5.5 %.

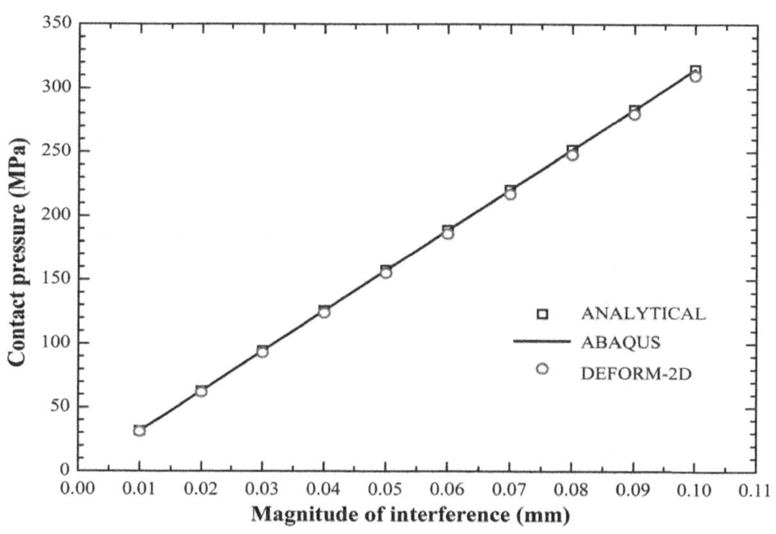

Fig. 4.7 Comparison of the variation of contact pressure with the magnitude of interference by simulation and analytical calculation

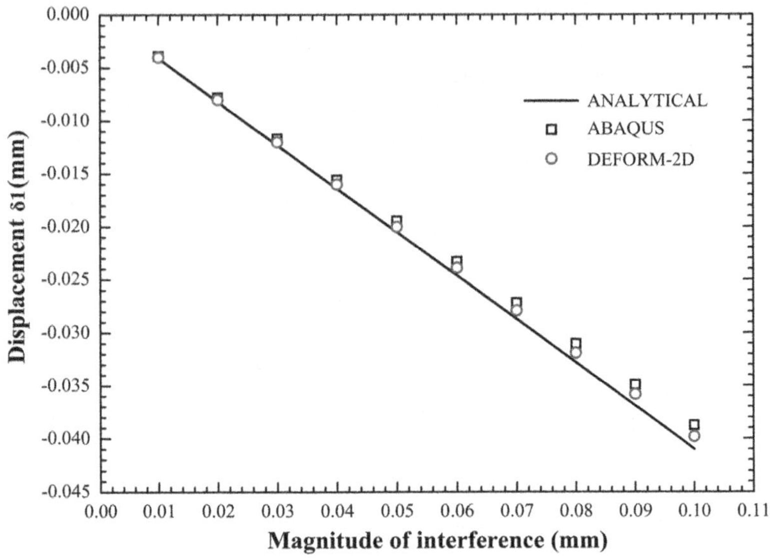

Fig. 4.8 Comparison of the interference displacement of die insert with the magnitude of interference by simulation and analytical calculation

On the other hand, Figs. 4.9, 4.10, and 4.11 show the stresses in radial and circumferential directions and the displacements of shrink ring in shrink fitting process determined by FE simulation and analytical approach. The agreement in between is similar to that of die insert. From these results, it can be seen that FE simulation can predict the stress and displacement well in both the die insert and shrink ring. Since the geometry of die insert is simple and there is a closed-form solution available, its analytical solution can be used to validate the FE simulation and verify the simulation result of shrink fitting scenarios. For many engineering problems, however, the cavity geometry of die insert is very complex, nonlinear, and there is no analytical solution available, the FE simulation can thus be directly used to determine the stress and displacement of fitting die components as long as the FEM-related models are accurately established and they can represent the real shrink fitting well. In addition, the implementation of these models is also accurate enough. In this a way, the FE simulation can be well to support the real design of die insert and shrink fit.

4.3.2 Die Working Stress

The working stress of die is a physical parameter which can be represented by different stress components or parameters, such as effective stress or maximum principal stress, in analysis of different deformation phenomena and working

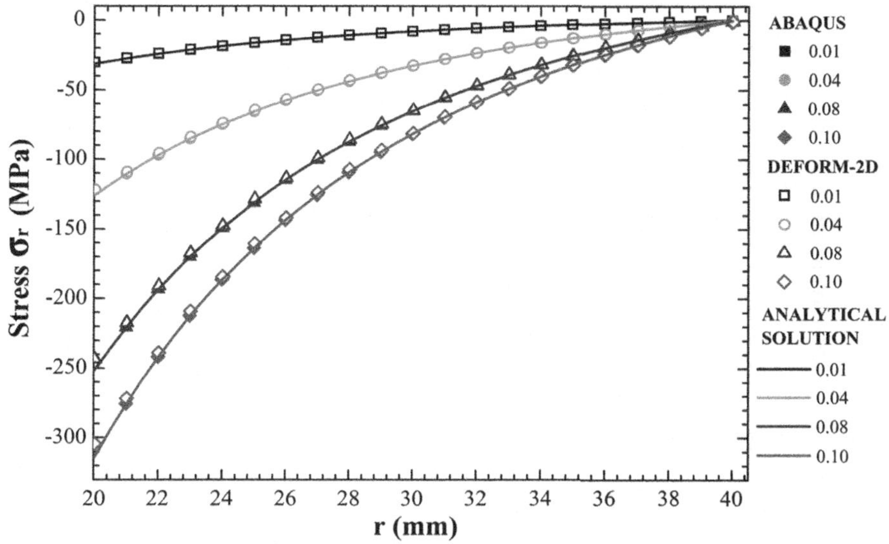

Fig. 4.9 Comparison of σ_r and its distribution in shrink ring between analytical solution and FEM calculation with different magnitudes of interference

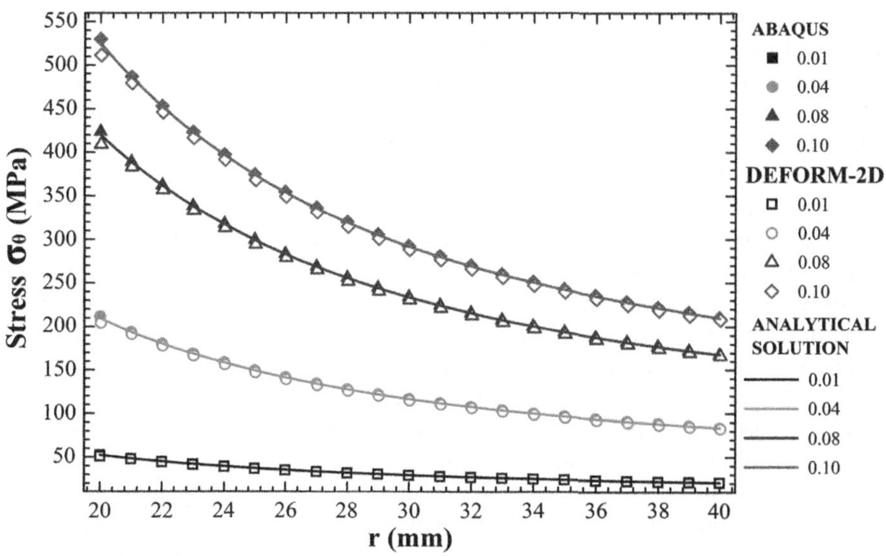

Fig. 4.10 Comparison of σ_θ and its distribution in shrink ring between analytical solution and FEM calculation via different magnitudes of interference

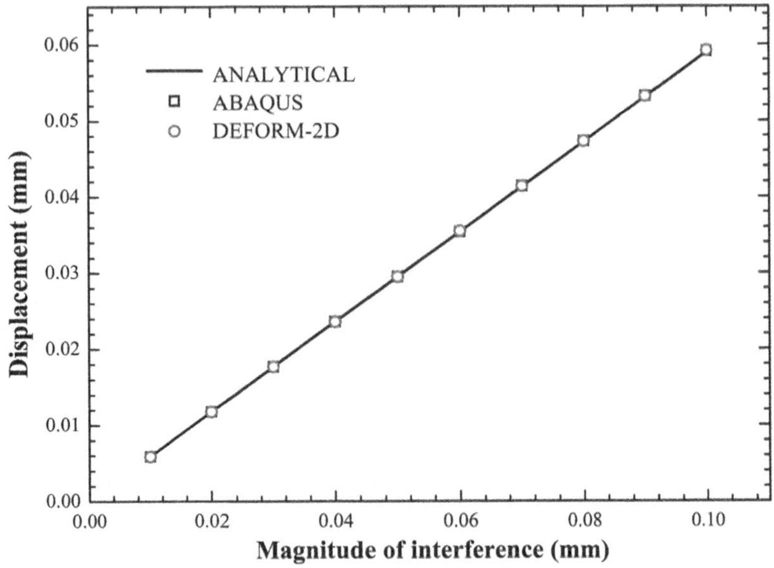

Fig. 4.11 Comparison of the interference displacement of shrink ring with the magnitude of interference by simulation and analytical calculation

performances of die in working conditions. To determine whether the die undergoes elastic or plastic deformation, the effective stress is generally used based on yield criterion. In analysis of die failure such as fracture, the maximum principal stress is usually employed to analyze the initiation, propagation, and growth of fracture in the die as well as the failure mode and root cause when the failure is related to the premature fracture of die, such as low-cycle fracture. In addition, the detailed stress components are also used in analysis of the detailed fracture scenarios. If the fracture occurs in the circumferential surface of die component, the circumferential stress component σ_θ needs to be considered. On the other hand, if the fracture happens along the axial direction of die component, the stress in the axial direction has to be considered in this analysis. When the fracture occurs at the maximum shear stress band, which is generally located at the plane with 45 degrees to the loading direction, the maximum shear stress is considered in assessing whether it exceeds the shear strength limit of die material. In die design, what representative stress needs to be used in analysis and prediction of die fatigue life, the detailed die working condition and the potential fracture scenario need to be considered first.

In die working stress analysis, the effective stress is generally used as a main parameter to represent the stress severity of die in working condition. The magnitude and distribution of the effective stress affect die performance and service life. In addition, a subtle increase of die working stress can significantly decrease die service life. In other words, die service life can be increased by a few folds via reducing the working stress a little bit. Furthermore, the effective stress helps pinpoint the root causes of die failure and identify the most potential failure location

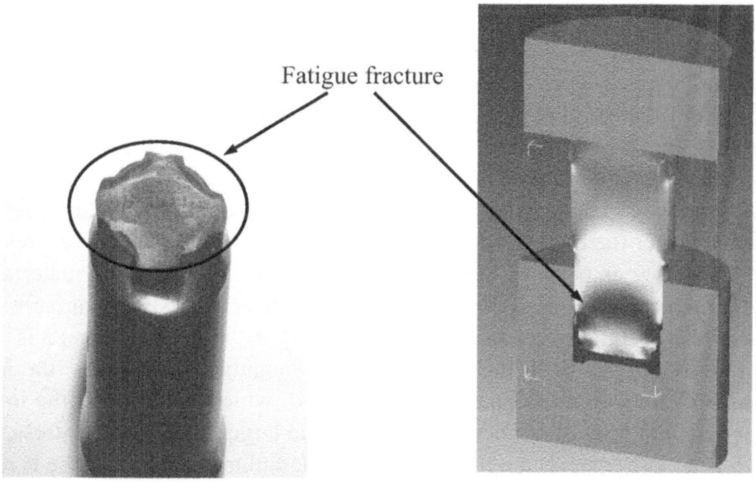

(a) Fracture occurrence at the location with the most severestress predicted by FEM

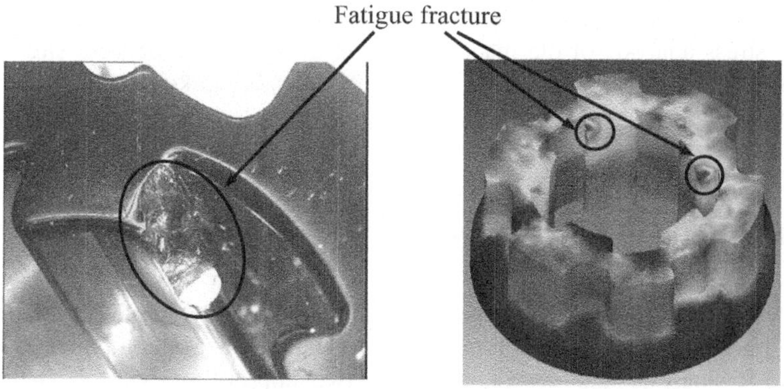

(b) Fracture located locally attheplace with themost severe stress in a die component

Fig. 4.12 Fatigue fracture occurrence at the place with the most severe stress in die

in the die in up-front design stage. Figure 4.12 shows the most severe stress locations predicted by FE simulation and their corroboration by the real fracture die components shown in the figure.

In addition, the effective stress cannot exceed the yield stress of die material. If it approaches the yield stress, some local yielding may happen, which is detrimental to die performance and service life. First of all, it could cause the change of dimensions and shape of die and further affect the dimensional accuracy of the deformed parts, and eventually the die cannot produce qualified products from dimensional accuracy and geometry tolerance perspectives. Secondly, the local yielding could lead to a large scale of plastic deformation of die such that the die components cannot work properly as the deformed die components may badly interact with other die components due to the mismatching of the geometries and dimensions of die components. Therefore, the effective working stress should reasonably be designed based on the yield stress of materials. However, there is no general guideline. Higher the working stress approaches the yield stress of die materials, shorter the die service life would be. Therefore, the effective stress of die should be as low as possible compared with the yield stress of die material.

Furthermore, mean stress plays a role in die stress analysis. Mean stress or the average stress designated as σ_m and represented by $\sigma_m = (\sigma_x + \sigma_y + \sigma_z)/3$ is relevant to die fatigue failure. With the decrease of the negative mean stress, the absolute magnitude of the compression stress is increased, which could resist the initiation, propagation, and growth of microcracks. With the larger compression stress, die can undertake the larger stress amplitude and thus the failure potential of die is reduced.

For the maximum principal stress, on the other hand, the higher tensile principal stress facilitates the propagation and growth of microcracks and fracture failure of die. Some materials such as carbide materials are very sensitive to the tensile principal stress, but it can undertake a larger compression stress. Therefore, in detailed stress analysis of die made of carbide materials, the detailed stress components need to be analyzed and figured out how they are linked to the detailed root cause of die failure.

4.4 Die Fatigue Life Analysis

As shown in Fig. 4.2, die fracture is a dominant failure mode. In die fracture failure, fatigue fracture is the most common fracture failure in die components. In die fatigue life analysis and design, there are two approaches, viz. stress-life and strain-life approaches. The stress-life approach employs the correlation between die fatigue life and its cyclic stress to assess die fatigue life; while the strain-life approach uses the total strain the part undergoes to assess its fatigue life. The two approaches have their own application scenarios. The stress-life approach is mostly used in the cases where the die stress is primarily within the elastic deformation range of die materials, and the die fatigue life is high-cycle fatigue, generally more than one thousand cycles. But for the strain-life approach, it is more appropriate in low-cycle fatigue scenarios where the die fatigue life cycle is less than one thousand times, and the applied strain has a significant plastic component. From forming process point of view, it is difficult to say which approach is more suitable for a given forming process as the same forming process could have different deformation modes, and thus, whether the die stress or strain is dominant is also dependent on the detailed forming scenario concerning metal-deformed part design, die material selection, material properties, tooling design, and process parameter configuration.

4.4.1 Stress-Life Approach

In stress-life approach, four characteristic stresses are defined for articulation of this approach. They are maximum, minimum, mean, and amplitude stresses. Figure 4.13 illustrates their definitions. In the figure, the loading cycles of the die in forming process are shown. In a loading cycle, a metal-deformed part is produced. In this process, the die undergoes a forming loading cycle with the occurrence of the above-mentioned four characteristic stresses in the die. Among these stresses, σ_{max} and σ_{min} represent the maximum and minimum principal stresses of the die, respectively. σ_m designates the mean stress during the loading cycle, which is equal to the average of σ_{max} and σ_{min}. σ_a, on the other hand, is the stress amplitude in the loading cycle. They are designated as follows:

$$\sigma_a = (\sigma_{max} - \sigma_{min})/2 \tag{4.7}$$

$$\sigma_m = (\sigma_{max} + \sigma_{min})/2 \tag{4.8}$$

To establish the relationship of die fatigue life with die stress and the material properties of die, the following Goodman equation is used [15]:

$$\frac{\sigma_a}{S_e} + \frac{\sigma_m}{S_u} = 1 \tag{4.9}$$

The Goodman equation represents the relationship of die fatigue life with σ_a, σ_m, and the material properties, viz. endurance limit S_e and ultimate strength S_u. Based on this equation, a diagram representing the correlation of cyclic stress and fatigue cycle life, viz. S-N diagram, can be constructed. From the S-N diagram, the die fatigue life can be determined based on the given amplitude and mean stresses. Figure 4.14 illustrates the S-N diagram for a given mean stress.

By using the stress-life approach for die fatigue life design and analysis, the S-N diagram of die material needs to be constructed. The experiment-based fatigue test is an approach to construct the S-N diagram. It is a costly and time-consuming

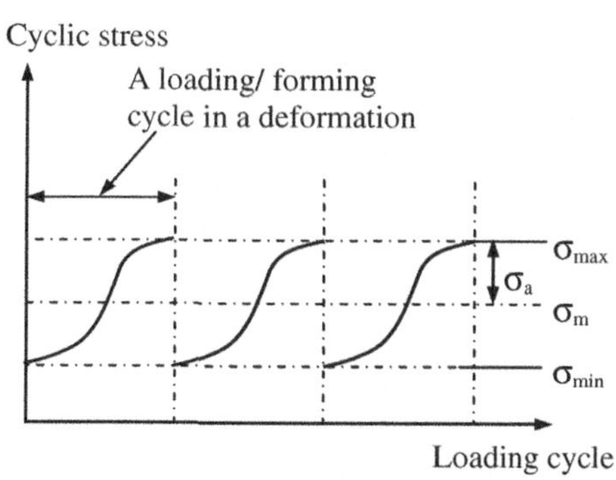

Fig. 4.13 The characteristic stresses in cyclic loading cycles [22]

Fig. 4.14 S-N diagram [22]

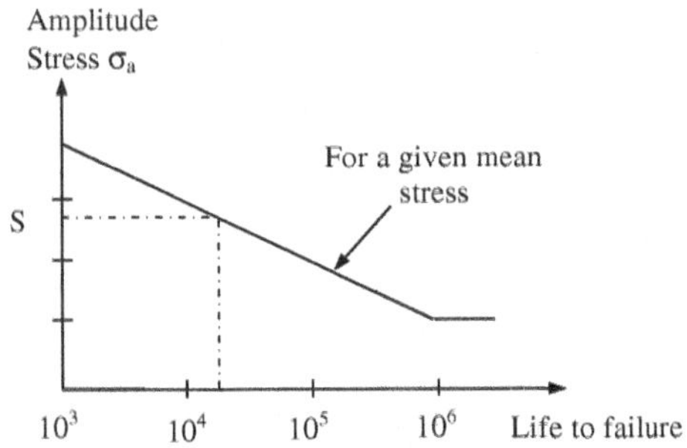

experimental work and many different experiments need to be designed and conducted to cover all the deformation scenarios with diverse stress conditions with different stress amplitudes and mean stresses, and in such a way, the constructed S-N diagram can be applied in different deformation scenarios and conditions.

In terms of determination of the most potential fatigue stage, location, and the four characteristic stresses in a forming cycle, FE simulation provides an efficient approach to identifying the stage at which the most severe stress occurs and the most potential fatigue location and determining the instantaneous values of the four characteristic stresses at the most potential fatigue occurrence stage and location in a forming cycle.

4.4.2 Strain-Life Approach

In the strain-life approach, the stress–strain curve is first used to analyze the fatigue durability of die with cyclic loadings in forming process. The physical response of die material to the cyclic loadings can be represented in the form of hysteresis loop, as shown in Fig. 4.15. The total strain range $\Delta\varepsilon$ and the total stress range $\Delta\sigma$ in the hysteresis are used for definition of strain and stress amplitudes as follows, respectively:

Fig. 4.15 Hysteresis loop and its stress and strain decomposition [22]

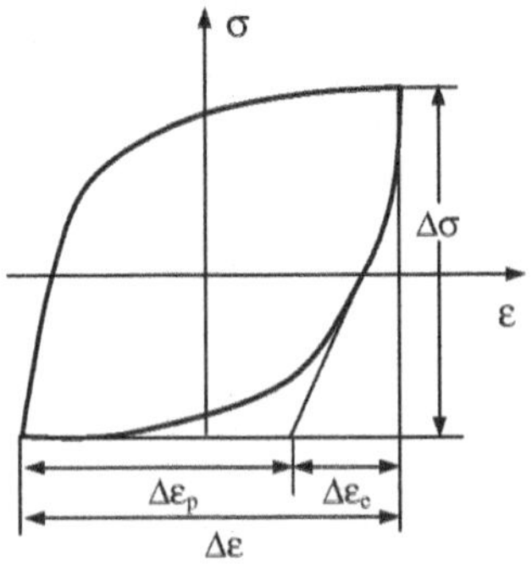

$$\varepsilon_a = \Delta\varepsilon/2 \tag{4.10}$$

and

$$\sigma_a = \Delta\sigma/2 \tag{4.11}$$

As the total strain is the sum of elastic and plastic strains, therefore,

$$\Delta\varepsilon = \Delta\varepsilon_e + \Delta\varepsilon_p \tag{4.12}$$

It can be further represented in terms of strain amplitude as:

$$\Delta\varepsilon/2 = \Delta\varepsilon_e/2 + \Delta\varepsilon_p/2 \tag{4.13}$$

By employing Hooke's law, Eq. (4.13) can be further expressed in the following

$$\Delta\varepsilon/2 = \Delta\sigma/(2E) + \Delta\varepsilon_p/2 \tag{4.14}$$

From the context of material fatigue, on the other hand, Basquin [15] presented the stress-life data in the following equation

$$\Delta\sigma/2 = \sigma_f'(2N_f)^b \tag{4.15}$$

where $\Delta\sigma/2$ is the true stress amplitude. $2N_f$ is the number of cycle to failure. σ_f' is the fatigue strength coefficient, which is approximately equal to the true fracture strength. b is the fatigue strength exponent or Basquin's exponent with a range between −0.05 and −0.12.

Furthermore, Coffin [16] and Mansion [17] found the relationship between plastic strain and fatigue life can be designated as:

$$\Delta\varepsilon_p/2 = \varepsilon_f'(2N_f)^c \tag{4.16}$$

where $\Delta\varepsilon_p/2$ is the plastic strain amplitude. ε_f' is the fatigue ductility coefficient, which is approximately equal to the true fracture ductility, and c is the fatigue ductility exponent ranging from −0.5 to −0.7. Using Eqs. (4.15) and (4.16), the total strain in Eq. (4.14) can be represented as:

$$\Delta\varepsilon/2 = (\sigma_f'/E)(2N_f)^b + \varepsilon_f'(2N_f)^c \tag{4.17}$$

Equation (4.17) provides a basis for strain-life approach for die fatigue analysis and is schematically illustrated in Fig. 4.16. In the strain-life diagram as shown in Fig. 4.16, the total strain-life diagram is the summation of elastic and plastic lines. With the large strain amplitude, the diagram approaches the plastic line, while it approaches the elastic line at the low amplitude [14]. For different materials, the

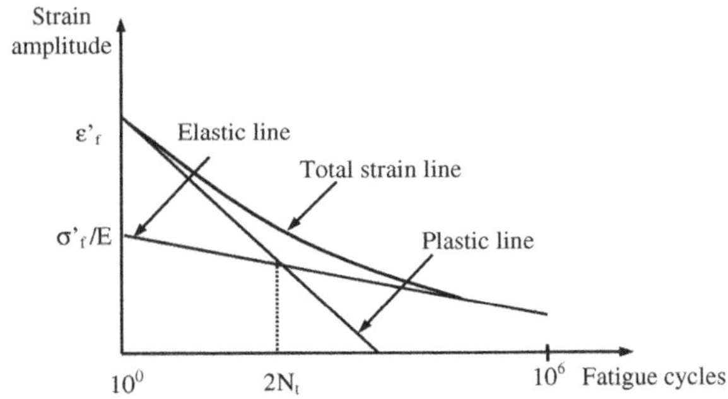

Fig. 4.16 Strain-life diagram [22]

strain-life diagram could be different. The construction of the strain-life diagram is generally based on the experimental data of the fatigue test of die materials.

4.4.3 Die Fatigue Life Assessment

In the previous sections, the stress-life and strain-life approaches for fatigue life assessment are formulated and articulated. By combining the approaches and simulation technologies, the die fatigue life can be estimated via CAE simulation. Figure 4.17 presents an implementation framework. Under the framework, the entire forming system is firstly modeled by physical, mathematical, and computational models. As mentioned before, physical model idealizes the real engineering problems and abstracts them to comply with certain physical theory with assumptions. For a metal-forming system to be simulated for die fatigue life design, the deformation of workpiece is assumed to be plastic deformation and its elastic deformation is generally ignored, while the tooling is assumed to have elastic deformation only. These assumptions are actually within the content of the physical model. For the mathematical model, it specifies the mathematical equations such as the constitutive equations the physical model should follow. In the current off-the-shelf and domain-specific FE simulation systems, the mathematical model is generally built into the systems and users generally need to specify some detailed data and information such as boundary conditions (BCs) and constraints and input into the simulation system. For the computational model, on the other hand, users need to choose elements type and specify mesh density and solution parameters. From the point of view of configuration of solution parameters, the detailed calculation tolerance, error bound, iteration control, and convergence criteria are needed to be provided for FE simulation to ensure the efficient convergence and the accuracy of the simulation results.

To conduct an integrated simulation of the plastic deformation of workpiece and the elastic deformation of die and its component, in addition to the above-described model-related data and information, the data preparation in the preprocessing stage further includes material models and properties, deformation process parameters

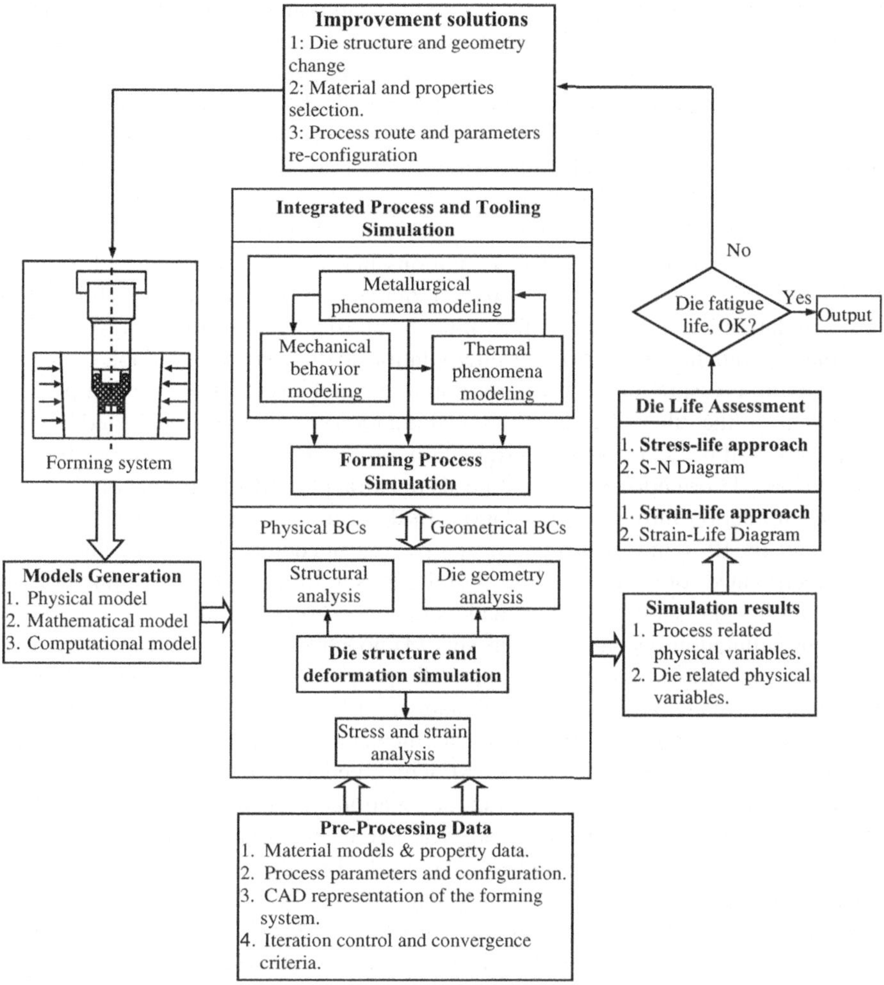

Fig. 4.17 Framework for die fatigue life assessment [22]

and configuration, and CAD models for representation of the entire forming system. In this process, the mechanical deformation behaviors, metallurgical phenomena, and thermal phenomena in the forming process should be well represented by the input data and information. The interaction and interplay of these physical behaviors needs to be considered and well represented by the input information. On the other hand, the simulations of tooling deformation and the plastic flow of workpiece in forming process are conducted simultaneously, which is the so-called coupled simulation.

In die fatigue life design and analysis, die structure and deformation simulation is the kernel. In Fig. 4.17, die structural analysis studies the structure relationship of die insert, casing, punch and other working components, and the interference relationship such as shrink fit between die insert and casing. Die geometry analysis determines and verifies the effects of the shapes and geometries of die components such as die angle, fillet, corner and dimensions on die deformation, and stress status. Stress and strain analysis, on the other hand, reveals the stress and strain distribution in the die set. The identified strain and stress provide the basic data and information for die life assessment.

In the integrated simulation of forming process and die deformation, the dynamic physical and geometrical boundary conditions between the simulation of forming process and die structure and deformation analysis are dynamically established in each simulation-loading step. To determine the physical and geometrical BCs in this integrated simulation scenario, the interpolation of the physical variables such as stress, strain, pressure, or temperature from one simulation kernel (forming process simulation) to the other (die deformation simulation) needs to be conducted in each simulation-loading step. The determination of the boundary conditions for die structure and deformation simulation needs to extract the BCs' information from the process simulation through interpolation. In deforming process simulation, on the other hand, the BCs' information from die structure and deformation simulation needs to be interpolated for the establishment of the BCs in forming process simulation. In the current off-the-shelf and tailor-made simulation software, all of these are done inside the simulation system. The users should know how they are done in the system.

Upon completion of the simulation, all the process and die related physical information is available. The stress-life and strain-life approaches can be employed for die life assessment. In a forming cycle, the stage at which the deformation load has the maximum value is identified first, which generally happens at the last stage of the forming cycle. The potential fatigue locations in the die are then determined under the maximum deformation load. The cyclic maximum principal strain and stress in these locations are further calculated. The die fatigue modes and the potential root causes need to be figured out. The amplitude stress and strain are eventually computed and the die fatigue life is thus assessed based on the developed approaches articulated previously. If the die life is not satisfied, the die structure and geometries, material selection and material properties configuration, and process route and parameters configuration can be re-examined for die life improvement. Furthermore, new simulations will be conducted for the modified forming system design until the satisfactory solution is obtained. The case studies presented in the following section provide an efficient approach for die life assessment and die design evaluation in metal-forming product and system development.

4.5 Case Studies

In this section, three case studies are employed to illustrate the above-described stress-life and strain-life approaches for die fatigue life assessment and analysis. The first case is to use the strain-life approach to studying the fatigue service life of a punch. The second one is to use the stress-life approach to exploring the fatigue life of a die insert. The last case study is to use the stress-life approach to designing die fatigue life in such a way to further rationally design the forming system.

4.5.1 Case Study 1

To illustrate the developed approach and the implementation framework for die life assessment, the first case study is conducted, which is an assessment and analysis of cold forging die life, as shown in Fig. 4.18. In the figure, the punch, punch holder, billet, die insert, casing, and counterpunch to be considered in the simulation are illustrated. The integrated simulation of the plastic deformation of workpiece and the elastic–plastic deformation of the die in forming process is conducted. The punch and die insert of the die assembly are considered to be elastic–plastic body, while the punch holder and counterpunch are supposed to be rigid body. The billet is the taken as a plastic body. The material of punch and die insert are M2, which is one of the popular tooling materials in cold forging industry. The billet material is SCM 415. Its material property is defined via the following stress–strain relationship:

$$\sigma = 150\varepsilon^{0.1} + 465 \tag{4.18}$$

where σ is the effective stress (MPa) and ε is the effective strain.

The above stress–strain relationship is determined through experimental compression test. To reduce the friction effect on the two contacting surfaces in between the die and billet in the compression test, a cycle groove on the top and bottom surfaces of the billet is machined to hold and contain the lubricant efficiently and in such a way the friction between the die and billet is reduced significantly. The experiments show that it is very efficient to provide a good lubrication on the contacting surfaces by employing the cycle groove in the contacting surface of the billet.

The integrated simulation is conducted by DEFORM™ 3D [19], which is a FE simulation software for analyzing metal-forming processes, die deformation, and further the design solution generation and verification in metal-forming product design and development. The system is developed and marketed by Scientific Forming Technologies Corporation based in USA [19].

Figure 4.19a shows the potential fatigue locations in the punch. There are two locations where the maximum principal strain is greater than other areas nearby. One is located at the radius corner of the first shoulder of punch and the other is at the radius corner in the second shoulder. This is caused by the stress concentration in these two locations. From the colored contour of the effective stress distribution, it can be found that the most potential fatigue area is located at the radius corner of the second shoulder, viz. the low shoulder of the punch. To investigate the detailed strain variation in a forming loading cycle, three tracking points are defined at this most potential fatigue area in the punch. The relative locations of the tracking points

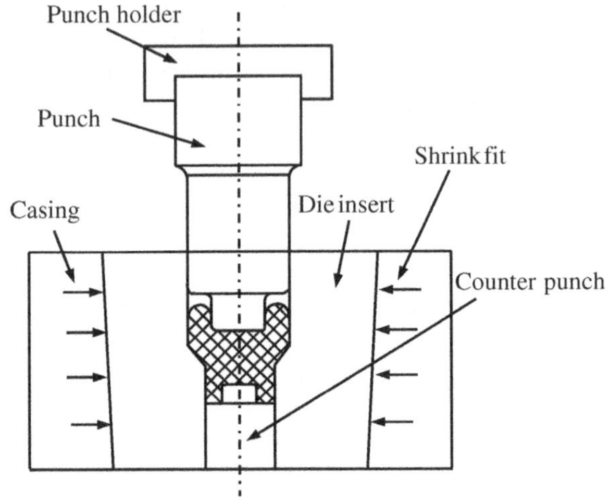

Fig. 4.18 Schematics of die structure in a combined extrusion process [22]

defined in the θ symmetry plane, which passes through the central axis of the part in Z direction, and their coordinates in the θ symmetry plane are shown in the figure.

The variation of the maximum principal strain for the tracking points is shown in Fig. 4.19b, and it can be seen that Point 1 has the maximum principal strain with 0.00814. The second tracking point is read with 0.00607, and the third point is 0.00520. The data show that the maximum principal strain decreases inwards. Therefore, the fatigue mode could be in such a way that the fatigue fracture may initiate first at the place where the punch has the maximum principal strain of 0.00814 and then propagates and grows inwards until the punch fails finally.

Since there is no initial strain in the punch, the strain amplitude can be determined as 0.00407. According to the strain-life diagram for this punch material M2, as shown in Fig. 4.20 [20], the designed die life can be estimated as 750 cycles. In other words, the punch could produce 750 parts under the current configuration of forming process, tooling and product design, and the material properties the simulation adopted. The predicted die life is in good agreement with the real scenario.

4.5.2 Case Study 2

The second case is shown in Fig. 4.21, which is to assess the service life of a gear forging die. The die structured is similar to the first case, except for the difference of billet material, which is AISI 1017. The flow stress of this material is shown in Fig. 4.22. From the figure, it can be found that the strain rate does not affect the flow stress much. The flow stress almost keeps the same when the strain rate changes from 1.6 to 40. The coupled analysis of die deformation and the simultaneous simulation of gear billet deformation are conducted by a CAE simulation software DEFORM. Figure 4.23a shows the potential fatigue location and the

Fig. 4.19 The potential fatigue locations and the variation of the maximum principal strain [22]

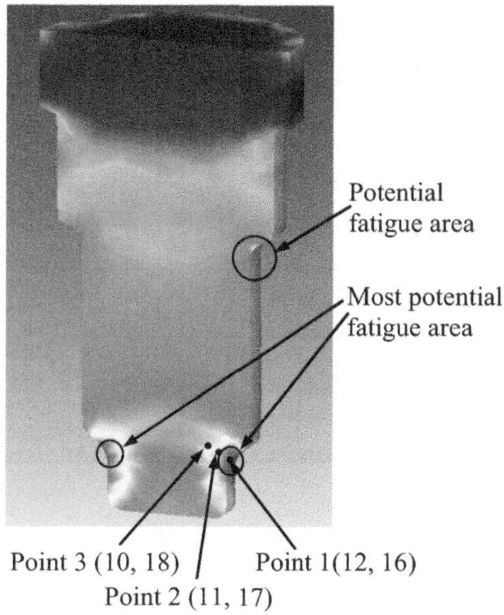

(a) Potential fatigue areas and the tracking points in the punch

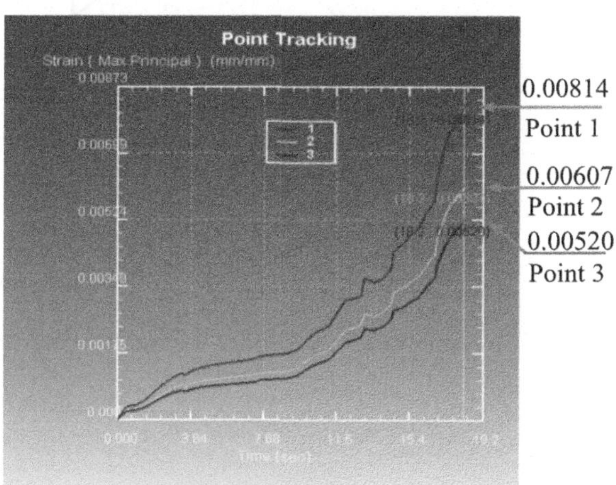

(b) Maximum principal strain at the tracking points

maximum principal stress distribution of the gear forging die. For the defined tracking point at the most potential fatigue areas as shown in Fig. 4.23a, the variation of the maximum principal stress is illustrated in Fig. 4.23b, which is read with 1600 MPa at the last stage of forging process. Since the initial maximum principal stress is zero, both the mean and amplitude stresses are 800 MPa. From

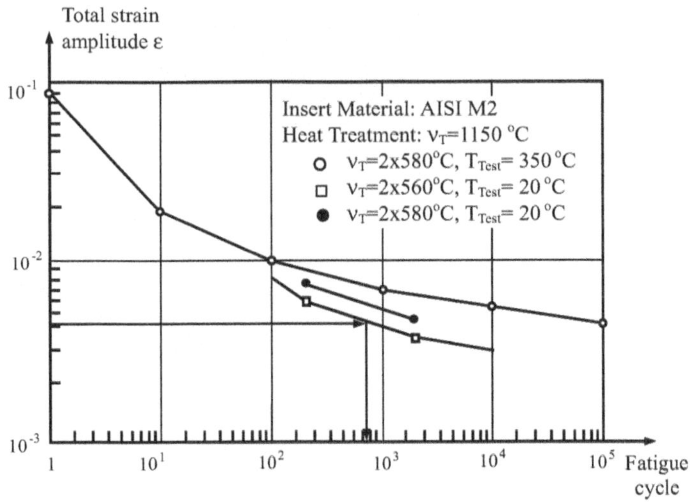

Fig. 4.20 Strain-life diagram for M2 [20]

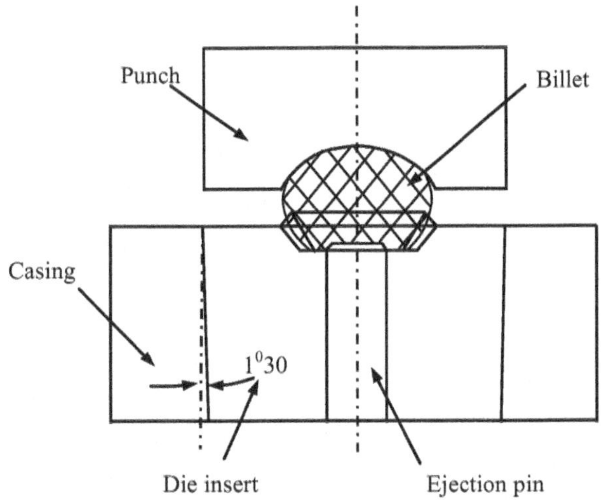

Fig. 4.21 Die structure schematics in forging of a gear [22]

the S-N diagram of M2 as shown in Fig. 4.24 [21], the assessed die fatigue life for this case is 3000 cycles, which means the die can produce 3000 parts before it fails.

4.5.3 Case Study 3

The third case study part is used to illustrate the above proposed methodology for die fatigue life optimization via the rational design of deformed part and its

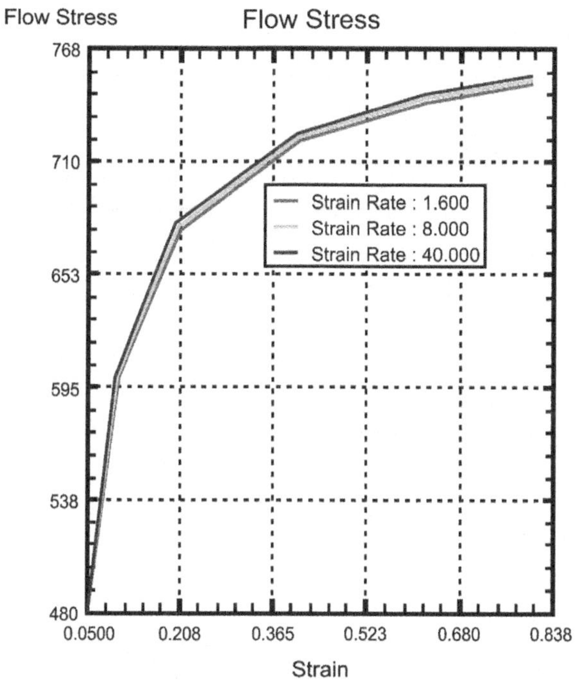

Fig. 4.22 Flow stress of AISI 1017 [22]

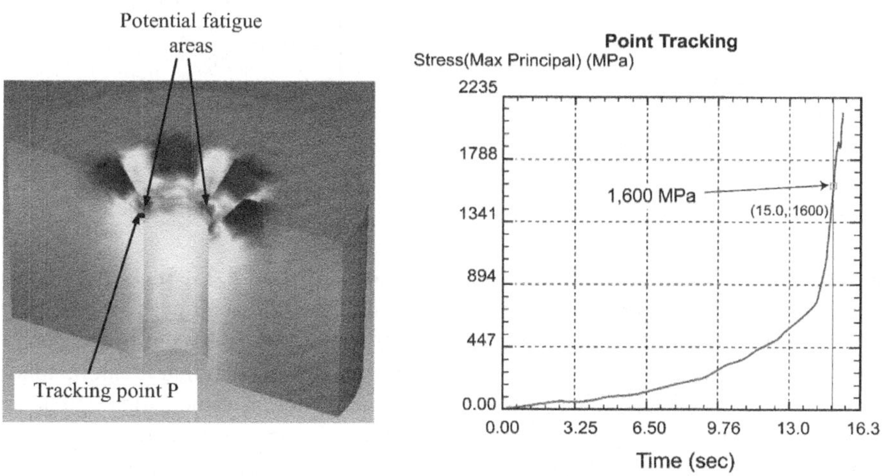

(a) Potential fatigue location (b) Maximum principal strain at Point P

Fig. 4.23 Potential fatigue areas and the maximum principal strain at the gear forging die [22]

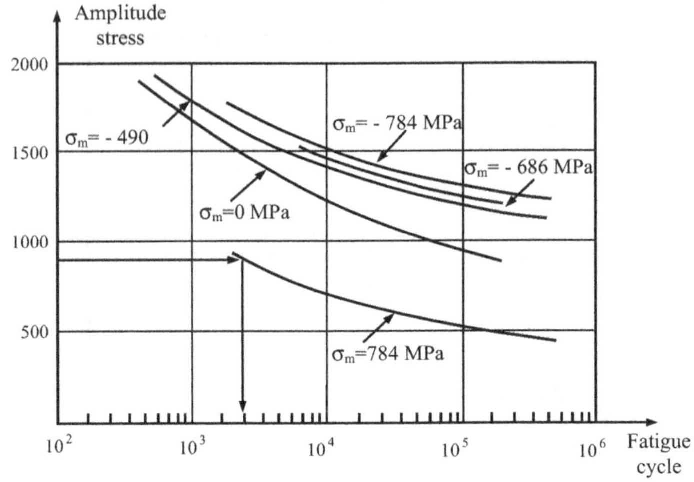

Fig. 4.24 S-N diagram of M2 [21]

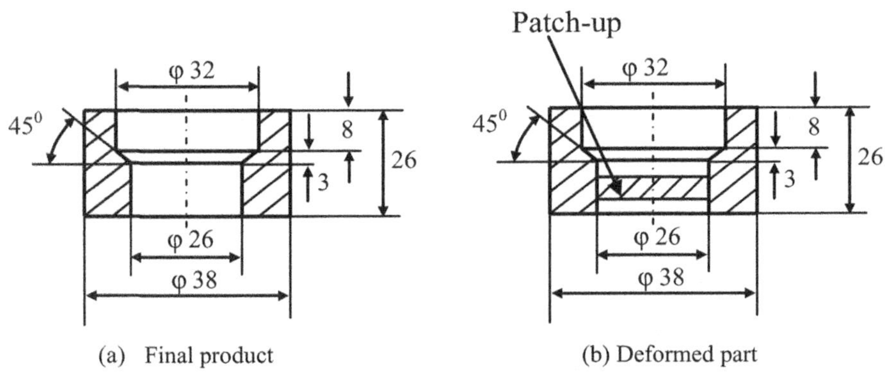

Fig. 4.25 Case study 3 [23]

corresponding forming system. The part is shown in Fig. 4.25. In Fig. 4.25a, it shows a final product geometries and this product can be formed by metal-forming process. To convert the final product geometry to the plastically deformed part, a patch-up is introduced into the metal-deformed part design such that it can be made by one-stroke metal-forming operation. However, different dimensions and locations of the patch-up constitute different metal-deformed part designs and their corresponding different forming system designs, which have different die service lives and process performances. The focus of this case study is to investigate the best design of die and forming system from die service life perspective.

4.5.3.1 Design of Metal-Forming System

To fully explore the potential design alternatives for forming the product, three factors are considered in this case study. The first one is the patch-up dimension of the deformed part as the forming system cannot form a product with a through central hole, as shown in Fig. 4.24, by one-stroke operation. The patch-up is thus needed in extrusion of a cylinder billet to the desired shape of the product. The second one is patch-up location and the third is punch geometry and shape. The shape of punch could be flat punch or bevel one. Based on these three factors, different design scenarios about in terms of metal-formed part, die structure, and the cylindrical billet and its dimensions can be figured out. Table 4.2 presents the twelve design scenarios through the configuration of the above different design variables. For the first design scenario as shown in Table 4.2, the dimensions of the cylindrical billet are determined as $\phi 37.6 \times 14.7$ based on the volume constancy of the billet and the deformed part.

4.5.3.2 Integrated Simulation of Forming System

To reveal the die stress magnitude and distribution in different design scenarios, the simultaneous simulation of the plastic flow of billet and the elastic deformation of die is conducted. First of all, the designed metal-forming system is modeled in a commercial CAD system and then converted into a data exchange format in such a way they can be imported into the FE simulation system for simulation.

Regarding the material behavior of the forming system, the punch material is M2 and is considered as an elastic deformation body during the forming process. The elastic module is 250,000 MPa and poison ratio v is 0.3. For the billet material, the following constitutive relationship of the material, viz. Equation (3.8) is employed

$$\sigma = 150\varepsilon^{0.1} + 547$$

where σ is effective stress or flow stress and ε is effective strain.

In addition, the friction between die and billet follows the shear friction model given by Eq. (3.9), which is represented as follows:

$$\tau = mK$$

where m is the friction factor, while K is the shear strength of billet material. m is equal to 0.1 in this case of cold forming process.

Table 4.2 Different design scenarios [23]

Design scenario	Formed part	Punch	Die structure
Scenario 1 $H = 6$ mm (bottom)	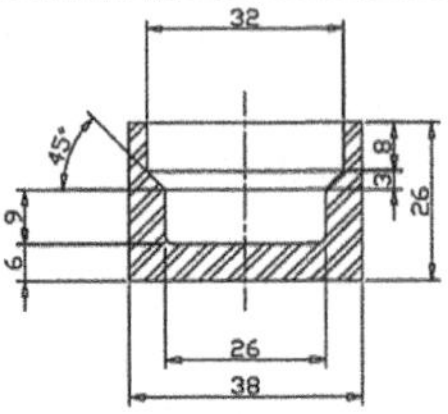	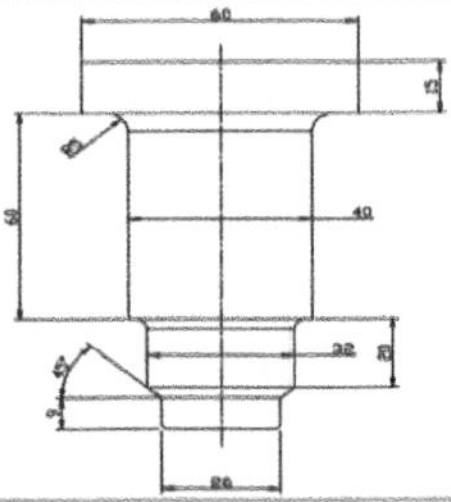	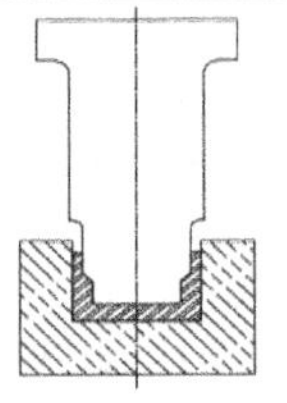
Scenario 2 $H = 6$ mm (middle)	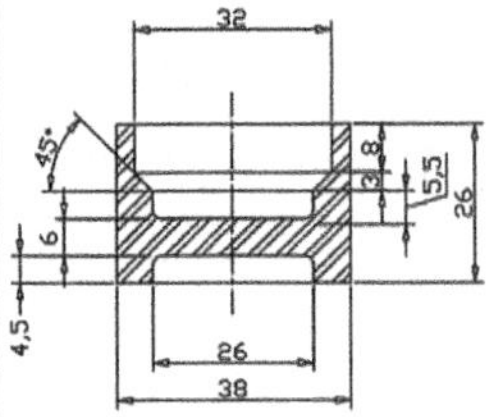	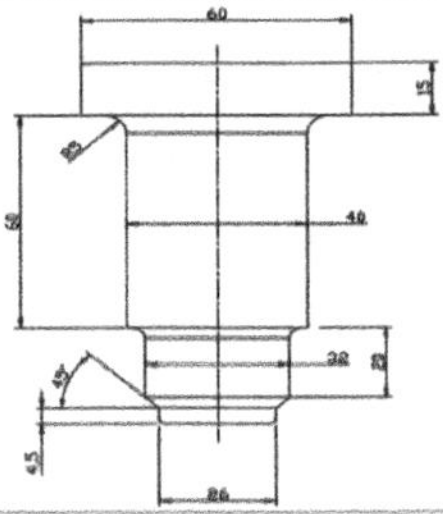	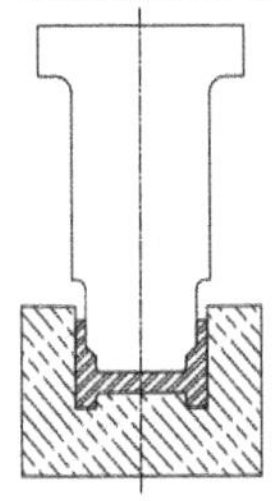
Scenario 3 $H = 6$ mm (top)	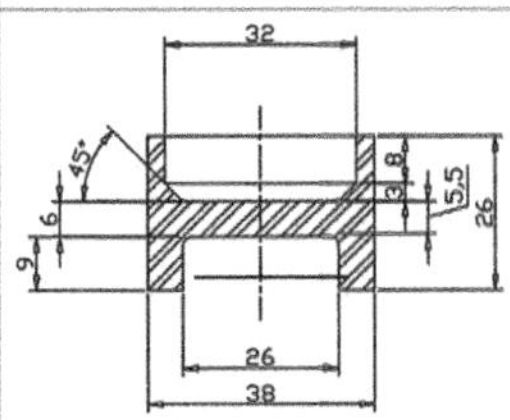	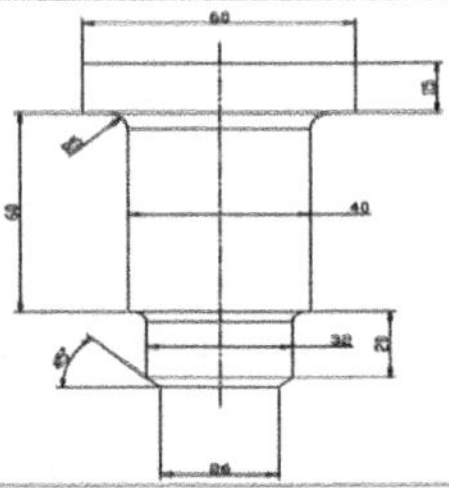	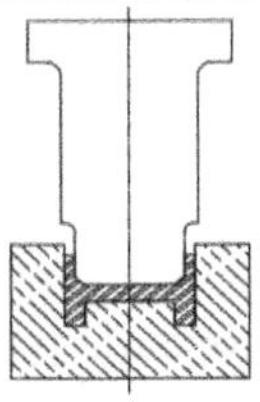
Scenario 4 $H = 4$ mm (bottom)	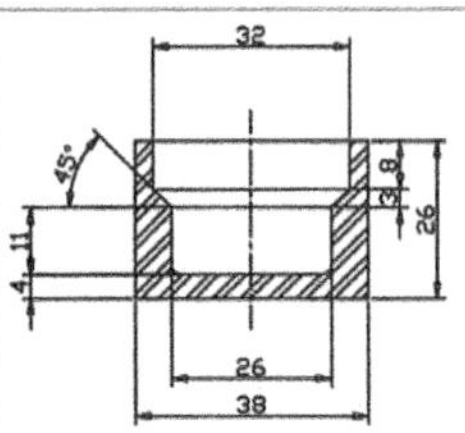	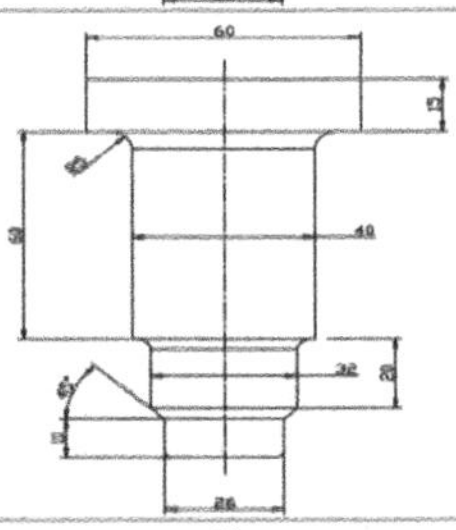	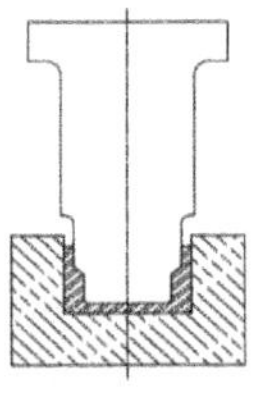
Scenario 5 $H = 4$ mm (middle)	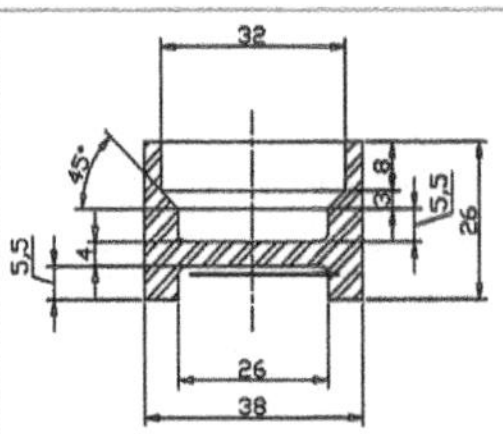	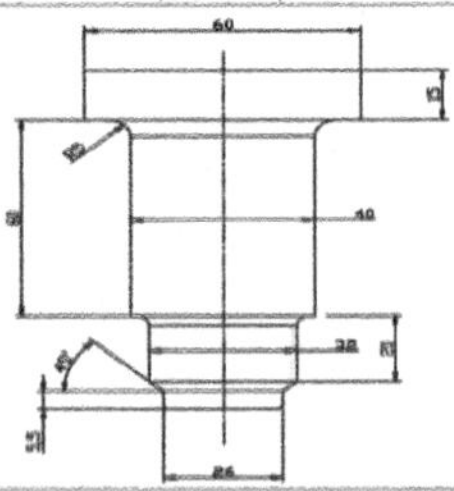	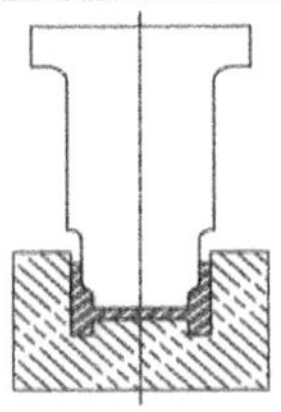

(continued)

Table 4.2 (continued)

Design scenario	Formed part	Punch	Die structure
Scenario 6 $H = 4$ mm (top)	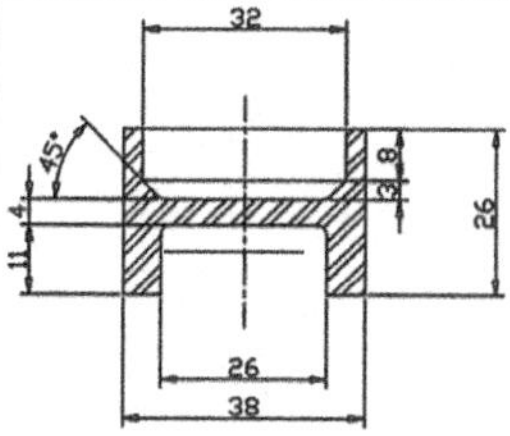	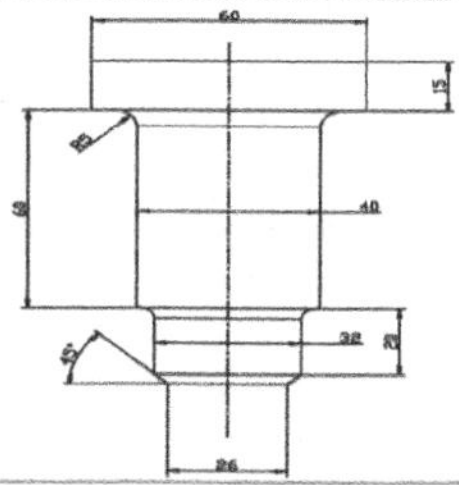	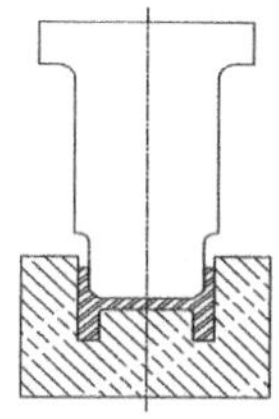
Scenario 7 $\theta = 10^{\circ}$ $H = 6$ mm (bottle)	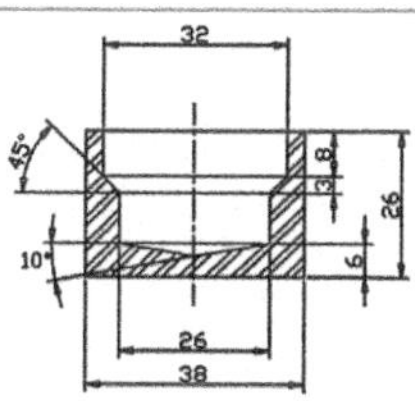	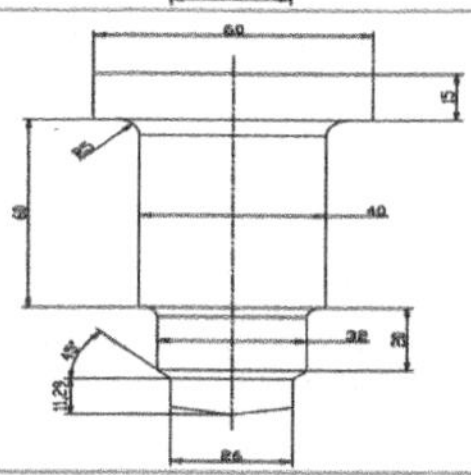	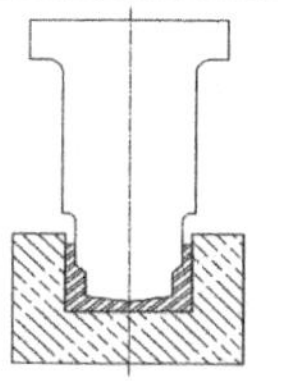
Scenario 8 $\theta = 10^{\circ}$ $H = 6$ mm (middle)	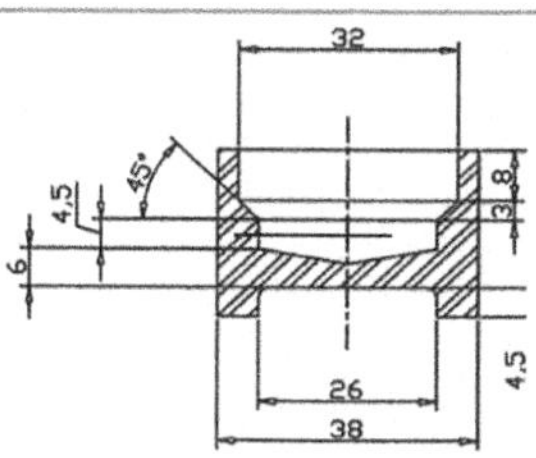	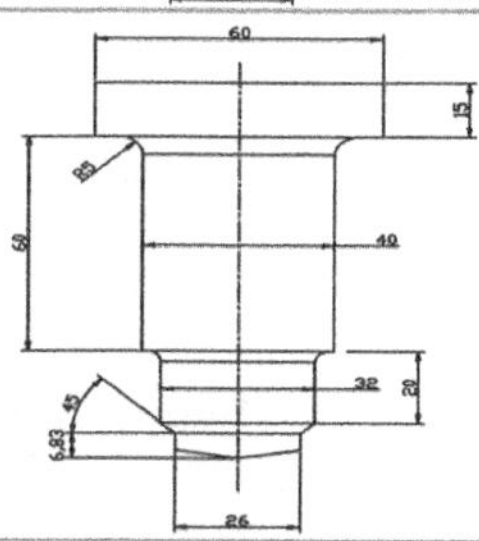	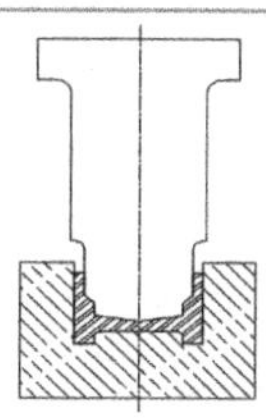
Scenario 9 $\theta = 10^{\circ}$ $H = 6$ mm (top)	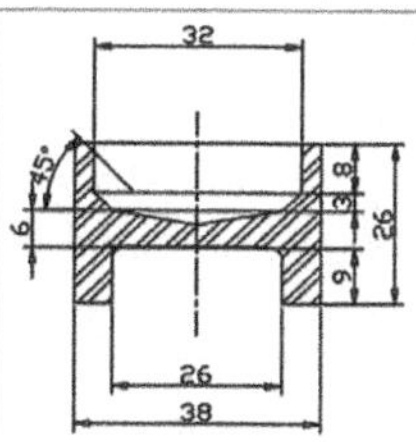	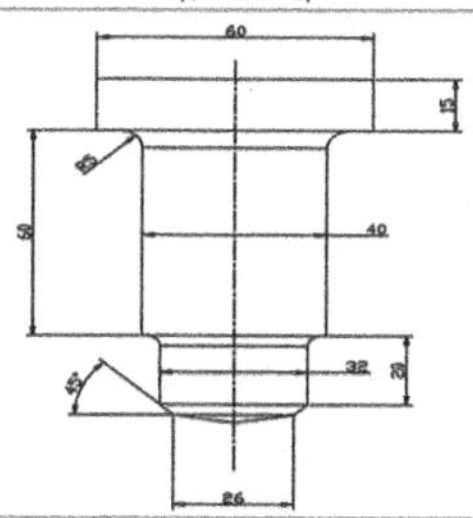	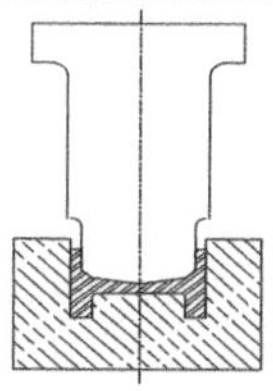
Scenario 10 $\theta = 10^{\circ}$ $H = 4$ mm (bottom)	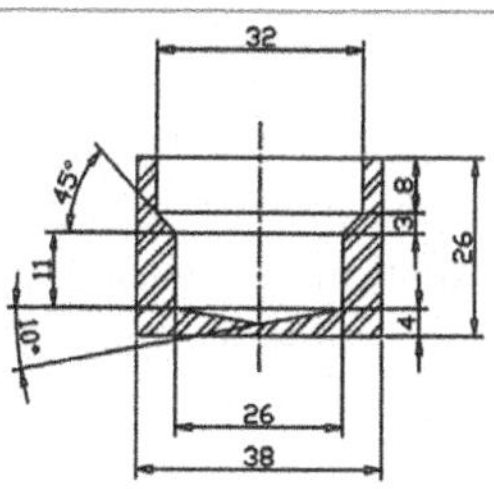	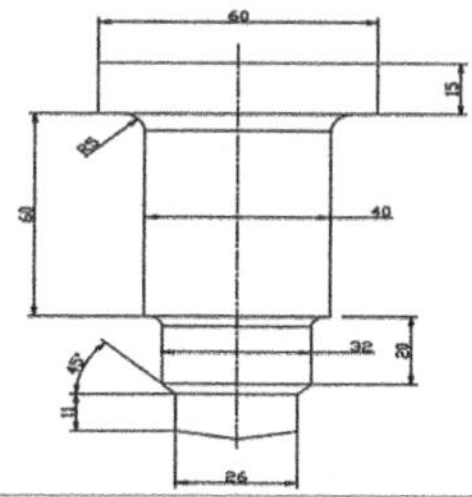	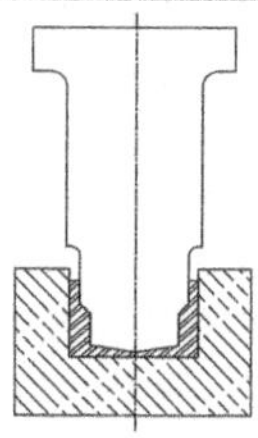

(continued)

Table 4.2 (continued)

Design scenario	Formed part	Punch	Die structure
Scenario 11 $\theta = 10^{\circ}$ $H = 4$ mm (middle)	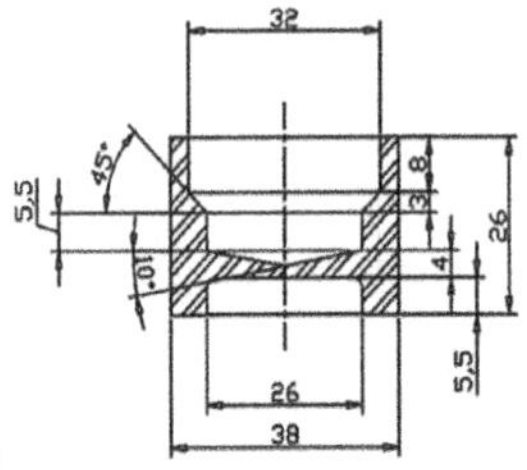	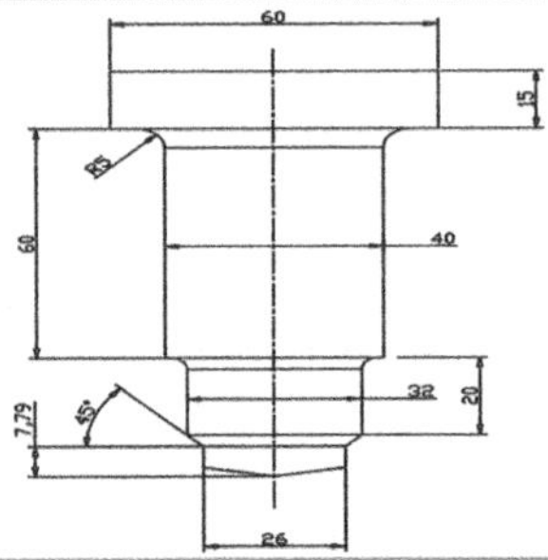	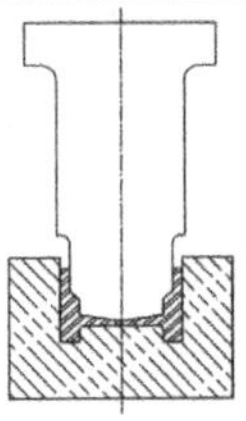
Scenario 12 $\theta = 10^{\circ}$ $H = 4$ mm (top)	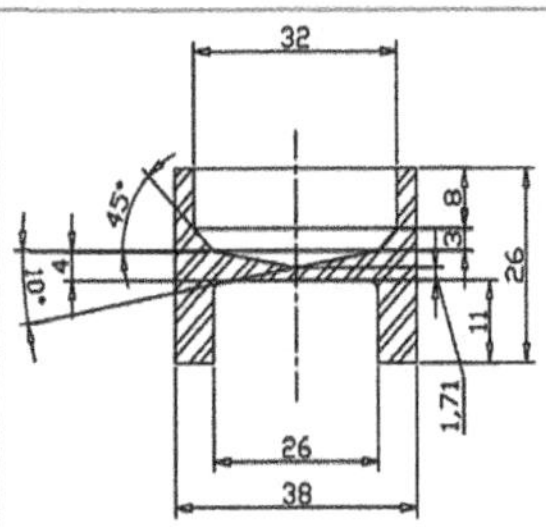	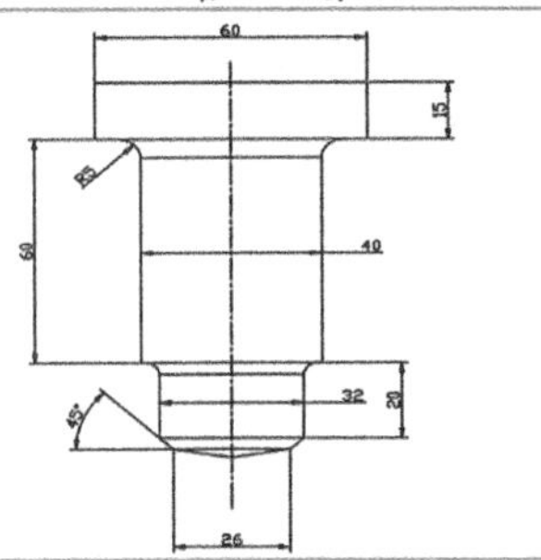	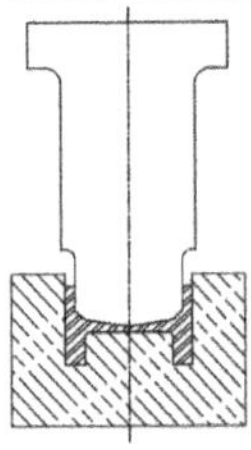

4.5.3.3 Procedure for Extraction of Simulation Results

A. Procedure for Extraction of Simulation Results

The integrated simulations are conducted for the above twelve design scenarios. Since the stress distribution of die is not uniform and the fatigue is assumed to be initiated at the most severe stress location, which is also the potential fatigue failure place. To determine this fatigue stress of die, the following procedure is adopted for extraction of the simulation results.

(1) Determination of the maximum deformation load and its appearance stage. In forming process, the deformation load varies in the process. Die service life is very sensitive to deformation load. A subtle reduction of die loading and die stress can increase the die service life significantly. The maximum deformation load and the deformation stage at which the maximum deformation load appears need to be determined. Through simulation of the plastic deformation of billet, the variation of deformation load in the entire forming process can be identified and the deformation stage where the maximum deformation load occurs can then be determined.
(2) Identification of the potential fatigue location. Under the maximum deformation load, the die stress is the most severe scenario. Under this condition, the non-uniform die stress distribution is then identified. To do so, the effective

stress can be used as a parameter and the potential fatigue location can then be fixed based on the distribution of the effective stress. The potential location is located at the place where the effective stress has the maximum value.

(3) Extraction of the maximum effective stress. From the most severe stress location, the maximum effective stress is identified. When the die is used for forming, the original stress state is zero as it does not have any prestress condition (no shrinking fit condition in this case). On the other hand, the fatigue is more related to the mean stress and the amplitude stress based on the Goodman's equation, the mean stress and amplitude stress of the effective stress are then determined and used for fatigue life assessment.

B. Results Analysis and Discussion

Figure 4.26 shows the deformed billet and the deformation load in Scenario 1. In the figure, "Step 158" is the simulation-loading step. In this case, each simulation-loading step is 0.1 mm displacement. "Step 158" would mean that the stroke of punch is 15.8 mm at this loading stage. Furthermore, from the figure, it can be seen that the maximum deformation load is located at the last deformation stage. From the simulation point of view, the simulation step is 158. The deformation load is 127 tons. In addition, Fig. 4.27 shows the simulation results of Scenario 2.

Based on the stage at which the deformation load has the maximum value, the punch stress is investigated. It is found that the potential fatigue location is located

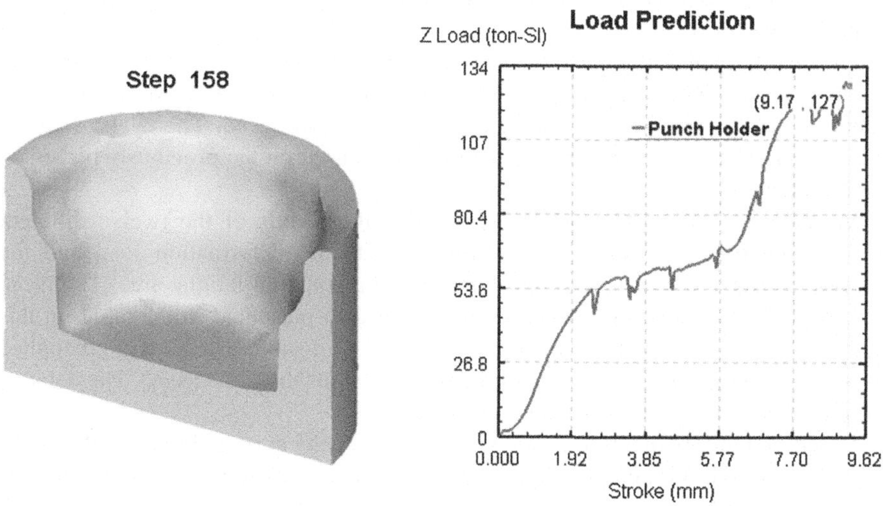

(a) Deformed workpiece (b) Deformation load

Fig. 4.26 Deformed workpiece and the deformation load in Scenario 1 [23]

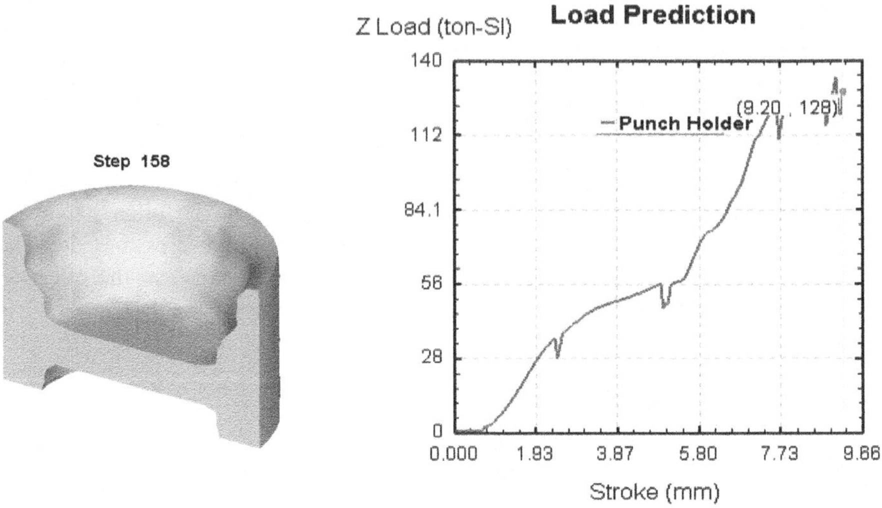

(a) Deformed workpiece (b) Deformation load

Fig. 4.27 Deformed workpiece and the deformation load in Scenario 2 [23]

at the radius corner in the second shoulder of punch. Defining a tracking point at point P1 at the most potential fatigue location, the effective stress at the tracking point can be identified. Figure 4.28 shows the potential fatigue location in the punch, the defined tracking point and the effective stress variation of tracking point P1 in the forming processes in Scenario 1. Similarly, Fig. 4.29 shows the similar results in Scenario 2.

On the other hand, Fig. 4.30 shows the cyclic stress variation in the course of die service. To produce a metal-deformed part, the die and metal-forming system undergo a cyclic loading and thus a cyclic stress is produced. The cyclic stress is used to determine the potential fatigue life of die based on the previously presented methodologies.

In addition, Tables 4.3 and 4.4 summarize the results of the twelve different design scenarios. The simulation results include the deformation load and the effective stress at the tracking point located at the potential fatigue area. The mean and amplitude stresses are then determined and also presented in the table. In this case, the die is supposed to be an elastic body. No residual stress and preapplied stress are considered in evaluation of die life. σ_a is thus the same as σ_m. The detailed values are listed in Tables 4.3 and 4.4.

Based on the S-N diagram as shown in Fig. 4.24, the expected fatigue life can be expected through the interpolation of the available curves in the figure and the die fatigue life can be determined. In Fig. 4.2, the vertical axis is amplitude stress σ_a and the horizontal axis is fatigue cycle, viz. fatigue life. For the given mean stress and amplitude stress, it needs to draw a horizontal line from the vertical axis and

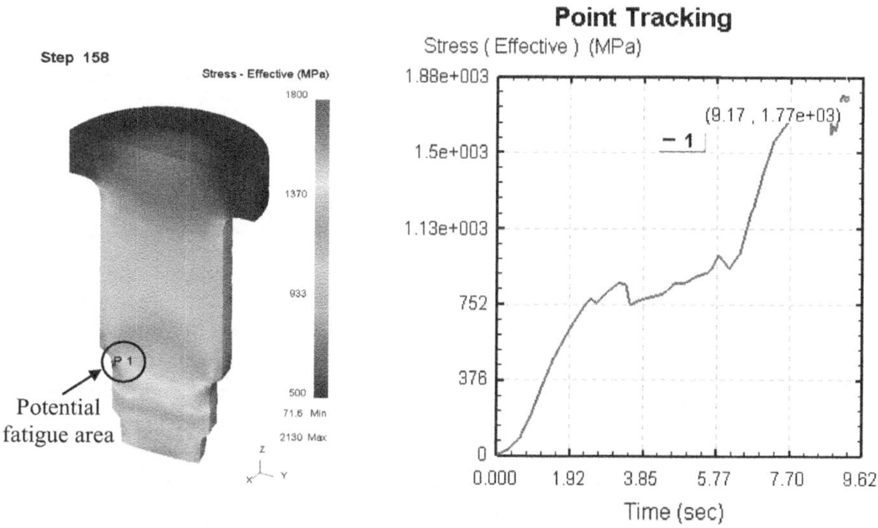

(a) Potential fatigue area (b) The effective stress at the tracking point P

Fig. 4.28 Potential fatigue location in Scenario 1 [23]

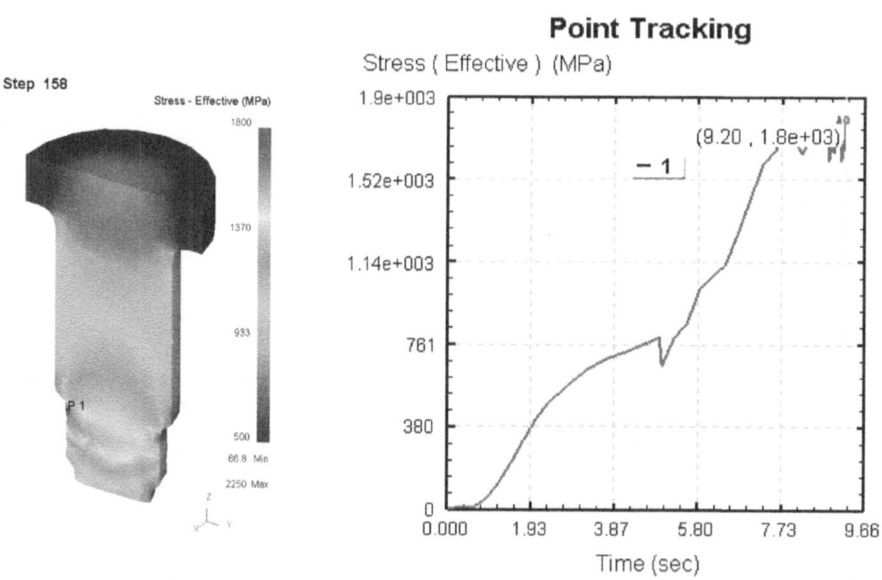

(a) Potential fatigue area (b) Effective stress at the tracking point P

Fig. 4.29 Potential fatigue location in Scenario 2 [23]

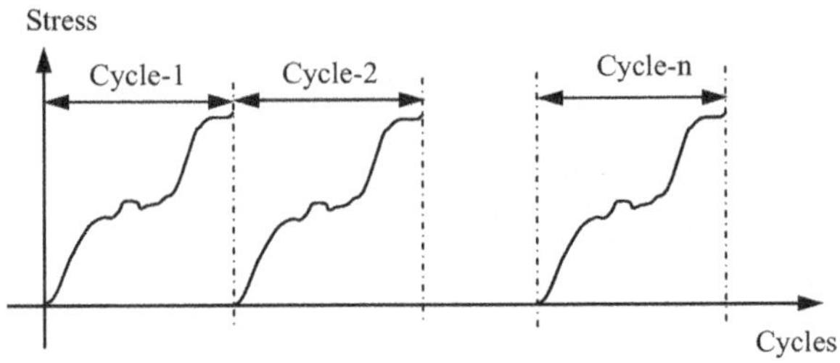

Fig. 4.30 The cyclic stress in forming process [23]

interpolate it based on the curves of the specific σ_m. The final fatigue life is determined based on the line from the interpolated stress point in the 2D plane.

From Tables 4.3 and 4.4, it can be found that the best design scenario is Scenario 9 where the expected die fatigue life is 9000. It means that the die can undergo 9000 cycles before it fails. From Table 4.3, it is found that the patch-up of 6 mm is located at the top of the smaller interior hole in the deformed part. This would make the material flow easily in backward and forward extrusion modes. In addition, the punch has a conic tip, which would facilitate the deformation of metal. All of these lead to the lowest stress in the punch and thus the punch has the longest service life.

Table 4.3 Simulation results for Scenarios 1–6 [23]

	Scenario 1 H6 (bottom)	Scenario 2 H6 (mid)	Scenario 3 H6 (top)	Scenario 4 H4 (bottom)	Scenario 5 H4 (mid)	Scenario 6 H4 (top)
Deformation load (Ton)	127	128	130	122	134	131
Effect stress at point P1	1770	1800	1630	1600	1810	1690
σ_a (MPa)	885	900	815	800	905	845
σ_m (MPa)	885	900	815	800	905	845
Expected fatigue life	2750	2300	4200	4800	2280	3600

Table 4.4 Simulation results for Scenarios 7–12 [23]

	Scenario 7 h6θ10 (bottom)	Scenario 8 h6θ10 (mid)	Scenario 9 h6θ10 (top)	Scenario 10 h4θ10 (bottom)	Scenario 11 h4θ10 (mid)	Scenario 12 h4θ10 (top)
Deformation load (Ton)	111	121	116	119	139	117
Effect stress at point P1	1510	1590	1450	1620	1880	1620
σ_a (MPa)	755	795	725	810	940	810
σ_m (MPa)	755	795	725	810	940	810
Expected fatigue life	7800	4700	9000	4500	1600	4500

4.6 Summary

Die performance and service life are affected by many factors related to die structure and component design, forming process determination, process parameter configuration, material selection and material properties configuration, and the entire forming system. Die design should be optimized such that the best die performance and the longest service life can be secured. To realize this goal, the above-mentioned affecting factors arising from different design perspectives need to be well considered. But how to consider so many affecting factors in design stage is a non-trivial issue. To address this issue, identification, prioritization, and classification of all the affecting factors are needed and the most important factors are then determined and fully considered in die structure and component design. To explore the whole panorama of their effect and explore the interaction and interplay among them, a systematic forming system design to consider all important factors is critical. In this chapter, some critical issues including die prestress design, die fatigue failure and die fatigue life design and analysis, and integrated simulation of metal-forming and die deformation are fully discussed such that the best die performance and service life are identified. Through case studies, the proposed methodologies for die fatigue life design, analysis, and assessment are validated and verified. The simulation-based design paradigm presented in this chapter can significantly shorten the design and development lead times of die and the entire forming system and further reduce the experimental tryouts in workshop in metal-forming and die making industries.

References

1. Falk B, Engel U, Geiger M (2001) Fundamental aspects for the evaluation of the fatigue behavior of cold forging tools. J Mater Process Technol 119:158–164
2. Knoerr M, Lange K, Altan T (1994) Fatigue failure of cold forging tooling: causes and possible solutions through fatigue analysis. J Mater Process Technol 46:57–71
3. Tong KK, Yong MS, Fu MW, Muramatsu T, Goh CS, Zhang SX (2005) CAE-enabled methodology for die fatigue life analysis and improvement. Int J Prod Res 43:131–146
4. Fu MW, Shang BZ (1995) Stress analysis of the precision forging die for a bevel gear and its optimal design using the boundary-element method. J Mater Process Technol 53:511–520
5. Lange K, Cser L, Geiger M, Kals JAG (1992) Tool life and tool quality in bulk metal forming. Ann CIRP 41(2):667–675
6. Fu MW, Yong MS, Tong KK, Muramatsu T (2006) A methodology for evaluation of metal forming system design and performance via CAE simulation. Int J Prod Res 44:1075–1092
7. Pedersen T (2000) Numerical modeling of cyclic plasticity and fatigue damage in cold-forging tools. Int J Mech Sci 42:799–818
8. Joun MS, Lee MC, Park JM (2002) Finite element analysis of pre-stressed die set in cold forging. Int J Mach Tools Manuf 42:1213–1222
9. Peder SH, Niel B, Jens G, Povl B (1999) Fatigue in cold forging dies: tool life analysis. J Mater Process Technol 95:40–48

10. MacCormack C, Monaghan J (2001) Failure analysis of cold forging dies using FEA. J Mater Process Technol 117:209–215
11. Lee YC, Chen FK (2001) Fatigue life of cold forging dies with various values of hardness. J Mater Process Technol 113:539–543
12. Nagao Y, Knoerr M, Altan T (1994) Improvement of tool life in cold forging of complex automotive parts. J Mater Process Technol 46:73–85
13. Lange K, Hettg A, Knoerr M (1992) Increasing tool life in cold forging through advanced design and tool manufacturing techniques. J Mater Process Technol 35:495–513
14. Bannantine JA, Comer JJ, Handrock JL (1990) Fundamentals of metal fatigue analysis. Prentice Hall Inc., New Jersey
15. Basquin OH (1910) The experimental law of endurance tests. Am Soc Test Mater Proc 10:625–630
16. Coffin LF (1954) A study of the effects of cyclic thermal stresses on a ductile metal. Trans ASME 76:931–950
17. Mansion SS (1953) Behavior of materials under conditions of thermal stress. In: Heat transfer symposium. University of Michigan Engineering Research Institute, pp 9–75
18. Boresi AP, Schmidt RJ, Sidebottom OM (1993) Advanced mechanics of materials. Wiley, New York
19. Scientific Forming Technoloiges Corporation (2014) DEFORM™ 3D manual, USA
20. Kocanda A (1990) Die steel for warm working—an evaluation of resistance to cyclic loading. Adv Technol Plast 1:349–354
21. Cooperation Research Group (1981) The fatigue strength of working forging tool steel of SKH 9. J Jpn Soc Plast Plast Work 22:158–165
22. Fu MW, Yong MS, Muramatsu T (2008) Die fatigue life design and assessment via CAE simulation. Int J Adv Manuf Technol 35:843–851
23. Fu MW, Chan WL, Lu J (2009) Die fatigue life improvement through the rational design of metal-forming system. J Mater Process Technol 209:1074–1084

Chapter 5
Flow-Induced Defects in Multiscaled Plastic Deformation

5.1 Introduction

In metal-forming industries, product quality is the most critical issue compared with other overriding issues such as productivity, product cost, and time-to-market. For metal-deformed parts, their quality can be represented by different factors from different aspects, which include geometrical and dimensional accuracies, surface quality, deformation defect, microstructure, phase composition and distribution, deformation and flow texture, and the mechanical properties of deformed parts. Therefore, the product quality assurance and control in the context of metal-deformed parts is a broad topic and also a non-trivial issue. How to design and develop defect-free and high-quality metal-deformed products aided by FE simulation to meet the required requirements and specifications is thus of concern and importance. It is also a critical, eluded, and tantalized issue in metal-formed product design and development.

Regarding the product quality from defect-free perspective, there are two main deformation defects to be considered in this book. One is stress-induced defect and the other is flow-induced defect. The stress-induced defect is more referred to as the stress-induced ductile fracture generated in the deformation process of ductile materials. The detailed forms of this category of defects include free surface, contacting surface, and internal fractures. The obvious root cause is the plastic deformation the material undergoes exceeds its deformation capacity at a given deformation condition. From stress point of view, the working stress in the

M.W. Fu, *Design and Development of Metal Forming Processes and Products Aided by Finite Element Simulation*, Engineering Materials and Processes,
DOI 10.1007/978-3-319-46464-0_5

deformation body exceeds the strength limit of material. For the flow-induced defect, it is caused by the irrational material flow in the deformation zones of the deforming materials. The common forms of this category of defects include folding, unfilling, and undercut. Among these defects, folding is one of the common flow-induced defects and also difficult to explore as it is often formed inside the deformation body and caused by the irrational flow inside the deformation material. It is thus the focus to explore and study in this book.

In this chapter, the flow-induced defects are extensively articulated, discussed, and analyzed in terms of flow behavior, formation mechanism, and avoidance method in multiscaled deformation processes from macro-, meso- to microscale by using experiments and simulations. In terms of the size scale of workpiece, macroscale refers to the parts with at least two dimensions larger than 10 mm, while mesoscale falls in the range of part dimensions from 1 to 10 mm. Microscale, on the other hand, is defined as the deformed parts with at least two dimensions in submillimeter. For the deformation processes for making of the macro-, meso-, and microscaled parts, they are thus called macro-, meso-, and microforming processes, respectively. In this book, the multiscaled plastic deformation processes are more focused on forging and extrusion in the defined three scales. The flow-induced defects occurred in forging and extrusion will be the kernel to be discussed and analyzed, and so will the stress-induced defects, which will be addressed in the next chapter.

5.2 Flow-Induced Defect in Forming Processes

Flow-induced defect is one of the common defects in metal-forming processes. By using the conventional metal-forming analysis approaches, it is very difficult to explore, analyze, and predict the flow-induced defects in the up-front design process as the defects are usually generated by irrational material flows, which could be different from time to time in forming process and thus difficult to predict and understand. In addition, they are affected by many factors, including the shape and geometry of the deformed parts, die structure and components, forming process route and process parameter configuration, and friction and lubrication in between workpiece and die. The typical flow-induced defects are folding and unfilling. The folding defect in bulk forming processes refers to the undesirable meeting of material flows in opposite directions and forming new contacting surfaces by merging of the unreasonable material flows in the deformation body. Furthermore, the newly merged material surfaces could be included inside the material in some occasions. In this scenario, the folding defects are difficult to be spotted out as there may not have obvious marks on the appearance of the deformed parts. Folding defect is detrimental to the properties and working performance of the deformed parts as the deformation texture and flowlines are not continuous or terminated at the folding defects such that part of the load-carrying capacity could be lost. In addition, the folding defect would be the source of stress concentration and initiation of microcrack when the deformed parts are in service. The unfilling, on the other hand, refers to the three-dimensional space between punch and die cavity, which is intended and desired to be filled up by material, but actually not in forming process, such that the designed part is not fully formed.

To show the folding defects in the parts of different size scales, Fig. 5.1a presents some axisymmetrically flanged parts in macroscale with folding defects. The defects can be clearly found on the appearance of the parts. In addition, it is quite interesting and surprising to find out that all the folding defects occur at the flanged features of the workpieces. To explore more on this type of flow-induced defect, Fig. 5.1b presents different types of flanged parts and the FE simulation results of material flow in forming process. It is found that all the parts with flanged features have the folding defect in the inner circumferential surface of the flanged feature. The results show that there would definitely be a common underlying mechanism to form the folding defects in these parts with flanged features or the so-called flanged parts in this book. Therefore, the mechanisms for the folding defect formation and avoidance in bulk forming processes need to be systematically studied via experiments and FE simulations.

In tandem with this goal, the study on forming of defect-free axisymmetrically flanged parts with one single forming operation is conducted. The axisymmetrically flanged part as shown in Fig. 5.2 is developed to study the defect formation mechanism in macroscaled forming process. The folding defect formation and avoidance mechanisms are investigated. A design of the dynamic change of tooling geometry and a FE simulation-based design framework for defect avoidance are presented. Through industrial case studies, the proposed design framework is validated and the findings are demonstrated.

In terms of mesoscale, a mesoscaled part with non-axisymmetric features, as shown in Fig. 5.3, is used as the case study to explore the flow-induced defect in mesoscaled forming process. The part is extruded via mesoforming process with flow-induced defects. The formation mechanisms of the flow-induced defects in this non-axisymmetric part are investigated using FE simulation and experiment. In addition, microscaled parts are also used as case studies to further explore and study the defect formation mechanisms and the size effect on the defect formation in microforming process is systematically investigated in the last part of this chapter.

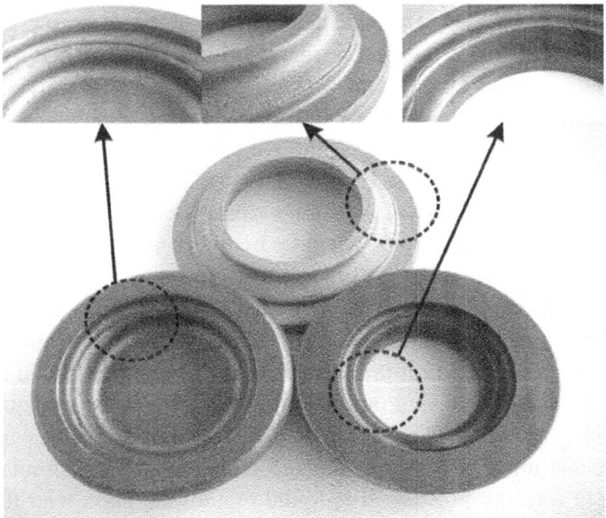

(a) Some axisymmetrically flanged parts in macro-scale with folding defects.

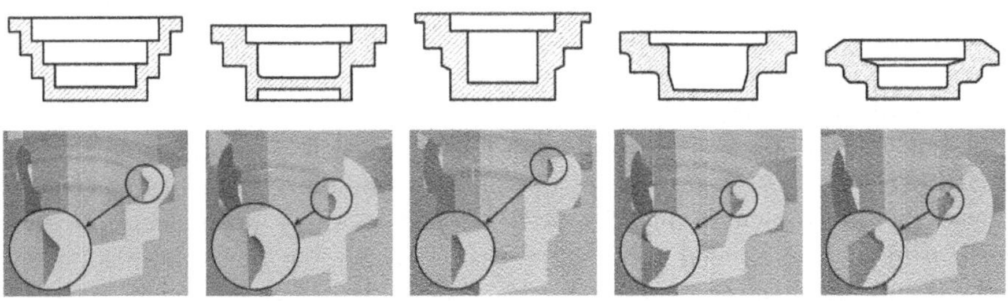

(b) Simulation results of different axisymmetrically flanged parts with folding defects [1].

Fig. 5.1 Examples of the axially symmetrical flanged parts in macroscale with folding defects [1]

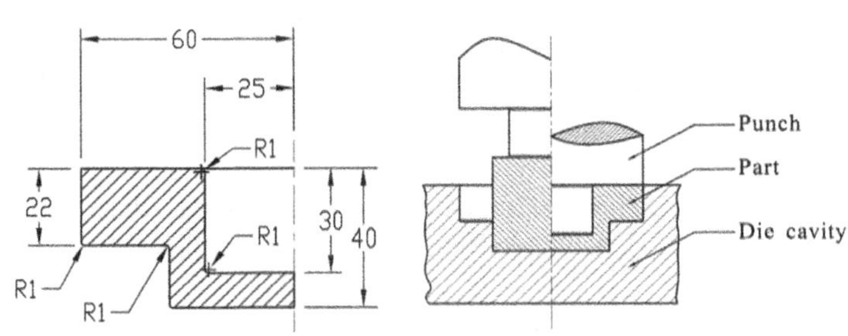

Fig. 5.2 Geometric parameters in axisymmetrically flanged parts in macroscale and illustration of die assembly [1] (dimensions in mm)

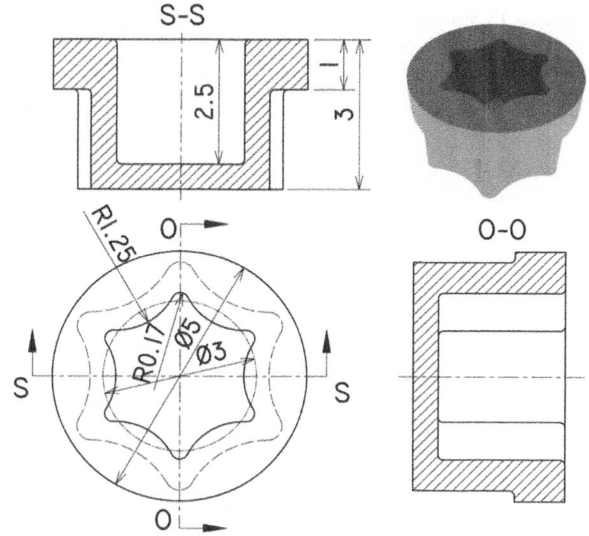

Fig. 5.3 A case study part in mesoscale with non-axisymmetric features [3]

5.2.1 *Flow-Induced Defect in Forming of Axisymmetric Parts with Flanged Features*

In forming the above-mentioned axisymmetrically flanged parts as shown in Fig. 5.1b, the punch in the die set applies an external deformation force onto billet material and the material flows into die cavity to form the flanged features. Generally, the flow patterns for all the flanged die cavities are similar for the studied cases. To conduct the FE simulation of the material deformation, DEFORM 3D [2], a commercial FE software system for simulation of material deformation and die stress and strain distribution, is employed. Since the focus of simulation is on the study of material flow, the billet is thus modeled as a rigid-plastic body, while the punch and die components are as rigid bodies. In hot forging process, tooling temperature is usually lower than billet, which leads to the heat transfer in-between. Taking heat transfer into account, FE simulation is thus needed.

In this study, two tooling thermal models, viz., constant temperature and heat transfer-enabled models, are used to compare the difference between material flow behaviors in material-forming process. The same geometry parameter configuration, as shown in Fig. 5.2, is used to study the difference by using the two temperature models. The simulation settings are presented in Table 5.1. The material properties represented by stress–strain curves are defined as a function of temperature and strain rate, as shown in Fig. 5.4. The workpiece material is AISI 1016 and is widely used in metal-forming industries. The information of preheating temperature and punch speed is obtained from industry, while the shear friction constant and heat transfer coefficient are determined based on the recommendation from the FE simulation system. Figure 5.5 reveals the material flow behaviors

Table 5.1 Simulation settings in simulation of folding formation process [1]

Billet (rigid-plastic model)	
Number of elements	25,000
Material	AISI 1016
Preheat temperature	1130 °C
Punch (rigid model)	
Number of elements	15,000
Speed	1000 mm/s
Material	AISI H-26
Preheat temperature	250 °C
Die cavity (rigid model)	
Number of elements	25,000
Material	AISI H-26
Preheat temperature	250 °C
Objects contact settings	
Shear constant	0.5
Heat transfer coefficient	5 N/s/mm/C
System settings	
Increment	0.2 mm/step
Convergence criteria	0.001

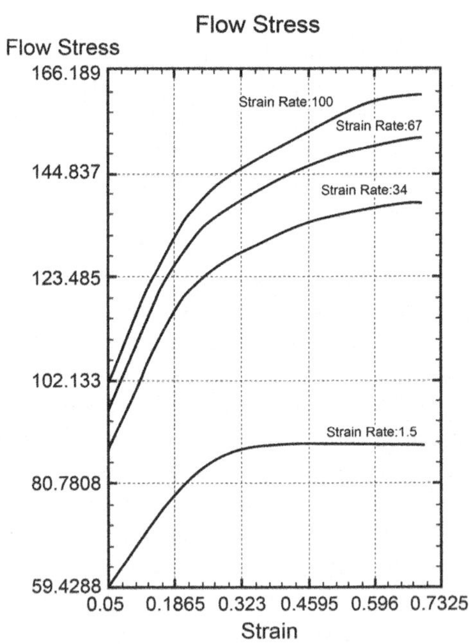

Fig. 5.4 Flow behaviors of AISI 1016 at 1130 °C [1]

predicted by both the tooling thermal models. It is found that the heat loss rate in workpiece is faster by using the constant temperature model. This is because the constant temperature of 250 °C is set for tooling, which is much lower than the

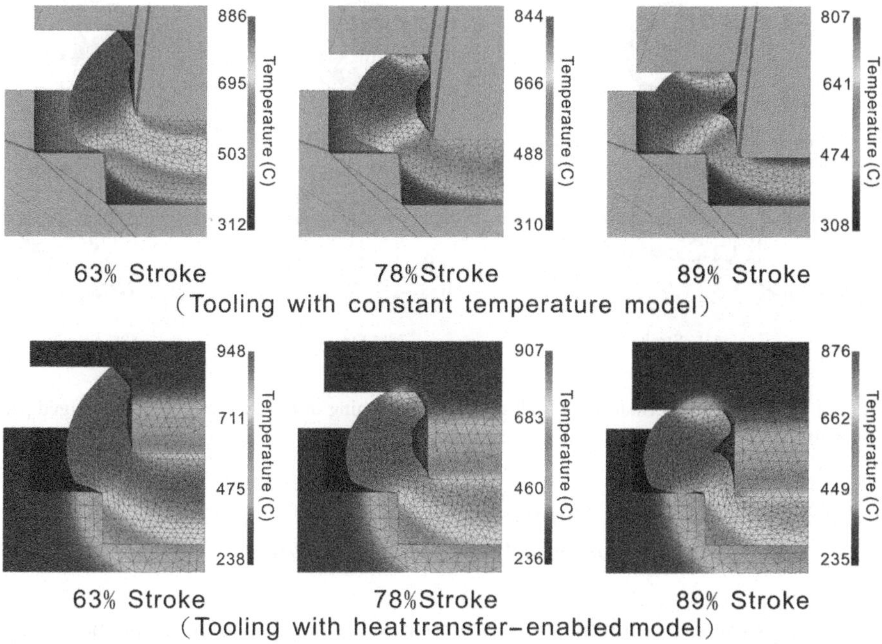

Fig. 5.5 Flow behavior comparison by using two different tooling temperature models [1]

workpiece temperature. For the tooling with the heat transfer-enabled model, the heat is transferred from workpiece to tooling, and thus, the temperature of workpiece at the contact region is higher than that based on the constant temperature model. From material flow behavior perspective, the material flow behaviors predicted by both the tooling temperature models are almost the same. Therefore, it is suggested that the modeling of tooling with the constant temperature model for study of the folding defect in forging of the axisymmetric components is reasonable.

Regarding the formation of folding defects in forging of the axisymmetrically flanged parts, Fig. 5.6 shows the two major stages in the formation process. In stage 1, when the punch presses billet material, the material under the punch shoulder flows up and a critical inner curved surface is formed, as shown in the figure with the deformation of 63 %. The curvature of the curved surface then becomes big and an empty space between the outer circumferential surface of punch head and the curved surface is formed and gets bigger. When the punch shoulder contacts the deformation material, stage 2 begins. In stage 2, punch shoulder presses the material in between the punch shoulder and die flange. The materials in between the punch shoulder and die flange have different flow directions. The materials under the punch shoulder flow downward and inward while the ones in between the punch head and above the die flange flow upward and outward. This flow pattern thus develops a folding defect, as shown in Fig. 5.6.

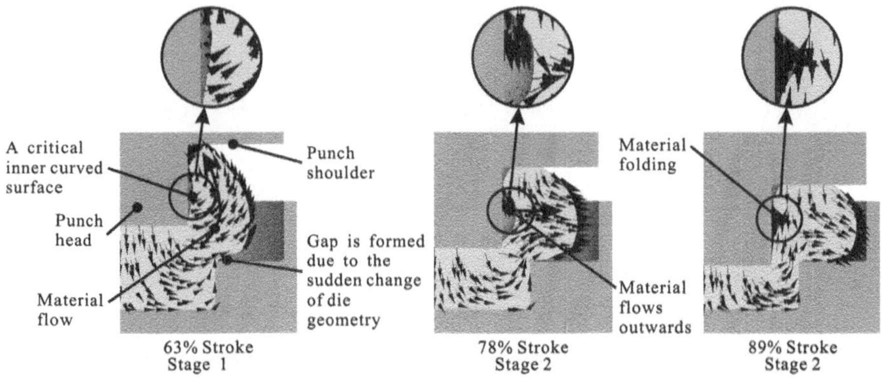

Fig. 5.6 Initiation and formation of folding defect in forming of the axisymmetrically flanged part in macroscale [1]

5.2.2 *Flow-Induced Defect in Forming of Non-asymmetrically Mesoscaled Parts*

Figure 5.3 shows a non-axisymmetrical part with flanged features in mesoscale. The part has non-axisymmetrically outer star contour and an internal star hole. The bottom thickness is 0.5 mm and the radius of the round corner is only 0.17 mm, which belongs to microscale. Considering the whole part geometry, the part is considered as a mesoscaled part, and thus, the part can be fabricated by mesoscaled forming process, viz., meso-extrusion. In this section, the material flow and the formation of flow-induced defects in meso-extrusion of this case study part are systematically discussed and investigated.

From deformation perspective, the plastic deformation of the workpiece includes two parts. The first one is the star contour feature, while the second one is the flanged feature. The die assembly developed for forming of the workpiece is shown in Fig. 5.7, in which the star contour feature is formed by bottom die, the flanged feature by die container, and the geometries of the inner star hole of the workpiece ensured by punch head. To facilitate the ejection of the deformed part from die cavity, the bottom die is designed into two parts. In addition, a punch guide is introduced in die structure to prevent the punch deflection in radial direction and further to ensure its perpendicular movement in vertical direction in such a way the vertical alignment of punch movement in forming process is secured.

5.2.2.1 Mesoforming Experiment

In forming of this case study part, pure copper is selected as the experimental material for its excellent plasticity and wide application in industries. Based on the volume constant principle in plastic deformation process, the cylinder billets with

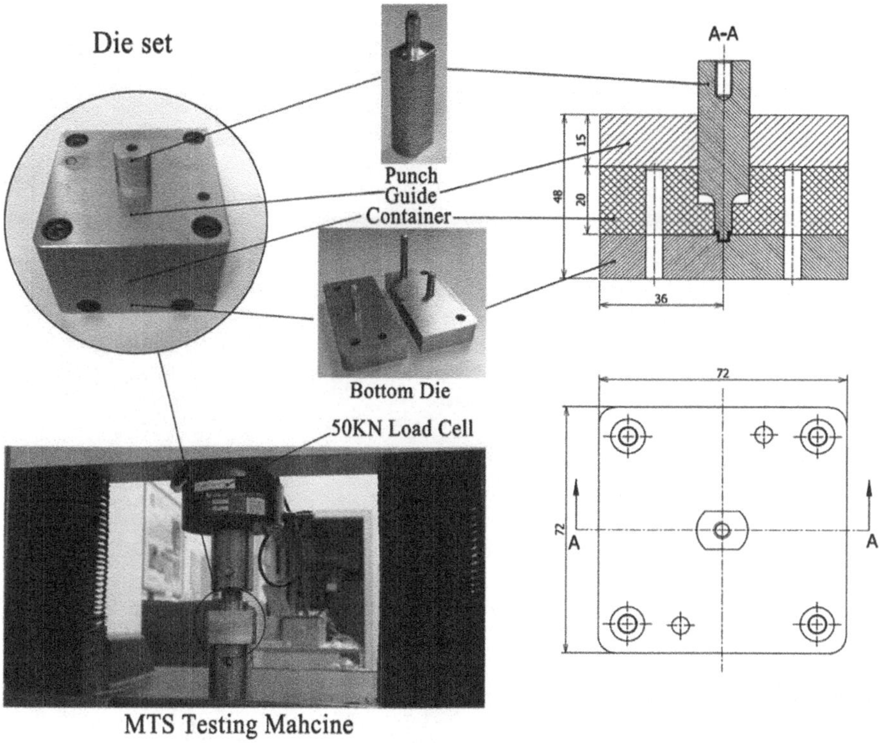

Fig. 5.7 Experimental die set and MTS testing machine [3]

the dimensions of Ø3.5 mm × 3 mm are fabricated. The part is deformed by one-stroke forming process. The punch stroke is 2.5 mm. The mesoforming experiment is conducted in a MTS material testing machine with a load cell of 50 KN. The testing platform is also shown in Fig. 5.7. The tooling and billet contact surfaces are lubricated by machine oil. The punch speed is 0.01 mm/s.

Figure 5.8 shows the deformed parts in different deformation strokes. It can be seen that there are visible folding defects in the inner star hole surface and the flanged feature of the parts. Therefore, the designed mesoforming process cannot produce defect-free parts and needs redesign. To avoid the occurrence of folding defects, the material flow pattern and defect formation mechanisms must be investigated. First of all, the deformation stages in the whole forming cycle are analyzed. In Fig. 5.9, it is found that the forming process could be divided into three stages. In the first stage, the billet undergoes deformation in both the radial and the axial directions simultaneously until the die cavity is completely filled up in radial direction. The second stage is the backward extrusion until the punch shoulder contacts the deformed workpiece. In this stage, the deformation load slightly rises. The last stage is the combined forward and backward extrusion until the entire die cavity is totally filled. At this stage, the deformation load increases

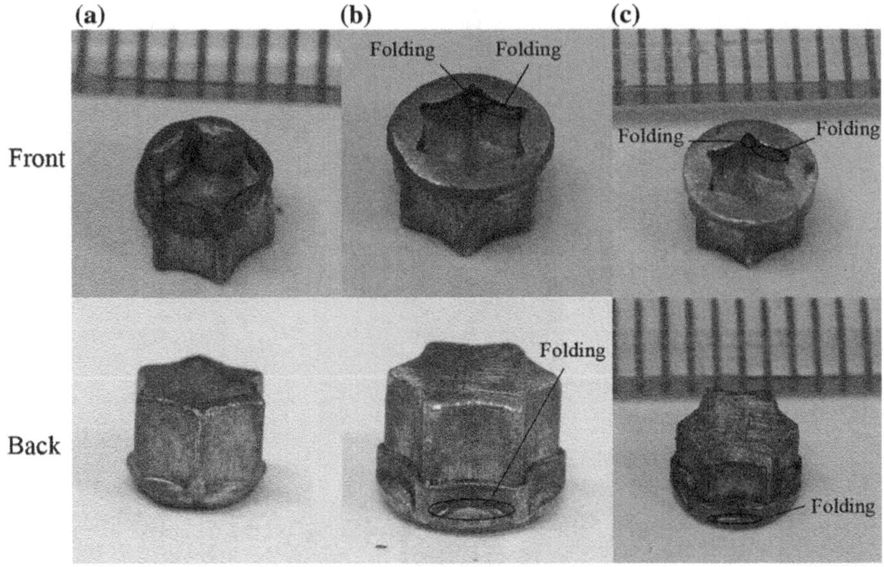

Fig. 5.8 Deformed parts with flow-induced defects in different strokes: **a** 65 %, **b** 75 %, **c** 85 % [3]

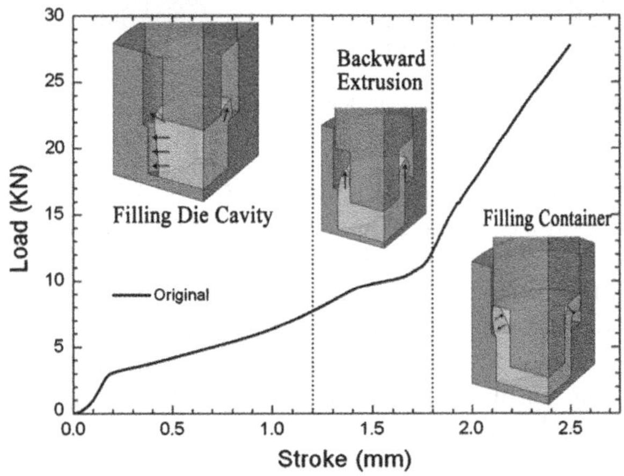

Fig. 5.9 Load–displacement curve and three different deformation stages [3]

sharply since the unfilled space in the cavity is very small, and thus, the final filling needs a great pressure compared with the preceding two stages.

In Fig. 5.8, it can be seen that there are two flow-induced defects, viz., folding defects, in the deformed part. At the stroke of 65 %, there is no defect formed. With the movement of punch, the two folding defects are formed at the stroke of 85 %. One is located at the inner hole and the other is located at the flange, as shown in Fig. 5.8b. In the end of deformation, the defects are more obvious and shown in Fig. 5.8c.

In this case, the first defect is an uneasily discovered folding defect appearing in the inner star surface. With the symmetry of mid-surface, a quarter of the part is analyzed to study the exact position and formation mechanism of the defect. The part is sectioned along the symmetrical plane as shown in Fig. 5.3. Figure 5.10 shows the O–O view with the strokes of 75, 90, and 100 %, respectively. Figure 5.10a shows the extrusion process in backward state and the inner surface with a little concave feature. From the cross section, it is found that an inner concave curve is formed at the top part of workpiece. This could be the origin of folding defect. Figure 5.10b shows that the material under the punch contacts punch shoulder and begins to flow downward. Meanwhile, the material moves upward continuously from the bottom and the opposite material flows meet up to form the vortex material flow. At the stroke of 100 %, as shown in Fig. 5.10c, the material fills the flange cavity and the materials flowing from the bottom further facilitate the formation of concave contour. The folding defect is thus inevitably formed.

In Fig. 5.10, it can be seen that the initiation of the concave surface is critical in formation of this type of flow-induced defect. The concave surface is the result of the irregular material flow, which is actually difficult to predict and control accurately, if not impossible, by the traditional metal-forming analysis approaches.

In addition, the S–S section shown in Fig. 5.11 reveals the defect initiation and formation process of another inner defect in the other side of the part. Figure 5.11a shows that there is no defect at the stroke of 75 %. With the stroke of 90 %, the opposite material flows are generated in the top section, as shown in Fig. 5.11b. In Fig. 5.11c, a slight folding defect is formed in the upper inner surface, which is the expansion of the vortex initiated in the previous stage and shown in Fig. 5.11b.

From the above analysis, the root causes of the folding defect in the flange and the slight folding defect in the up inner surface in the mesoscaled non-axisymmetric part are revealed. The common root cause for these defects is the opposite material flow, which could be further caused by the significant change of the cross-sectional areas in different features in the deformed part. The FE simulation later will explore more in terms of the occurrence of the opposite material flows.

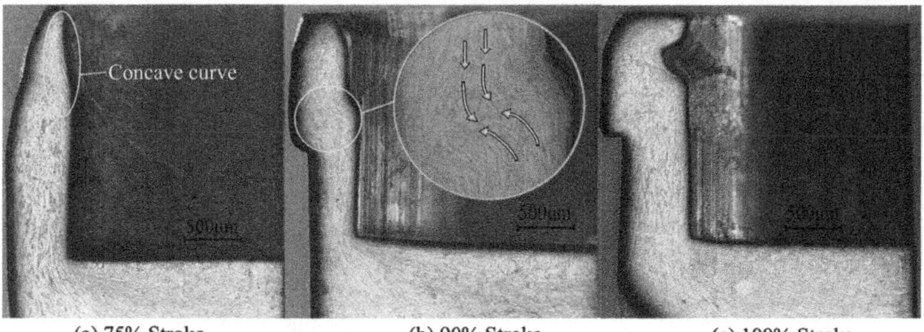

Fig. 5.10 Folding formation at the O–O section at different strokes [3]

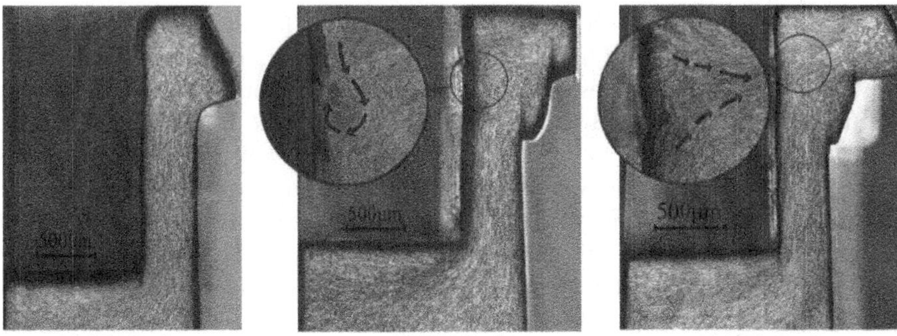

(a) 75% Stroke (b) 90% Stroke (c) 100% Stroke

Fig. 5.11 Folding formation in S–S section at different strokes [3]

On the other hand, the overall dimension of the investigated part is mesoscaled, but some features are microscaled. To clearly observe the defects, the deformed part is observed by SEM and shown in Fig. 5.12. In Fig. 5.12a, there is a V-shaped folding cavity in the O–O section and it gradually becomes smaller along the circumferential direction and eventually reaches the smallest level in S–S plane and shown in Fig. 5.11. The material flow texture near the cavity obviously shows the opposite material flow directions from the top to the bottom. The folding defect in the flange is shown in Fig. 5.12b. The appearance of the defect is also non-axisymmetric. In the connection region with the concave surface, the folding defect is very apparent and severe. But in the connection region with convex surface, there is no defect. Moreover, the degree of folding defect decreases with the increase of convex as the change of the sectional area decreases gradually. To avoid this defect, it is crucial to avoid the severe change of the sectional area in deformed part design, if possible.

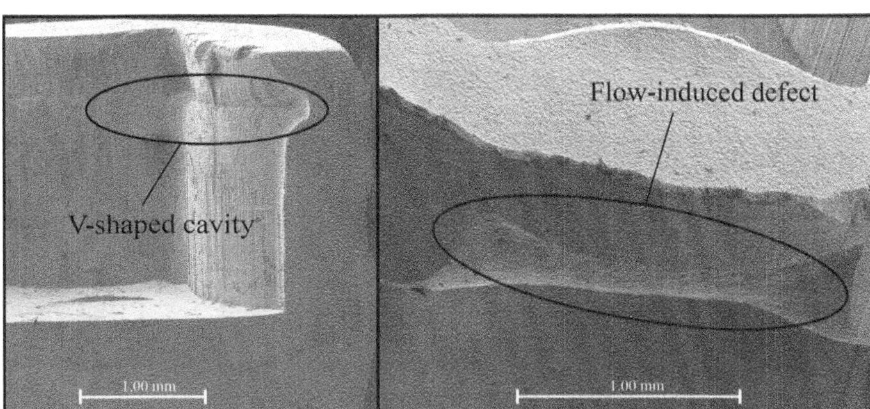

(a) Inner star hole (b) Flange section

Fig. 5.12 3D view of the inner folding defect [3]

According to the experimental results, three folding defects are found in the deformed part. The defect formation mechanism is analyzed based on the physical observation. In the actual mesoforming process, it is very difficult, if not impossible, to obtain the instantaneous material flow and to reveal the panorama of the defect formation process. FE simulation is thus conducted to address this issue.

5.2.2.2 FE Simulation

The case study part is quarterly axisymmetric and the deformation is represented and modeled by a quarter of the workpiece and the forming system. The simulation is focused on the material flow behavior and the elastic deformation of the billet material is thus neglected. The workpiece is considered as a plastic deformation body, while the tooling system is treated as a rigid body for representing the geometries of tooling components only. A commercial deformation FE simulation software DEFORM 3D is employed to conduct the simulation for revealing the formation of the flow-induced defects in the deformed part. In FE simulation, the speed of punch is 0.01 mm/s and the constant friction coefficient is 0.12 in this mesoforming process.

With the help of FE simulation, the material flow behavior and the effects of the forming process parameters involved in the process are revealed. The nodal velocity revealed by simulation is employed to investigate the material flow pattern. Figure 5.13 shows the flow behavior of the case study part at different strokes. Figure 5.13a shows that a critical curve is formed at the stroke of 75 % and the material contacts the punch shoulder and then begins to flow downward. In Fig. 5.13b, three opposite metal flow regions are discovered with the stroke of 90 %. The root cause of the folding defect is the opposite material flow, leading to the mutual contact of the free surfaces. In Fig. 5.13c, it shows the opposite material flows further carry on until the forming process is finished. Finally, the severe and reverse material flow is observed in the inner surface of the star hole feature where a folding cavity is formed. The simulation results also tally well with the experimental results and are further corroborated by the experiments.

The formation process of the folding at the flange of the mesoformed part is shown in Fig. 5.14. With the stroke of 40 % as shown in Fig. 5.14a, the forming process is more on die cavity filling stage and the part of the star outer contour and flanged feature is formed with different cross-sectional areas. Figure 5.14b is the backward extrusion stage and the features formed in Fig. 5.14a are further formed. At the stroke of 80 % in Fig. 5.14c, the process is the flange filling stage in which the material at the top is forced to flow downward. At the stroke of 90 %, the material begins to flow downward and moves in radial direction simultaneously to form the flanged feature. In the intersection part between the vertical wall and the flange, the folding defect is obviously observed, as shown in Fig. 5.14e. Through FE simulation, the folding defect mechanism in the flange section is also revealed.

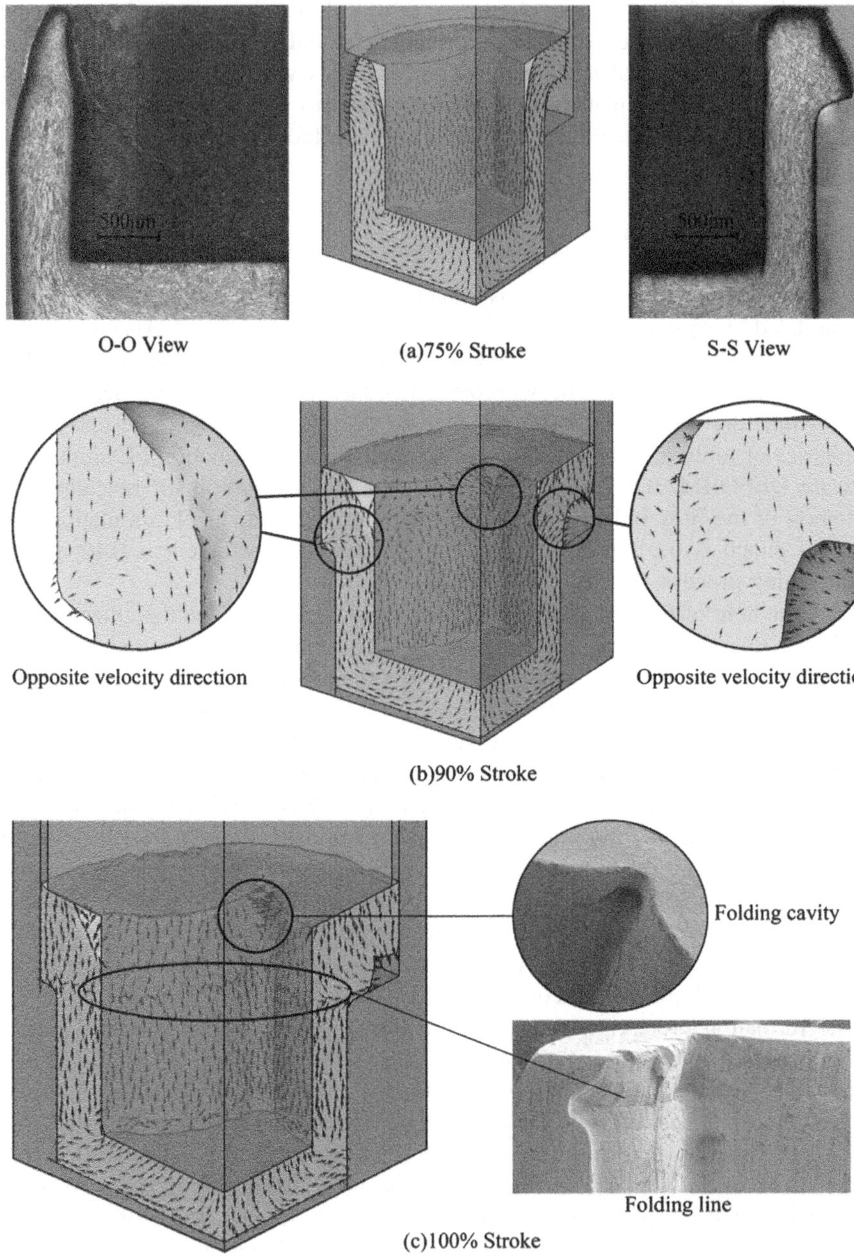

Fig. 5.13 Inner folding formation in forming process [3]

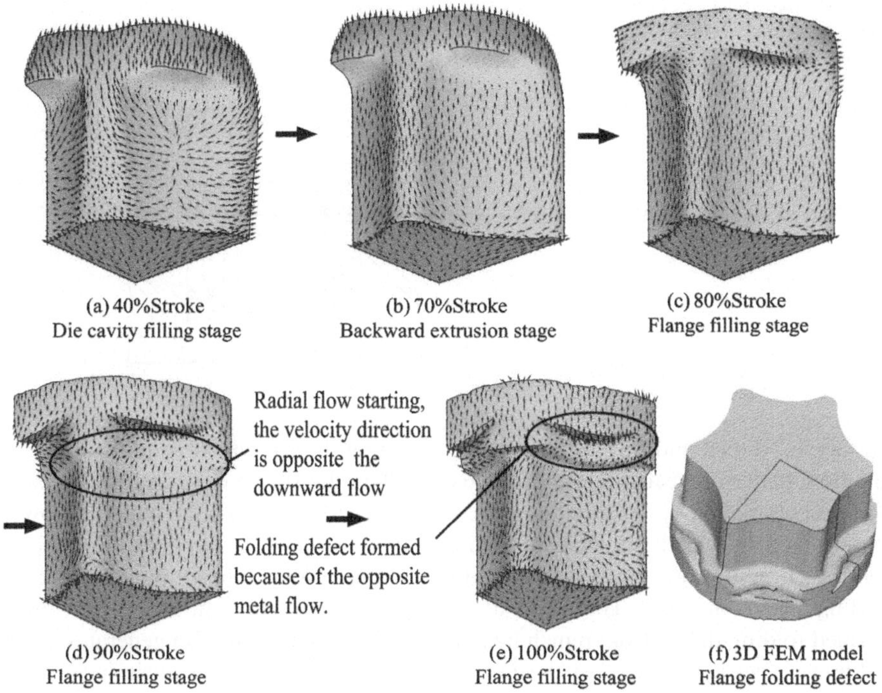

Fig. 5.14 Folding formation in the flanged feature [3]

5.3 Defect Avoidance in Forming Process

5.3.1 Employment of Spring-Driven Die Insert Structure

5.3.1.1 Sliding Die Insert Design and Material Flow Behaviors

Based on the revealed folding defect formation mechanism, it is suggested to introduce a spring-driven sliding die insert into the die structure to facilitate the material flow along the surface of punch head and further to avoid the formation of the critical inner curved surface. Figure 5.15 shows the schematic of the tooling structure with a spring-driven sliding die insert. The influences of the initial height and position of die insert (Parameter h in Fig. 5.15) are studied. The punch provides an external force to overcome the spring force so as to make the die insert slide along the inner surface of die cavity. To have a better quantitative comparison of the deformation force, the spring forces in the last forming stage at which the final part geometries are obtained for all the design scenarios are set to be the same for different heights and positions of the die insert. Figure 5.16 shows the simulation results of the material flow behavior with different heights and positions of the sliding die insert. When the height position is set to 10 mm, the inner curved

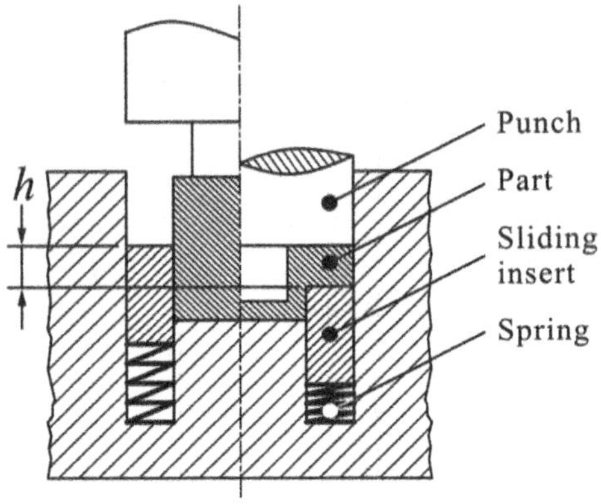

Fig. 5.15 Schematics of the tooling assembly with a sliding die insert [1]

surface is formed and further turns to be a folding defect, which is quite similar to the flow pattern revealed in the prior cases. When the stroke is 80 % (100 % is set for the completion of deformation), the flow pattern shows that the material flow is constrained by punch shoulder and sliding insert in such a way that the material tends to flow upward to fill the flanged space among punch head, punch shoulder, sliding insert, and die. This kind of flow pattern not only shifts the previously formed folding defect from punch head to punch shoulder, but also generates a new folding defect at the edge of die insert. When the insert height position is set to 20 mm, the material near the punch head is not trapped at the corner of punch head, but flows outward. Meanwhile, the inner curved surface does not become a folding defect. At the stroke of 66 %, however, a similar flow pattern is generated for the scenario with the insert height position of 10 mm at the stroke of 80 %. The folding defect occurs at the insert edge region. When the insert height position is set to 30 mm, the material flows outward smoothly to fill up the die cavity. The die insert moves downward and a S-flow pattern is formed. With the insert height positions of 40 and 50 mm, the forming system has the backward cup-like extrusion flow pattern and the material reaches the punch shoulder before the material is separated from the punch head. The space at the punch shoulder corner is significantly reduced. This condition prevents the formation of folding defect.

Figure 5.17 shows the load–stroke curves in different insert heights and positions. It is found that all the studied scenarios have almost the same maximum forming load at the end of forming. Figure 5.18 presents three forming stages in the deformation process based on the characteristics of the load–stroke curves. The first stage is the punch head pressing the billet and the billet material flowing backward. The higher the initial die insert position of die insert, the higher the load needed to deform the material and the earlier the material reaching the punch shoulder. Once the material contacts the punch shoulder, the forming load rises rapidly and stage 2 begins. In stage 2, the punch head and shoulder press the upfront material to flow outward and fill up the die cavity. Once the force applied on the sliding die insert is larger than the sum of friction force and spring recovery force, the sliding die insert

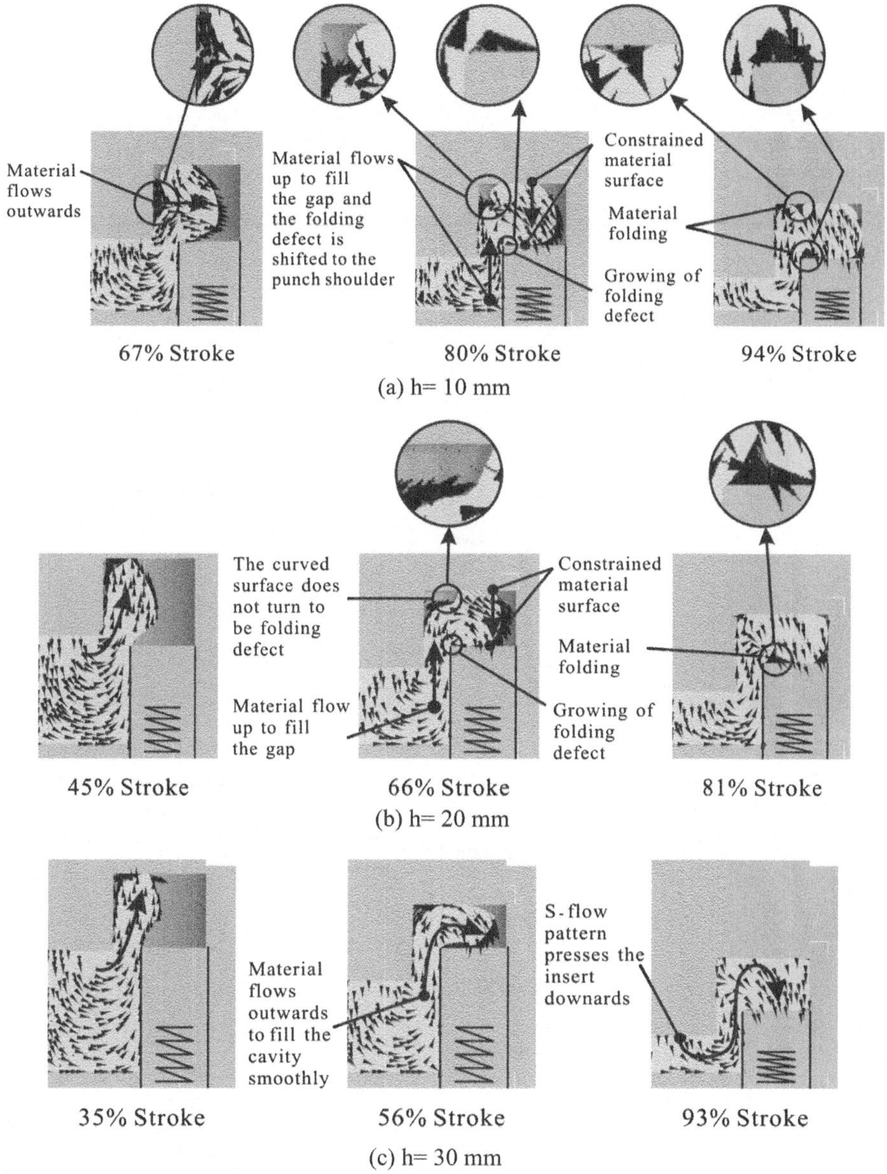

Fig. 5.16 Flow behaviors with different insert heights and positions [1]

moves downward and stage 3 begins. In stage 3, the die insert keeps moving downward to release the space for material flow.

The influence of spring stiffness on material flow and deformation load is also studied. The deformation behavior of material using the springs with different

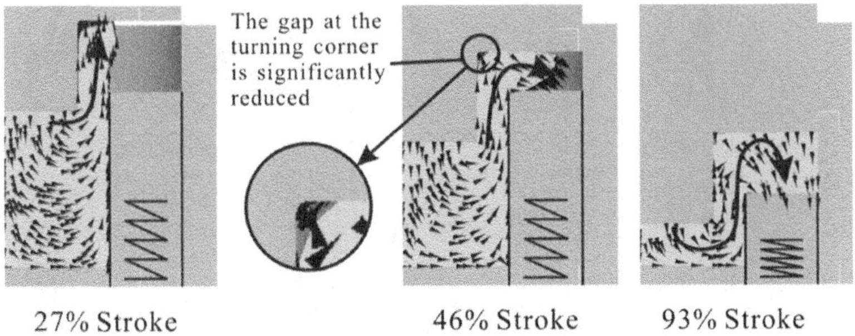

(d) h= 40 mm

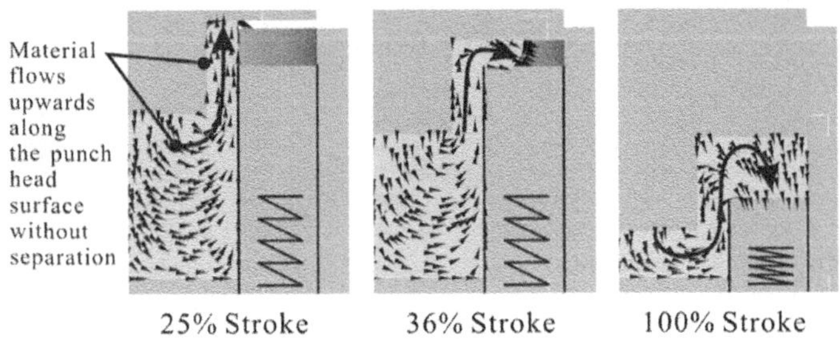

(e) h= 50 mm

Fig. 5.16 (continued)

Fig. 5.17 Load–stroke curves in different insert heights and positions [1]

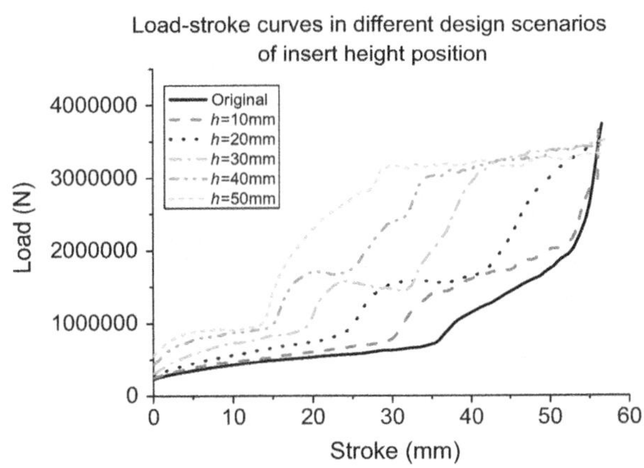

stiffness constants in the scenario with the initial insert height position of 40 mm is simulated. Figure 5.19 shows the load–stroke curves with different spring stiffness constants. It reveals that the spring stiffness does not play an important role in stages 1 and 2, but it affects the deformation load in stage 3. The figure further shows that the larger the spring stiffness constant, the steeper the load curve and the higher the deformation load at the end of stroke. Since the height position of sliding

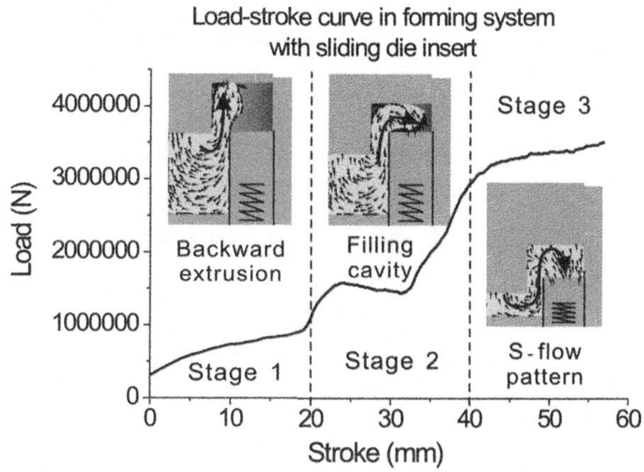

Fig. 5.18 Three forming stages in load–stroke curve (h = 30 mm) [1]

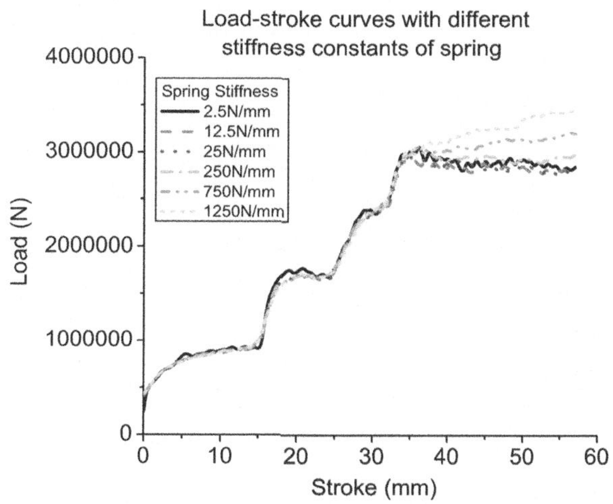

Fig. 5.19 Load–stroke curves with different spring stiffness constants [1]

insert is critical in controlling flow behaviors and avoiding folding defect, it needs to maintain a critical height and position of die insert to avoid the defect initiation and formation in stages 1 and 2. The variation in the height and position of sliding die insert in the process is studied. It is found that the variation in position or displacement of die insert is not affected by spring stiffness, as shown in Fig. 5.20. In addition, the influence of friction conditions on the insert movement is also investigated. It is found that the friction condition plays a role on delaying the movement of die insert, as shown in Fig. 5.21, but does not affect the displacement of die insert significantly.

Furthermore, when the punch moves down and the material flows upward, the flowing material provides an upward force to sliding insert so that the die insert could keep its height and position. If the die insert fails to maintain the effective height and position, the folding defect would be induced. Figure 5.22 shows the

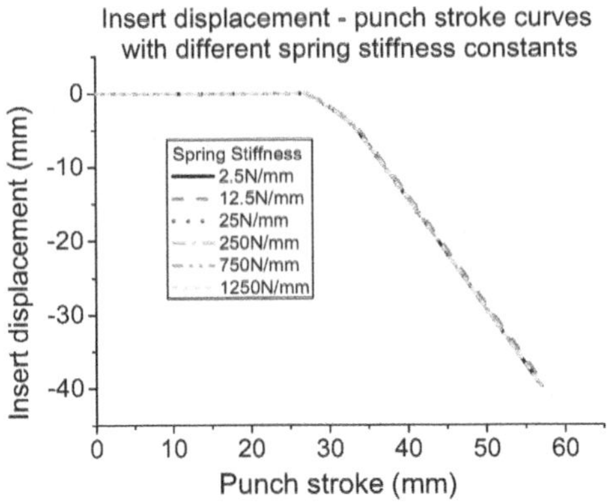

Fig. 5.20 Insert displacement–punch stroke curves with different spring stiffness constants [1]

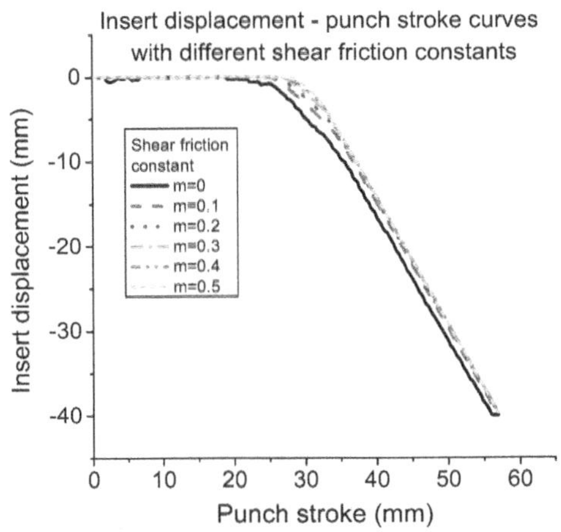

Fig. 5.21 Insert displacement–punch stroke curves with different shear friction constants [1]

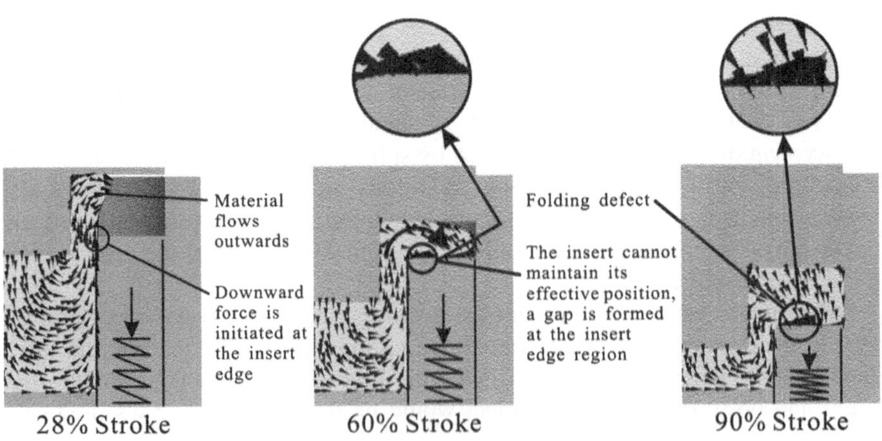

Fig. 5.22 Folding defect formation with a sliding insert in the scenario without friction [1]

flow behaviors in the scenario without friction on the surface of sliding insert. A downward force is applied at the insert edge region once the material flows outward. There is no upward force which can be generated without friction on the surface of die insert. In addition, the material cannot fill up the space near the insert edge and a gap is formed at the edge of die insert since the insert keeps moving downward, which leads to the formation of folding defect.

5.3.1.2 Spring and Die Insert Design Aided by FE Simulation

Based on the simulation results of the developed FE models in the previous sections, a spring and die insert design framework aided by FE simulation is presented in Fig. 5.23. The objective of this design framework supported by simulation is to reveal deformation behavior and flow pattern of the design configuration such that folding defect prediction and avoidance can be done. The first part of the framework is the modeling of forming system, which represents the geometries and shapes of metal-formed product, tooling geometry and structure, and billet geometries and dimensions using CAD systems. The generated CAD models are

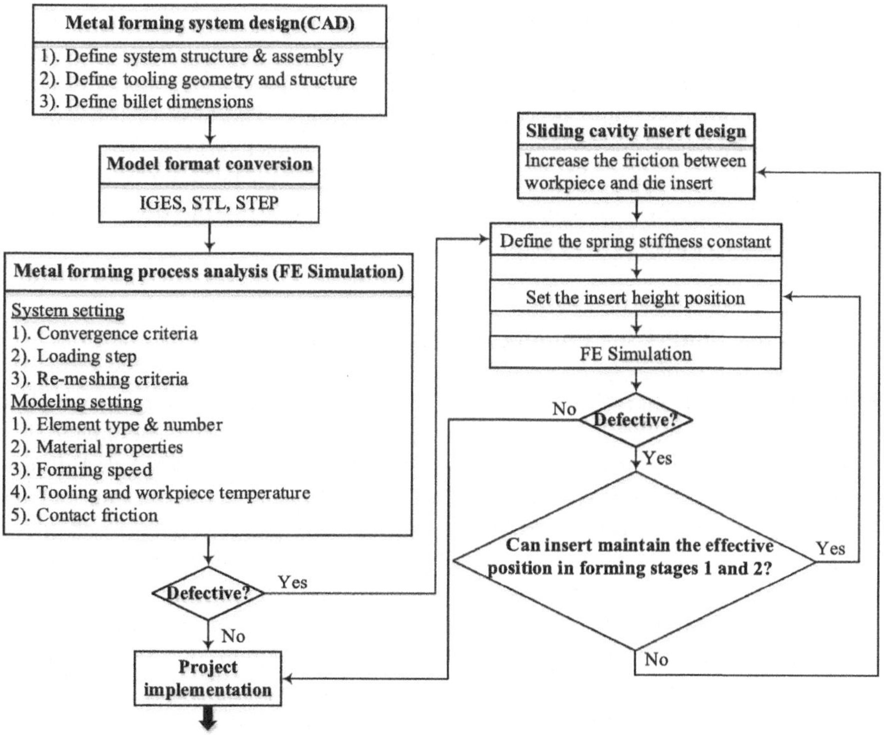

Fig. 5.23 Spring and die insert design framework aided by FE simulation [1]

then converted to a data exchange format such as IGES, STL, and STEP and then imported to a FEM system for simulation. To conduct the material-forming simulation, it is necessary to configure the setting of the simulation system, which includes convergence criteria, loading step, and remeshing criteria. From modeling setting perspective, element type and number, material properties, forming speed, tooling and workpiece temperature, and friction condition need to be given. If the simulation shows the occurrence of folding defect, a sliding die insert and a spring component are introduced into the die structure to avoid the defect.

To design the sliding die insert, it needs to define the initial height position of the insert and the spring stiffness constant. Since the initial height position is the most critical factor to control material flow and avoid folding defect, it needs to be optimized first. The height position of the insert should be defined based on the tooling structure and its influence on material flow behavior. The spring does not play a role on controlling the movement of insert in forming process, but it is used to return the die insert to its initial position after the deformed part is ejected. Therefore, the stiffness constant of the spring should be defined based on the contact friction of sliding surfaces and the weight of sliding module. Since the additional components are needed to set up sliding insert, the interference and die stress concentration problem may exist and should be considered. If the simulation result shows the occurrence of folding defect, it needs to investigate whether the defect can be avoided by increasing the height and position of die insert. If defect occurs and the insert cannot maintain its effective height position in forming stages 1 and 2, degrading the surface finishing of the sliding insert and increasing the friction constant in simulation could be considered accordingly. The upward force of die insert can be increased by increasing the friction of contact surfaces. By using the simulation results, the modification of tooling design can be done based on the folding formation and avoidance mechanism. The optimized die design to avoid folding defect can thus be identified.

5.3.1.3 Case Studies for Folding Defect Avoidance

An airplane component from literature [4] is selected as a case study part to explore the formation mechanism of folding defect and its avoidance. The material is Al-alloy and the part is formed by cold forging process. The detailed part dimensions and die assemble are shown in Fig. 5.24. The geometries of the intermediate preforms at different forming strokes are shown in Fig. 5.25. It is found that a folding defect occurs in the forming process. Through FE simulation, the formation process of the folding defect is revealed and illustrated in Fig. 5.26.

To avoid the folding defect at the flanged region, a sliding die insert is introduced in the die structure. The initial height and position of the insert, viz., $h = 10$ mm of the first case in Fig. 5.16, is firstly considered. The scenario with the sliding displacement of 10 mm is simulated and the results are shown in Fig. 5.27a. It is found the folding defect occurs, but shifts upward. Therefore, the initial height of the die insert needs to be increased. The second scenario with the sliding

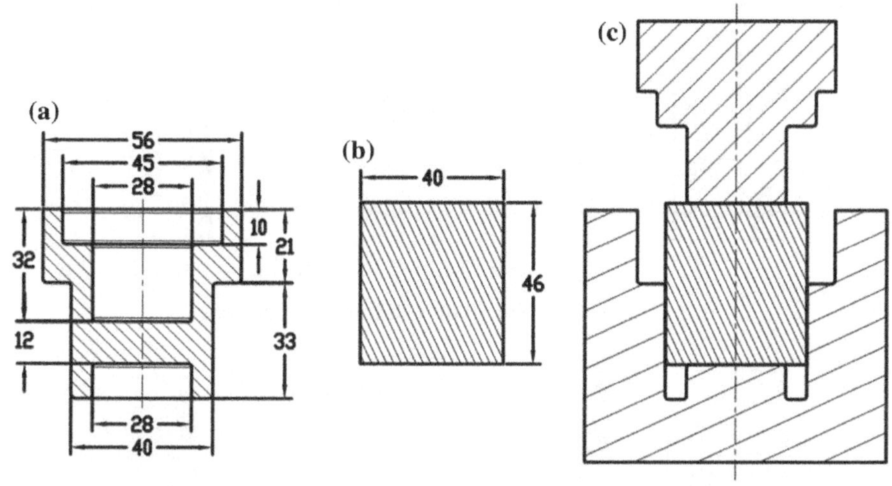

Fig. 5.24 A case study part and its die assembly. **a** Part dimensions, **b** billet dimensions, and **c** die assemble [1]

Fig. 5.25 Folding defect formation in the intermediate preforms in forming process [4]

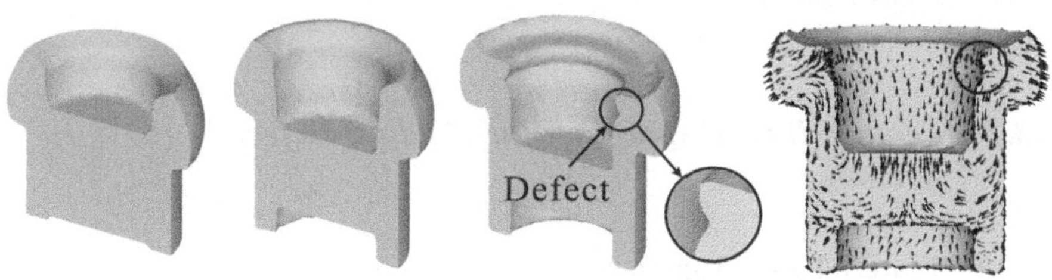

Fig. 5.26 Defect formation process revealed by FE simulation [1]

displacement of 20 mm is then simulated and the results are shown in Fig. 5.27b. With the stroke of 50 % as shown in the figure, the material flows without leaving the punch head before the material reaches the punch shoulder. The material then flows to fill up the flanged cavity and the die insert moves downward. An S-flow

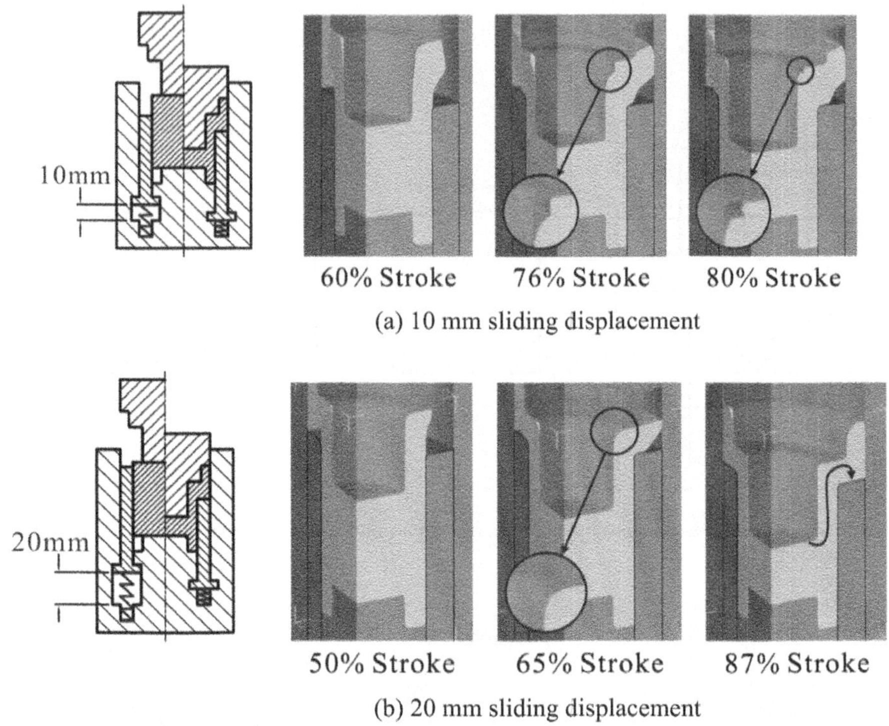

Fig. 5.27 Die assembly with a sliding die insert and the flow behaviors of material in forming scenarios with different sliding displacements [1]

pattern of material flow and movement is formed. In addition, the simulation shows that no folding defect occurs, and thus, the flow-induced folding defect can be avoided in this forming system design.

5.3.2 *Feature-Based Approach for Folding Defect Avoidance*

5.3.2.1 Feature-Based Approach

Flow-induced folding defect is caused by irregular material flow, which is closely related to forming process and sequence, and thus, their design is very critical in terms of control and change of material flow pattern and avoidance of irregular material flow. A slight change of forming process and sequence can significantly affect the material flow pattern. From metal-deformed part geometry perspective, a metal-deformed part is composed of different geometry features. Metal forming is actually the process to form the geometry features of the deformed part sequentially or simultaneously and thus, the design of forming process also includes the determination of forming sequence. Forming sequence determination is actually the determination of the operation sequence for forming of different features of the designed part. The traditional know-how-based design approach, however, is difficult to predict and reveal the flow pattern of the designed part in up-front design stage and FE simulation is an efficient approach to exploring the metal flow in different deformation stages. In this section, a feature-based approach aided by FE simulation is proposed to identify the most rational forming process sequence in which the flow-induced defects in mesoforming process can be efficiently avoided.

To illustrate the feature-based forming sequence design, the case study part shown in Fig. 5.3 is further used as the case study. The part consists of three features, namely flange, outer star contour, and inner star hole. The part can thus be formed mainly by three forming sequences. The first sequence is to form the flanged feature first, and then the outer star contour and followed by the inner star hole features. The second sequence is to generate the outer star contour feature first and then to form the flange and inner star hole features. The last sequence is to form the inner star hole first and then to form the last two features. The feature-based approach is to design forming sequence based on this rationale. In feature-based approach for defect avoidance, all the features in the deformed part need to be identified first and all the possible forming sequences are then determined. From the perspective of defect-free flow pattern, the best material flow pattern in which no flow-induced defect occurs is to be eventually identified. To achieve the objective of defect-free forming, the key features which have flow-induced folding defects should be firstly formed without defect and other features are formed subsequently. This design sequence is the most effective way to avoid the occurrence of flow-induced defects in forming process. It is also the kernel of this feature-based approach for defect avoidance design of forming system.

For the part as shown in Fig. 5.3, it cannot be successfully formed using the natural forming process in one stroke due to the occurrence of the irregular material flow pattern in forming process. To control the material flow, the forming is realized by a few steps based on the revealed defect formation mechanisms previously and the proposed feature-based approach. For the case study part, there are two main defects in the inner star hole and the flange section. The main reason is that the cross-sectional area of material flow in the flow path is different as the sectional area of flange is larger than the hole section of the part, which causes the material flowing in opposite directions. The critical feature is thus the flanged feature. Therefore, the best forming sequence is to form the flanged feature first. In order to prove and verify the effectiveness and the availability of this proposed feature-based approach, the above-presented three forming sequences are studied by FE simulation, and the simulation results are shown in Fig. 5.28.

In Fig. 5.28, it is can be found that forming Sequence 1 (S.1) successfully avoids the flow-induced folding defects, but Sequence 2 (S.2) and Sequence 3 (S.3) generate the folding defects in the flange section and the inner star hole section, respectively. Based on the simulation results of the different forming sequences

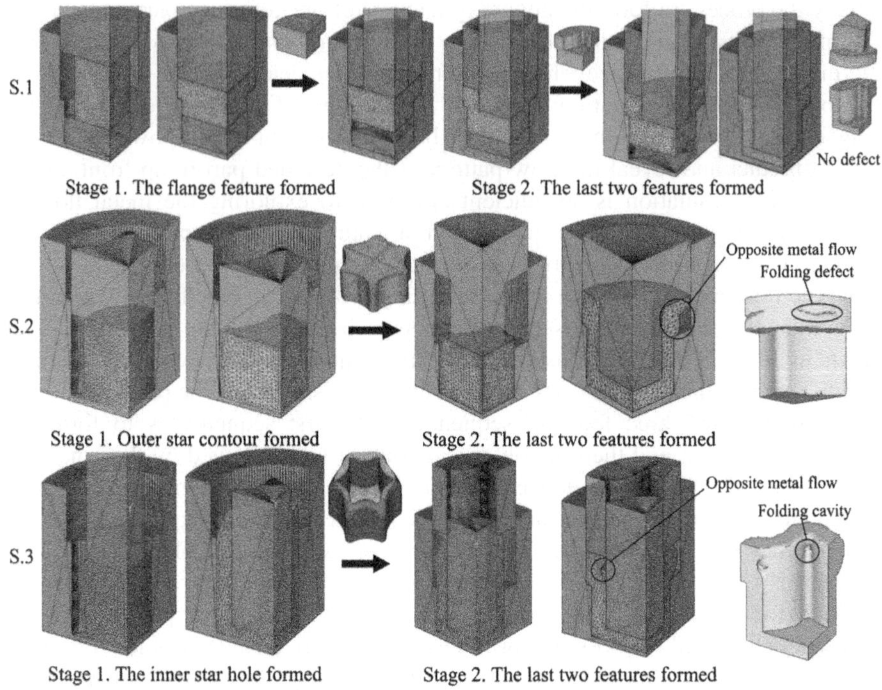

Fig. 5.28 Forming sequence determination and comparison for defect-free forming [3]

shown in Fig. 5.28, a feature-based approach to determining the defect-free forming sequence is proposed and shown in Fig. 5.29. In this approach, the first step is the metal-formed part design, which defines the geometries and features of the deformed part. The detailed features are then extracted out and a feature list is then determined. The potential forming sequences are then figured out by which the features are formed in different sequences. From forming system design perspective, it covers the design of tooling geometry and structure and billet design. From simulation perspective, it needs to address the system setting and modeling setting. The former needs to input convergence criteria, loading step, and re-meshing condition. For the latter, it needs to specify the details related to selection of element type and number, specifying material properties including stress–strain curve of material and the deformation condition including deformation speed and friction conditions. As for the model conversion from the generated CAD models to the models which can be recognized by CAE simulation systems, the conversion to data exchange standards, such as IGES, STL, and STEP, is needed to realize the model conversion. Through FE simulation, the detailed plastic deformation and the deformation behavior of the specific forming sequence can be revealed and the occurrence of any defect in each forming sequence can be identified.

In the feature-based approach for determination of defect-free forming sequence as shown in Fig. 5.29, once the FE simulation of the original forming sequence

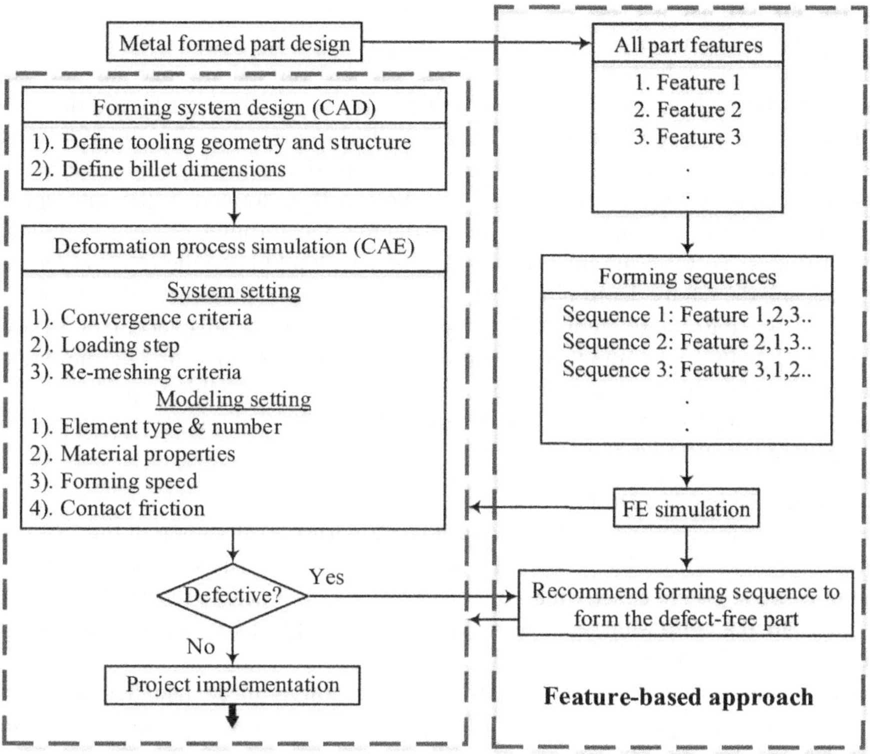

Fig. 5.29 Feature-based approach for defect-free forming sequence design [3]

shows the existence of flow-induced defects, the critical features with defects will be identified. Combining all the identified features of the designed part, all the potential forming sequences which form the critical features first are determined. Through FE simulation, all the potential forming sequences are investigated from defect-free perspective and the best forming sequence which produces the defect-free-deformed part is finally identified.

5.3.2.2 Detailed Approaches to Avoiding the Flow-Induced Defects

Based on the revealed formation mechanism of folding defect and the determined forming Sequence 1 by using the feature-based method and the FE simulation as shown in Fig. 5.28, two process designs, viz., four steps extrusion design and backward extrusion, are proposed. However, the detailed process parameters are not yet determined. In this section, the detailed process parameters will be generated with the aid of FE simulation and the feasibility of the identified process sequence will be investigated and verified through physical experiments.

Four-step Extrusion Process

According to the material flow behavior and forming sequence, the flanged feature should be formed firstly. Therefore, in the first deformation stage, the flanged feature is formed using the punch with the diameter of Ø5 mm. In addition, the depth of die cavity is 0.75 mm and the punch displacement is 1.25 mm. After stage 1, the preformed part 1 is formed completely as shown in Fig. 5.30. Once the flanged feature is formed, the star contour hole will be formed in the subsequent step. Considering the plastic deformation degree, the forming process is divided into 4 stages. To form the hole feature, the punch should have the star contour. In stage 2, the die cavity depth is 1.25 mm and the punch displacement is 1.0 mm. In stage 3, the die cavity depth is 1.75 mm and the punch displacement is also 1.0 mm. Finally, the component is formed completely in stage 4. The die cavity depth is 2 mm and the punch displacement is 0.5 mm. After this step, the desired defect-free component is made, as shown in Fig. 5.30. The process variables are summarized in Table 5.2.

Backward Extrusion Design

Another forming process, namely the backward extrusion, as shown in Fig. 5.31, is studied in this section. This backward extrusion process consists of three steps, and the first two steps are exactly the same as the four-step process presented in the last section. In the third step, the die cavity depth is still 1.25 mm and the punch displacement is 0.75 mm.

After the second stage, the intermediate preform is made and the third step configuration is shown in Fig. 5.31. In this scenario, the flange and star contour

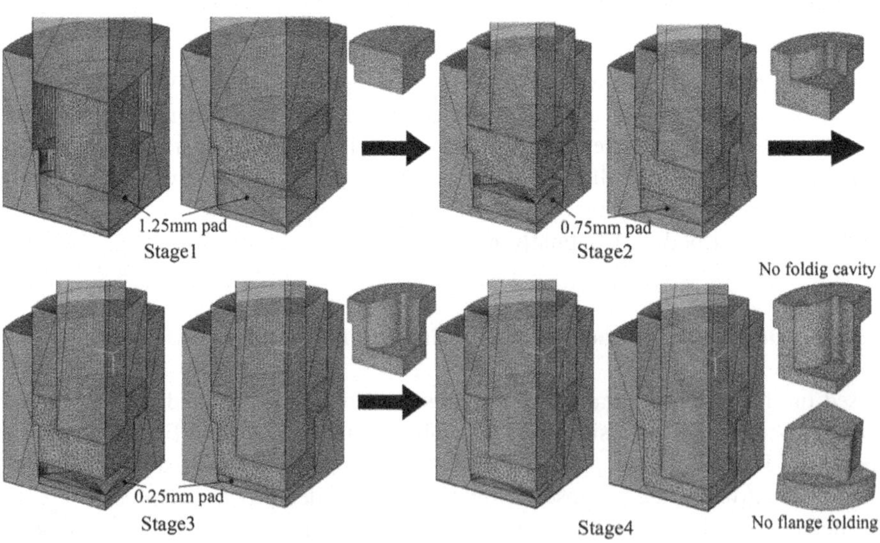

Fig. 5.30 Four steps extrusion and the simulation results [3]

Table 5.2 Simulation setting of the four steps extrusion [3]

	Die cavity depth (mm)	Punch displacement (mm)
Stage 1	0.75	1.25
Stage 2	1.25	1
Stage 3	1.75	1
Stage 4	2	0.5

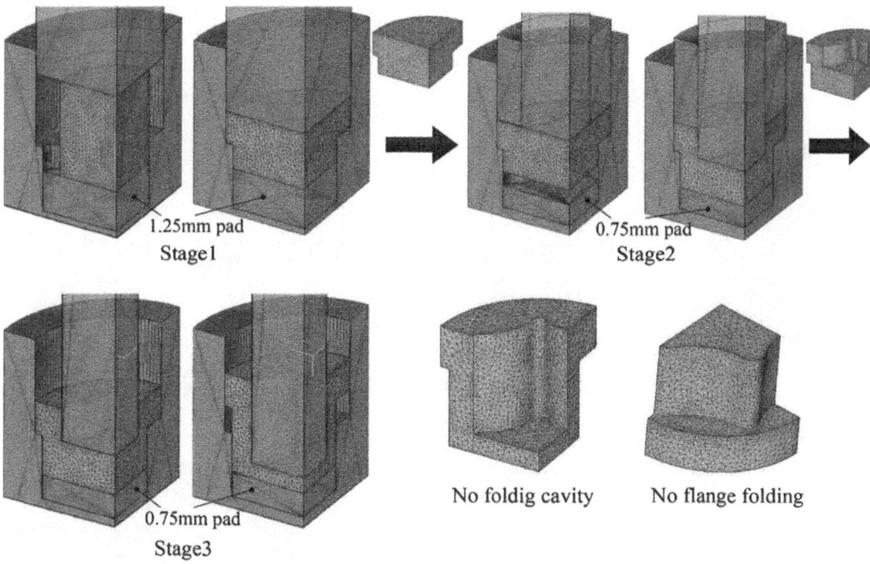

Fig. 5.31 Backward extrusion and the simulation results [3]

container is no longer needed. From the simulation results, it is found that there is no any defect in the flange section or in the inner surface of the star hole. The backward solution is thus a feasible method to produce the defect-free component.

5.3.2.3 Physical Implementation of the Designed Forming Processes

Four-step Extrusion Experiment

In order to test and verify the designed process and the accuracy of the simulation results, a four-step extrusion experiment is conducted. The tooling set is shown in Fig. 5.32. This tooling set has four layers in vertical direction. The first and second layers are punch guide to make sure the punch movement is vertical to bottom die. The third layer is a container, which is different in different stages. In the first step,

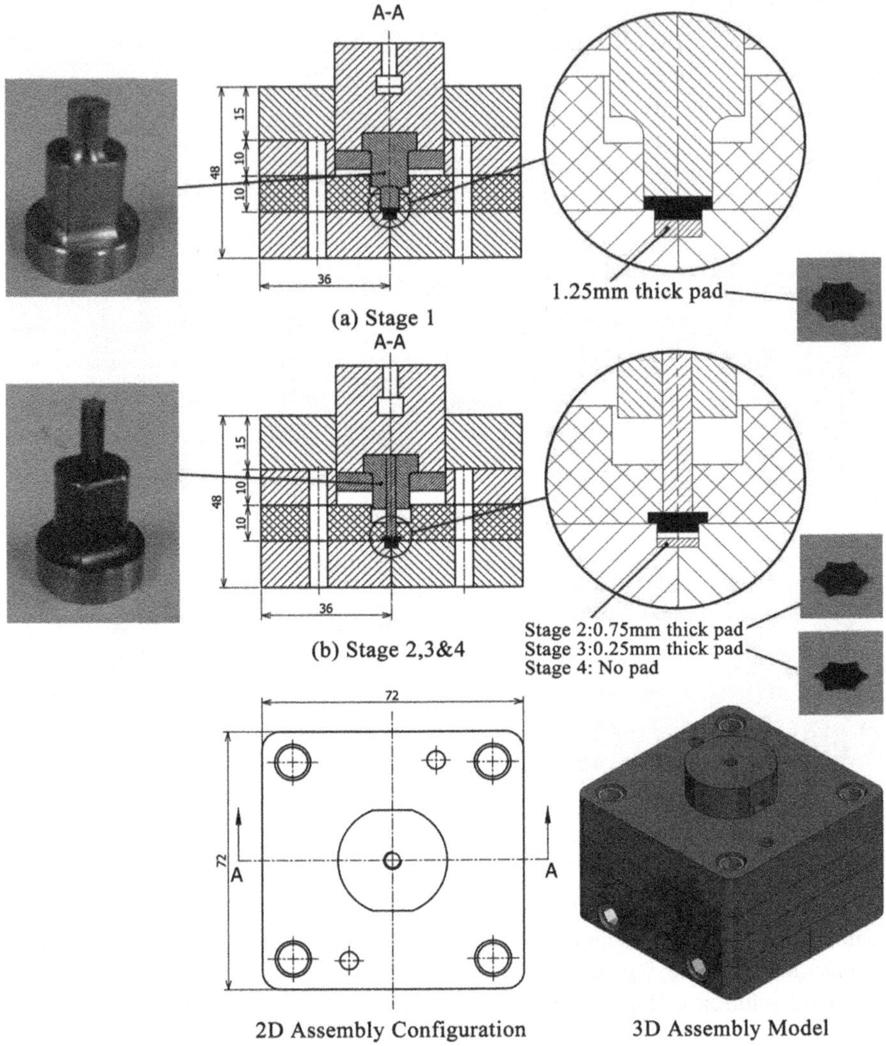

Fig. 5.32 Die set of four steps extrusion [3]

the circle inner hole in the container is to form the flanged feature, and in the next steps, the function of third layer is to protect the formed flanged feature and guide the star punch movement. The fourth layer is the bottom die to form the star contour. In this scenario, there are two types of punches and four different die cavity depths. One punch is cylindrical used to form the flanged feature, and the other is star contour punch used to form the star blind hole. Regarding the die cavity depth, a pad is introduced to adjust the depth of cavity. The thicknesses of the pad are 1.25, 0.75, and 0.25 mm, respectively, as shown in Fig. 5.32.

The preformed part 1 is shown in Fig. 5.33. From the figure, it can be seen that the preformed part in the experiment has the expected shape which is consistent with the simulation results.

In the second step, the flange container is replaced by the star feature container. The die cavity depth is 1.25 mm and the punch displacement is 1 mm. In the third step, the die has the same configuration with the second step. The preformed parts 2 and 3 are shown in Fig. 5.34. There is no defect in the preformed parts.

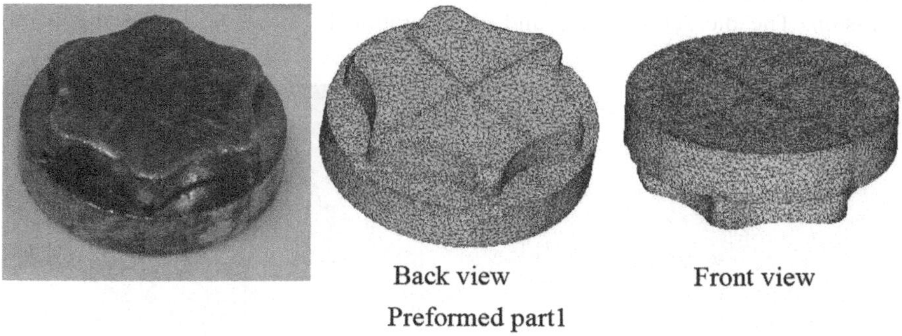

Fig. 5.33 Experimental results in stage 1 [3]

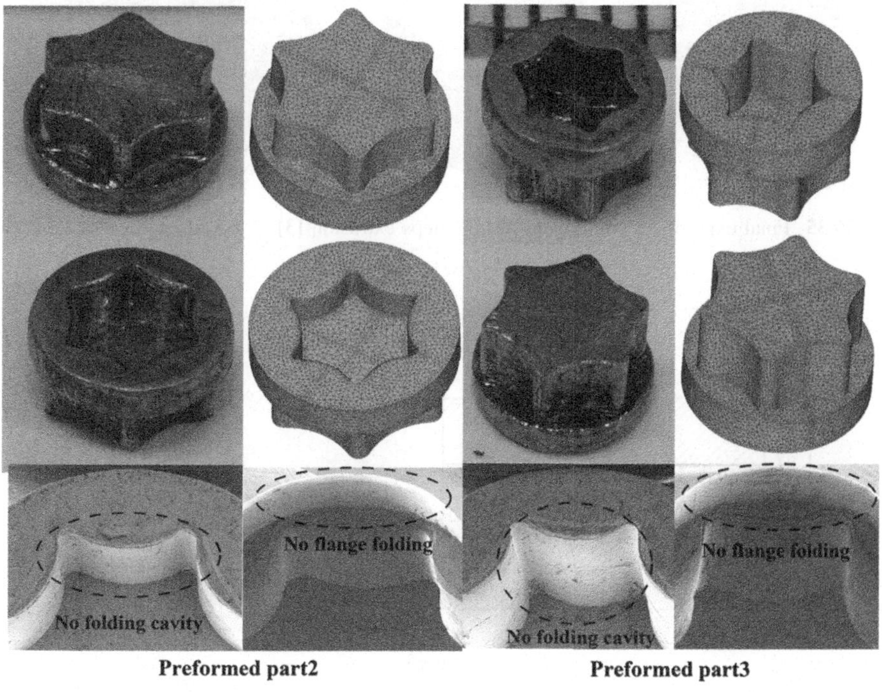

Fig. 5.34 Experimental results in stages 2 and 3 and the comparison with simulation results [3]

In the fourth stage, the final component is completely formed without any defect and shown in Fig. 5.35.

Backward Extrusion Experiment

A backward cold extrusion experiment is also conducted. Its die set has four layers. The first, second, and fourth layers have the same function and structure as the previous die set, as shown in Fig. 5.32. In this experiment, the third layer has three different structures. The inner diameter of the container must be larger than the flange diameter for facilitating the material flow and movement in backward extrusion. The die set structure and configuration is shown in Fig. 5.36. In this

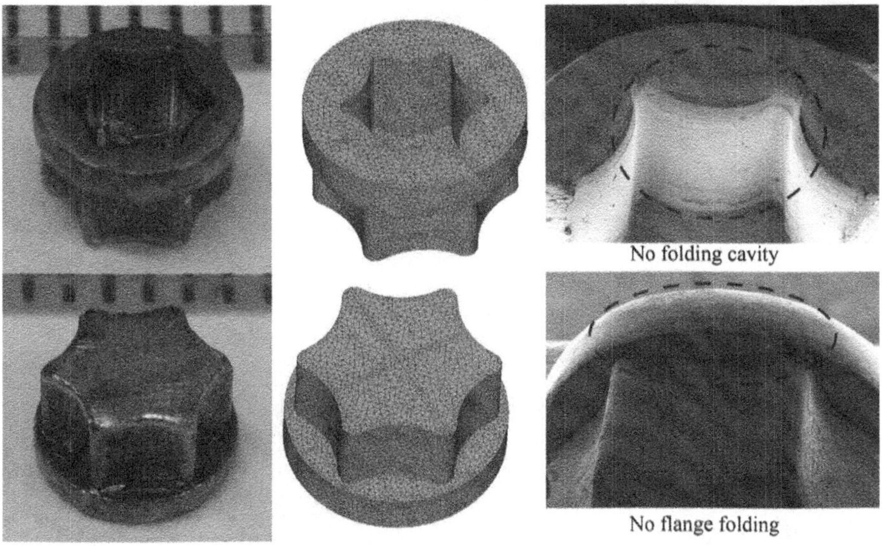

Fig. 5.35 Final experiment component of four steps extrusion [3]

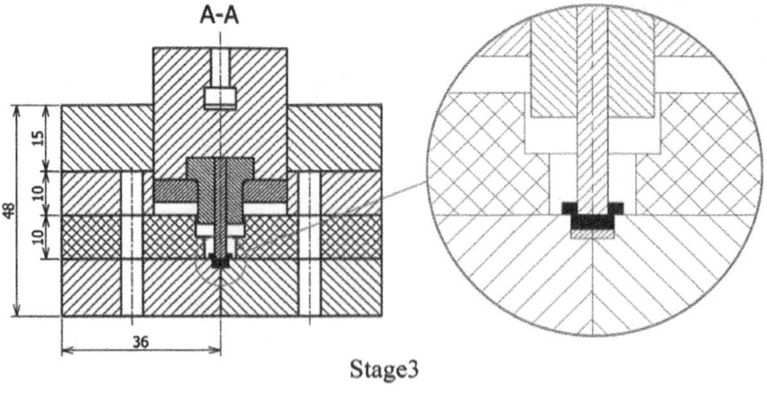

Fig. 5.36 Die set of backward extrusion [3]

experiment, the first two steps are the same as the four-step scenario. In the third step, the preformed part is extruded. The punch displacement is 0.75 mm. The final part as shown in Fig. 5.37 does not have any defects, which tallies with the FE simulation results and the proposed feature-based approach is validated for its efficiency for defect avoidance in forming process.

5.4 Flow-Induced Defect and Size Effect in Meso- and Microforming

The extensive exploration and study on flow-induced defect is mainly focused on macroforming [1, 4] and the meso-extrusion presented in Sect. 5.3 [3]. In micro-forming, however, this type of defects is still lack of in-depth understanding and systematic investigation despite of fact that more investigation and exploration have been conducted recently [5, 6]. With the decreasing size of the metal-deformed parts, individual factors such as workpiece geometry size, grain size, orientation, properties, and the interfacial conditions between workpiece and tooling all demonstrate their significant effects on the flow behaviors of material, the formation of flow-induced defect, the microstructure of the deformed parts, and the

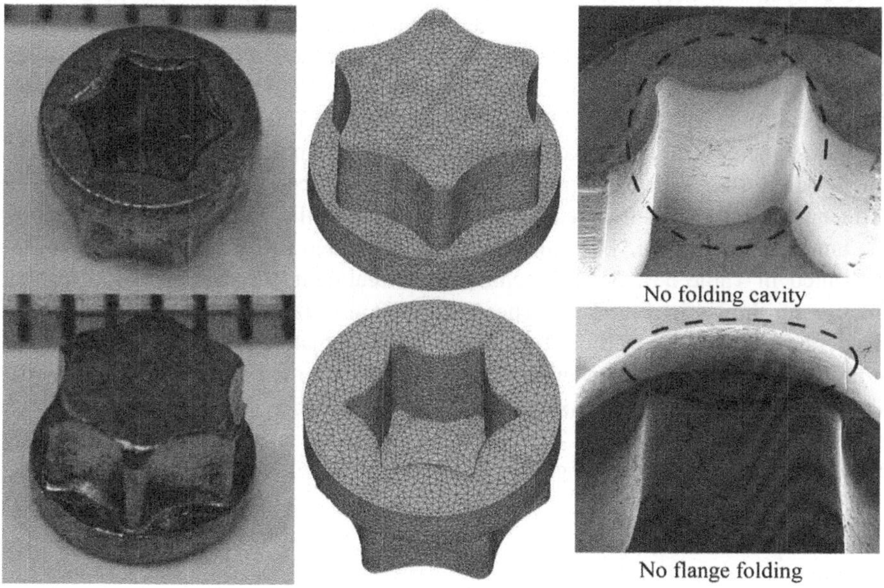

Fig. 5.37 Final experiment component of backward extrusion [3]

process-related behaviors and performances in microscaled plastic deformation process.

In tandem with this research status, this section summarizes the research outcome in terms of the formation of flow-induced defect in microcold extrusion process and microstructure evolution of the microextruded parts. This is the continuous effort on the flow-induced defect in mesoscaled forming process presented in the previous section. In this section, the focus is on how the size effect affects the flow-induced defect and what main factors such as size effect, grain, or geometry size effect greatly influence the flow-induced defect. All of these have not yet been fully and systematically reported. Furthermore, how the flow-induced defect evolves with the material microstructure and the workpiece geometry size in microforming process is also addressed in this section. To illustrate all of these easily, two different scale parts, one non-axisymmetrical part and the other axisymmetrical part, as shown in Fig. 5.38, are used as the case study parts made by microcold extrusion experiment. As for the grain size effect, the heat treatment of the billet material is conducted to get three different grain sizes. The formation mechanisms of the flow-induced defects in the two different types of parts revealed by experiment are also presented in this section.

5.4.1 Experiments of Meso- and Microscaled Parts

Pure copper is widely used in electrical and electronics industries for its excellent electrical conductivity, reasonable mechanical properties, good manufacturability, and plastic formability. Two kinds of pure copper billets with the diameter of 1.8 and 0.9 mm are used for non-axisymmetrical parts. Three types of pure copper billets with the dimensions of Ø2.6 × 3, Ø1.3 × 1.5, and Ø0.65 × 0.75 mm represent meso- (the large and small mesoscaled in the first two cases), and microscaled (the last case) axisymmetrical parts, respectively. To explore the influence of grain size effect, each type of billet with different grain sizes is obtained via annealing treatment. The annealing conditions and the obtained grain sizes are presented in Table 5.3. After annealing heat treatment, the microstructures of the billets are shown in Fig. 5.39. The average grain size increases with annealing temperature and holding time. For the as-received materials with different dimensions (case 1), the average grain size is about 10 μm. For case 2 as shown in Table 5.3, the average grain size is about 20 μm, which is almost twice the average grain size in case 1. For case 3, the average grain size is around 30 μm. In the last case, viz., case 4, the average grain size is 50 μm, which is much larger than that of other cases.

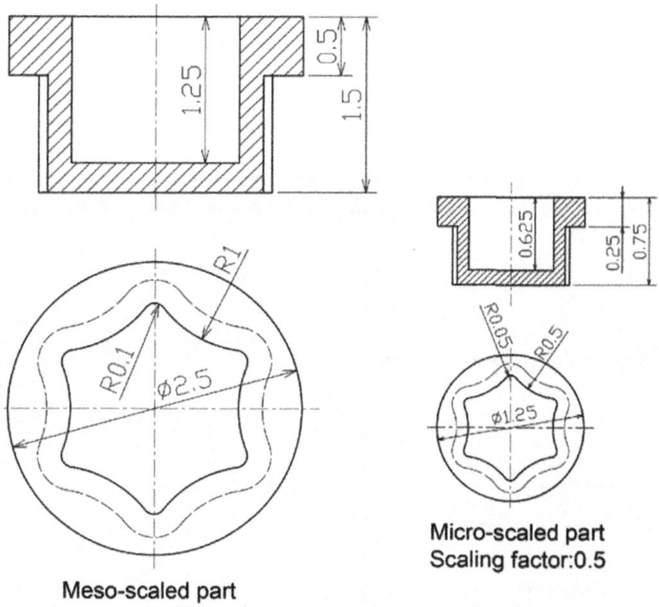

(A) The non-axisymmetric parts with two different scales [5]

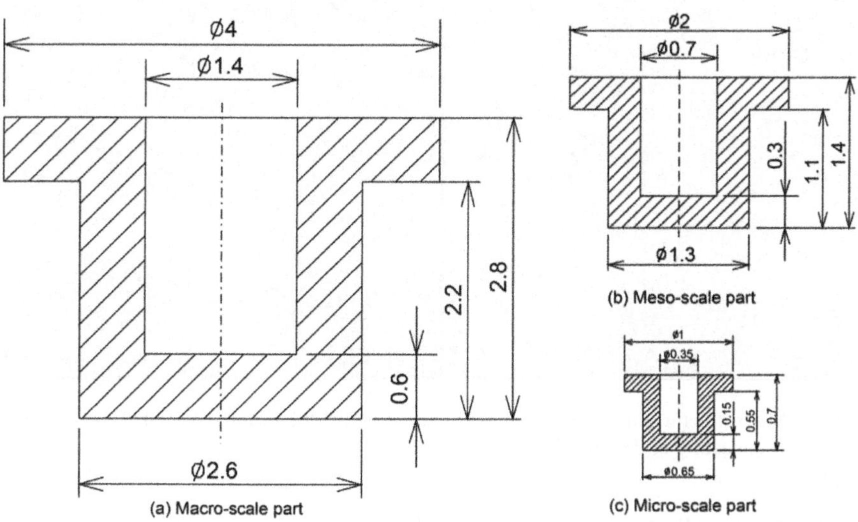

(B) The axisymmetric parts with three different scales: (a). Scale factor (large meso-scaled): 1, (b). Scale factor (small meso-scaled): 0.5, and (c). Scale factor (micro-scaled): 0.25. [6]

Fig. 5.38 Case study parts [5, 6]

Table 5.3 Heat treatment parameters [6]

Annealing conditions	Case 1 as-received	Case 2 450 °C, 2 h	Case 3 600 °C, 2 h	Case 4 750 °C, 3 h
Grain size	10 μm	20 μm	30 μm	50 μm

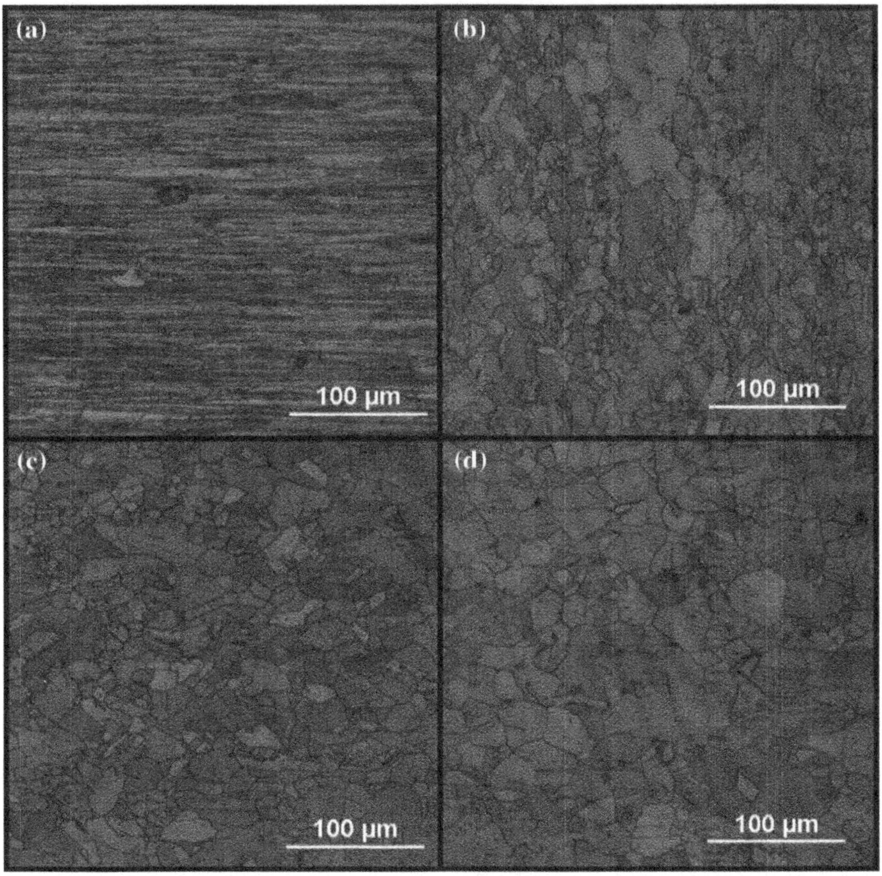

Fig. 5.39 Microstructures of the pure copper. **a** As-received, **b** annealing at 450 °C, **c** annealing at 600 °C, **d** annealing at 750 °C [5]

5.4.1.1 Microcylindrical Compression Test

The mechanical properties of the annealed treatment metal billet are determined by the microcompression test conducted in a MTS material testing machine with a load cell of 30 KN. A microcompression tooling set is employed as the experimental platform. The specimens are lubricated with machine oil in order to minimize the frictional effect. For each category of metal billet, three tests are conducted to minimize testing error. The crosshead velocity of the testing machine is set to be 0.02 and 0.01 mm/s for meso- and microscaled billets, respectively. All the specimens are compressed by 60 % of the corresponding specimen height at room temperature. The dynamic loading and displacement data are recorded via a built-in data acquisition system. The true stress–strain curves for the two kinds of billets with different grain sizes are therefore obtained.

The true stress–strain curves for the five kinds of metal billets with different annealing conditions and dimensions are shown in Fig. 5.40. It is found that the flow stress decreases with the increase in grain size for the same billet. The flow stress of the materials with different grain sizes is different significantly for the meso-scaled billet. This indicates that the grain size effect obviously affects the flow stress of these materials and also in tandem with the Hall Petch equation. In addition, the stress–strain curves have a good repeatability and consistency for the same metal billet. However, for the microbillet, the flow stress of the billet heat-treated at the temperature of 600 °C is slightly larger than that of the billet treated at the temperature of 750 °C and in the large strain region, the flow stress of the billet treated at the temperature of 600 °C even smaller than that heat-treated at the temperature of 750 °C, which could be caused by the friction effect between the interfacial surfaces of workpiece and tooling. For the workpiece with the small volume to grain size ratio, single grain plays an important role in the entire deformation process. The scattering of flow stress is thus also caused by the friction between random grain and tooling surface.

According to Fig. 5.40, it is found that flow stress varies with grain size and the grain size effect thus exists in microforming process. These five types of billets with different grain sizes are used to conduct microforming. The deformed parts with flow-induced defects are influenced by grain and geometry sizes.

5.4.1.2 Microforming Experiments

The microcold extrusion experiments are conducted to investigate the influence of size effect on the formation of flow-induced defects. All the microforming experiments are done in a MTS test machine with a load cell of 30 KN. The tooling sets are shown in Fig. 5.41. As shown in the figure, two sets of dies are designed and fabricated for making two non-axisymmetric and axisymmetric parts with different scales, respectively. In addition, the die structure can be used for general purpose in nature and the parts with different scales can be deformed using this die structure via changing the related punch and die. In Fig. 5.41A, the die set is used for forming of non-axisymmetric part. In Fig. 5.41B, the different punch and die are designed for forming of the axisymmetric parts with different scales.

To avoid the influence of strain rate, the velocity of the crosshead of the test machine is set to be 0.01 mm/s. Based on the volume constant principle in plastic deformation process, the cylinder billets are prepared. The parts are deformed by one-stroke process. The tooling and billet contact region is lubricated by a machine oil.

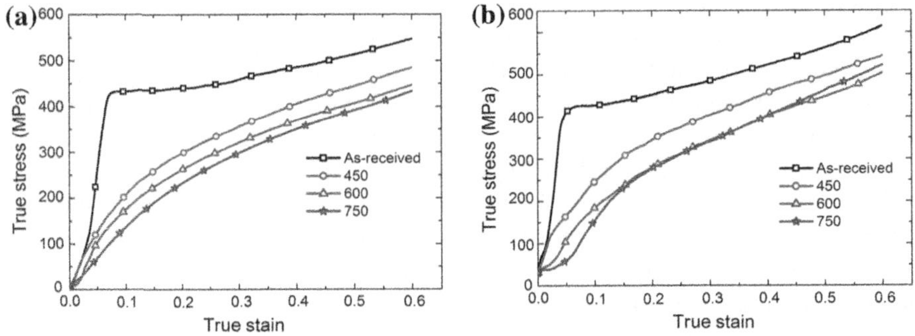

(A) Two types of billets with different grain sizes: (a) Dimension = Ø1.8×1.3mm, (b) Dimension = Ø0.9×0.65mm [5]

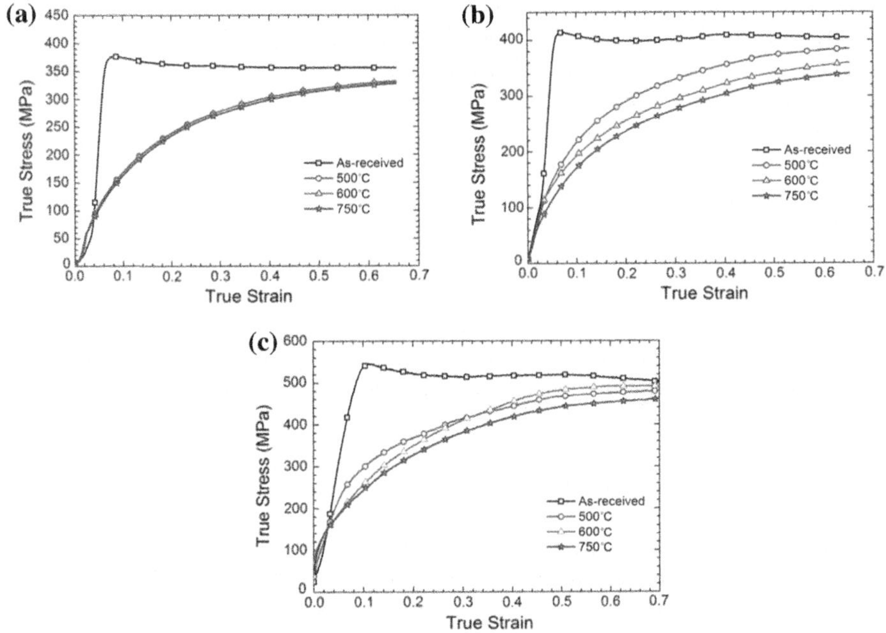

(B) Three types of billets with different grain sizes. (a) Dimension=Ø2.6×3 mm, (b). Dimension=Ø1.3× 1.5 mm, and (c). Dimension=Ø0.65×0.75 mm. [6]

Fig. 5.40 True stress–strain curves of billets [5, 6]

The load–stroke curves of the two non-axisymmetric and axisymmetric parts with two and three different size scales, respectively, and using the materials using different heat treatment conditions are shown in Fig. 5.42. In Fig. 5.42, it is found that the entire forming process can be divided into three stages. In the first stage, the

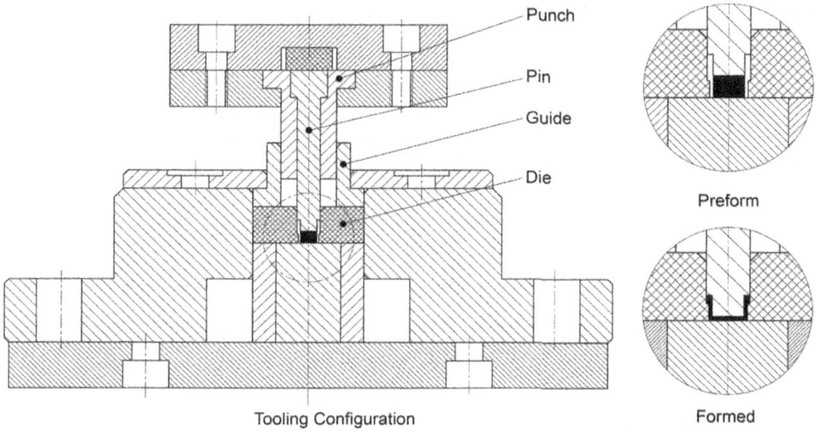

(A)Die set for forming of non-axisymmetric parts [5]

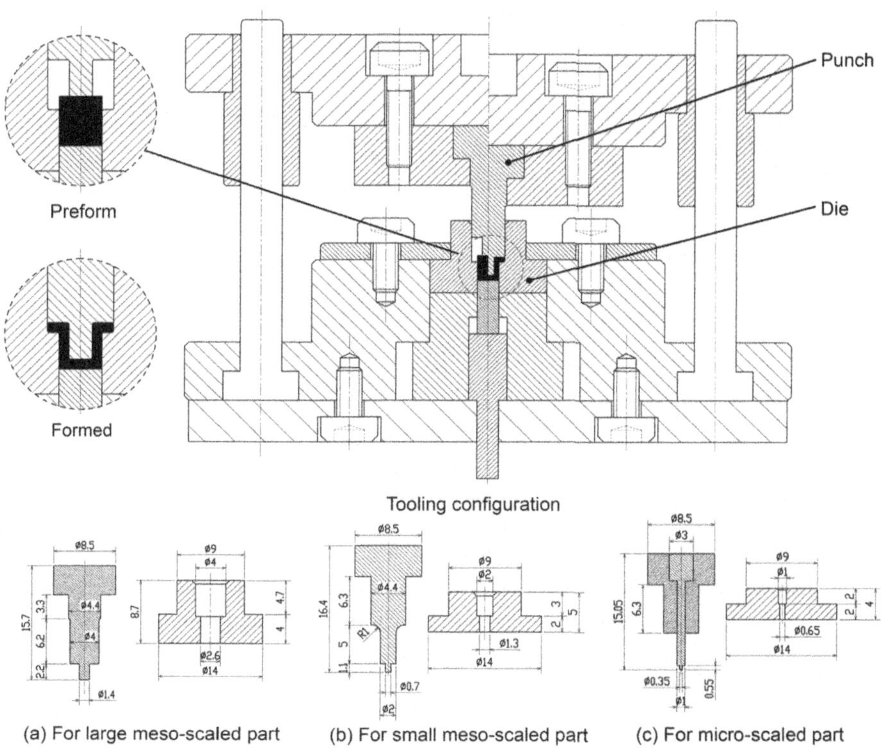

(B) Die set for forming axisymmetric parts [6]

Fig. 5.41 Experimental tooling set [5, 6]

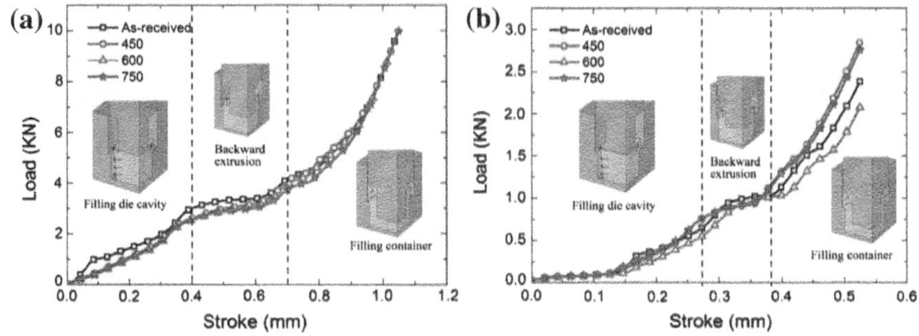

(A) The non-axisymmetric parts with two different scales, (a) Meso-scaled part, (b) Micro-scaled part [5]

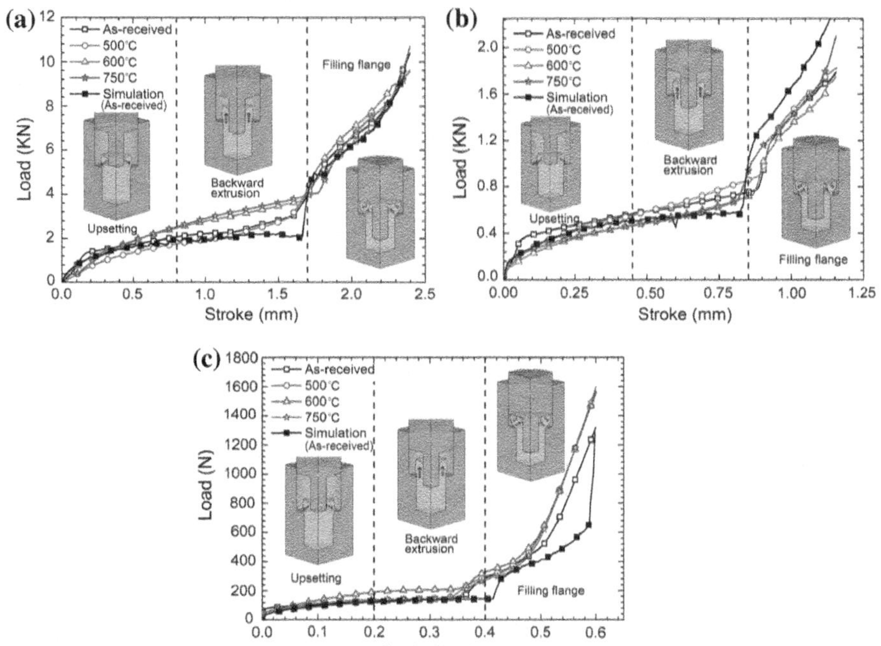

(B) The axis-symmetric parts with three different scales, (a) Large meso-scaled part, (b) Small meso-scaled part, and (c) Micro-scaled part [6]

Fig. 5.42 Load–stroke curves [5, 6]

billet undergoes deformation in both the radial and the axial directions simultaneously until the radial die cavity is completely filled. The second stage is the backward extrusion until the punch shoulder contacts the deforming workpiece. In this stage, the deformation load has a slight increase. The last stage is the combined forward and backward extrusion until the entire die cavity is totally filled up. At this stage, the deformation load sharply increases since the unfilled place in the cavity is very small and thus, the final filling needs a great pressure compared with the preceding two stages to drive the material to flow into the remaining empty space in the cavity.

For the non-axisymmetric part, the mesoscaled part is formed using three different materials prepared by different heat treatment conditions and the one with as-received condition. The load–stroke curves as shown in Fig. 5.42A-a, however, have almost the same variation trend and the deviation among them is very small. For the microscaled part as shown in Fig. 5.42A-b, despite of the fact that the same materials are used, the load–stroke curves have a great scattering and there is a big deviation among them, especially for the later forming stage. In addition, there is no an obvious variation pattern of the load–stroke curves by using four different materials. This is also mainly caused by the nature of microforming process in which the individual grain properties, grain orientation, and behaviors all affect the deformation behavior of the whole part. In this case, the maximum diameter of the part is only 1.25 mm, which fully falls into the category of microforming, and thus, the scattering of the deformation loading is quite big. Moreover, there are three different stages shown in the load–stroke curves in Fig. 5.42A in both meso- and microscaled parts and the different material conditions do not affect the occurrence of the three different stages in the deformation process. For the detailed variation of the deformation loading, the deformation load increases with the decrease in annealing temperature in the first two stages. As for the third stage, the all deformation loads of the workpieces with different grain sizes have a significant increase due to the difficulty of filling the remaining space in die cavity in the last stage and more surface contact between workpiece and die in this stage and thus the big friction effect.

For the load–stroke curves as shown in Fig. 5.42B, there are three different scales and the materials with four different conditions for the axisymmetric part. For the large mesoscaled part as shown in Fig. 5.42B-a, there is no big difference in the load–stroke curves among the four differently prepared materials and the simulation results. The FE simulation is based on the as-received material. For the small mesoscaled part, the load–stroke curves are shown in Fig. 5.42B-b. There is a bigger deviation between the simulation result and the experimental one for the scenario of as-received material, and the simulation result is much larger than the experimental ones, especially in the last forming stage. For the microscaled part and the load–stroke curves as shown in Fig. 5.42B-c with four differently prepared materials, there is a biggest deviation between the simulation result and the experimental one with the as-received material. Similarly, it occurs in the last deformation stage. But in the small mesoscaled part, the simulated load is much larger than that of the experiment for the as-received materials. In the microscaled part, however, the loading predicted by simulation is much smaller than the real loading in the same as-received condition.

In addition, Fig. 5.43 shows the non-axisymmetric and axisymmetric deformed parts with different size scales. The parts have two flow-induced defects, viz.,

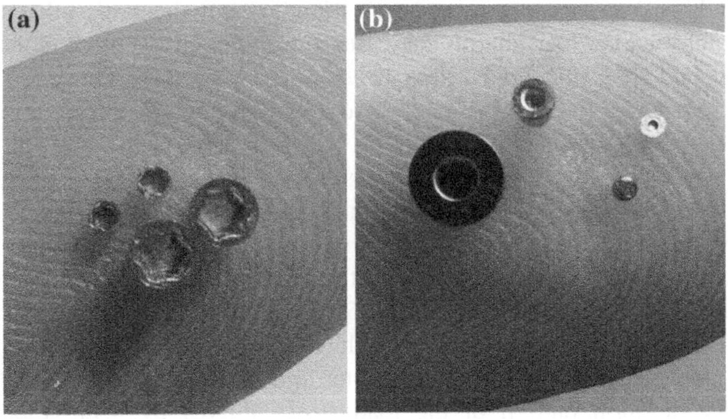

Fig. 5.43 Deformed parts: **a** the non-axisymmetric parts with two different scales: mesoscaled and microscaled parts [5], **b** the axisymmetric parts with three different scales: large meso-, small meso-, and microscaled parts [6]

folding defect at the inner star surface and the folding defect at the flange, as shown in Figs. 5.44, 5.45, 5.47, and 5.48. In this case, the first defect is not easily found as it appears in the inner star surface. The detailed analysis of the two types of defects is presented in the next section.

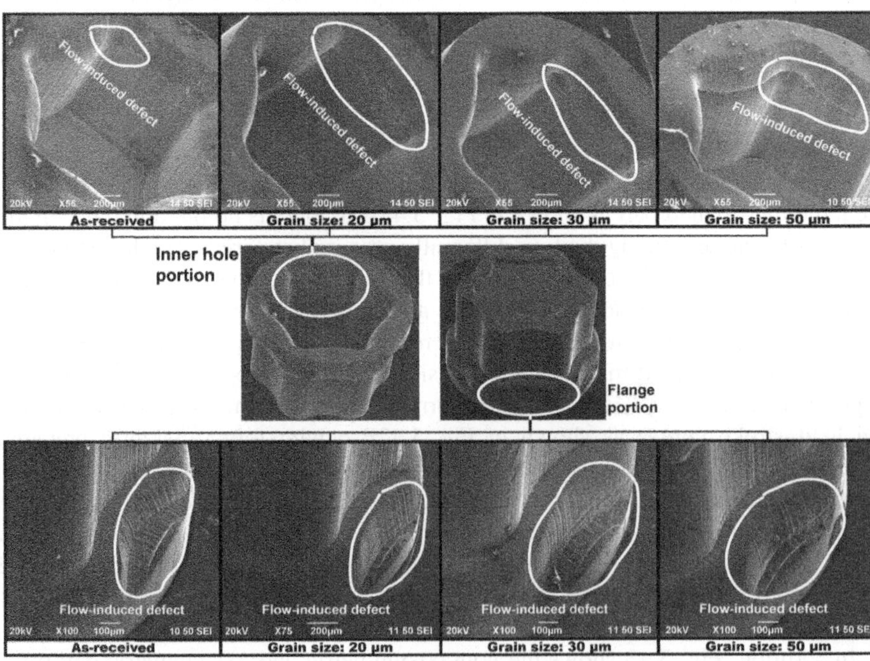

Fig. 5.44 SEM photographs of the mesoscaled non-axisymmetric part [5]

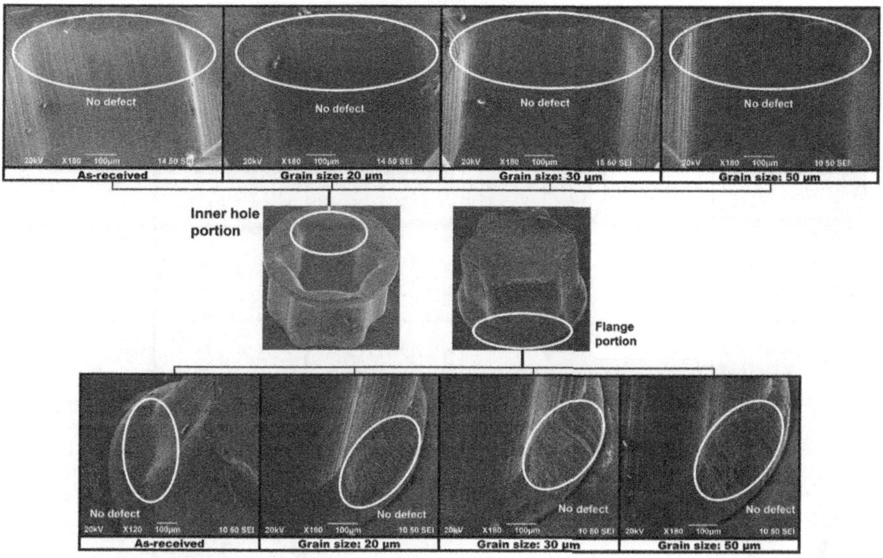

Fig. 5.45 SEM photographs of the microscaled non-axisymmetric part [5]

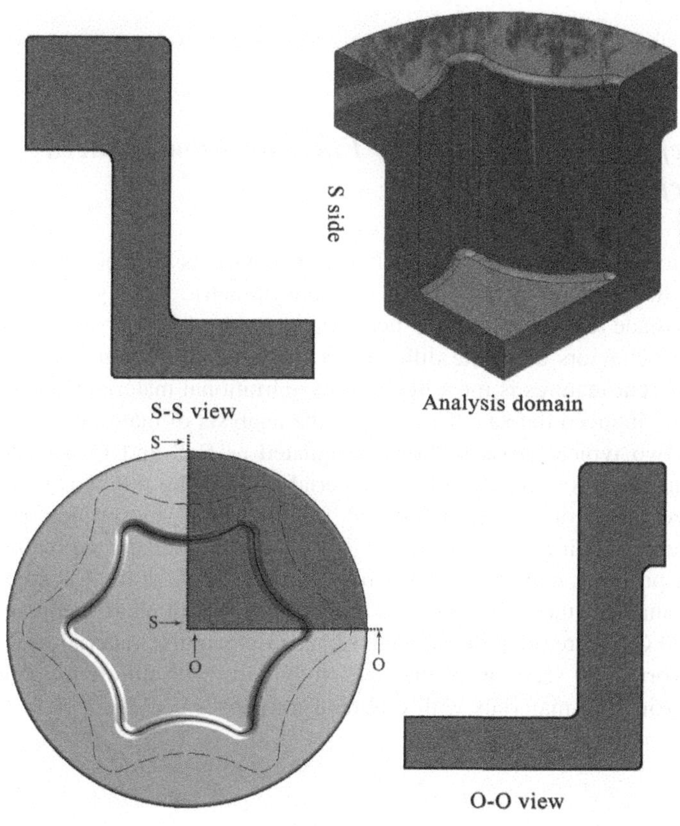

Fig. 5.46 Cross-sectional illustration in the non-axisymmetric part [5]

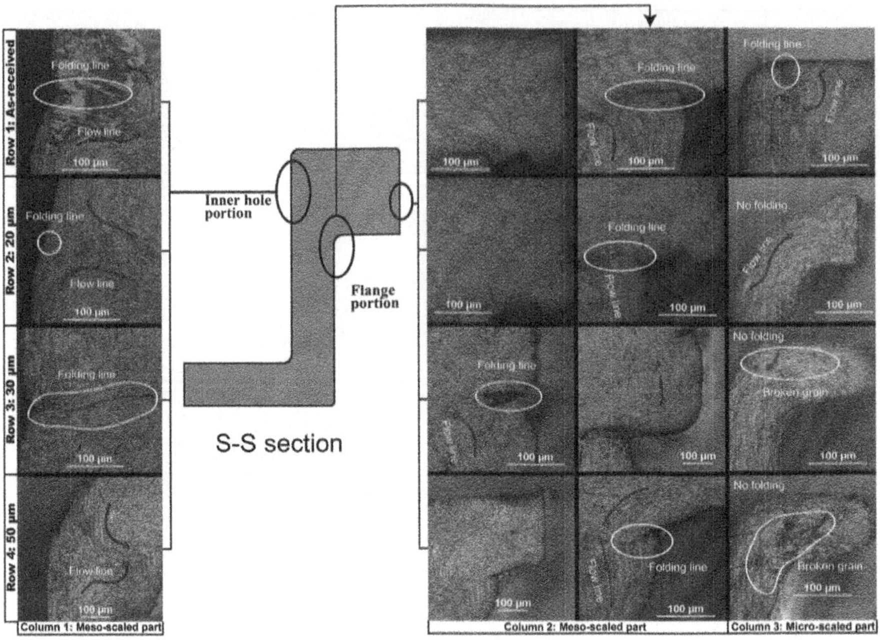

Fig. 5.47 Folding defect formation in the non-axisymmetric parts with meso- and microscale and different grain sizes in S–S section [5]

5.4.2 Defect Analysis and Size Effect on Flow-Induced Defects

In microdeformation process, material flow behavior is further affected by size effect. For the two non-axisymmetric and axisymmetric parts with different size scales and made of the materials heat-treated with different grain sizes, their deformation behaviors are quite different. It is found that the intersecting portion between different features is most likely to have irrational material flow and thus to form the flow-induced defects. To facilitate the analysis of material flow inside the workpiece, two typical cross sections designated as S–S and O–O, as shown in Fig. 5.46, are used for analysis of the material flow of the non-axisymmetric part and the size effect on the formation of folding defect in forming process. In Figs. 5.47 and 5.48, it can be concluded that the folding defects are mainly located at the upper portions, namely the inner hole and the flange of the workpiece. In the subsequent analysis, these two portions in the two different plane symmetry planes, viz., S–S and O–O, are the focus to analyze and compare the material flow behavior and defect formation scenario of the case study parts with different geometry size scales and using the materials with different grain sizes.

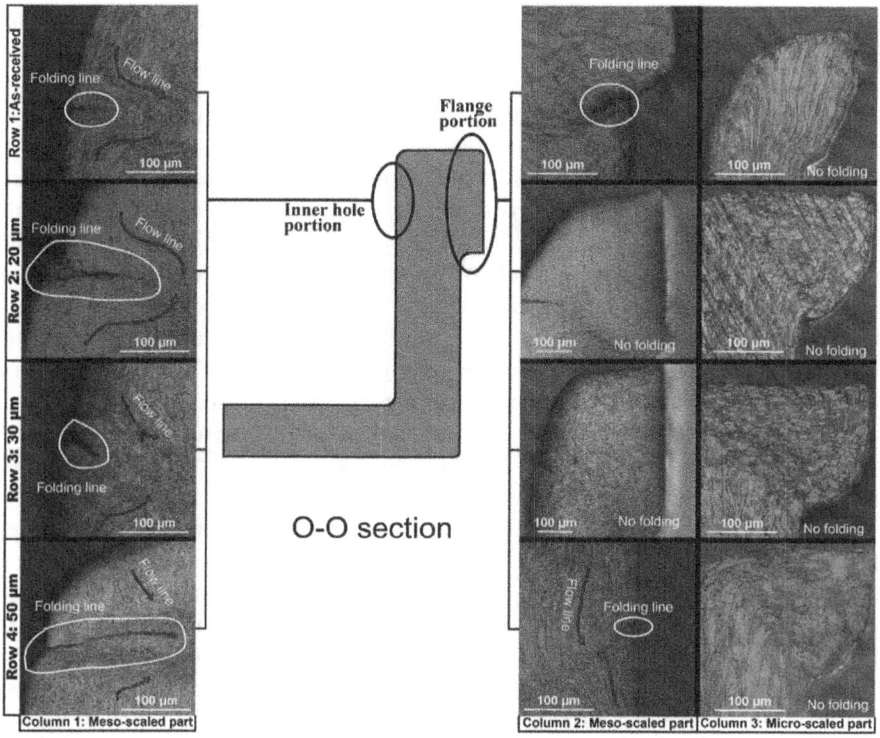

Fig. 5.48 Folding defect formation in the non-axisymmetric parts with meso- and microscale and different grain sizes in O–O section [5]

5.4.2.1 Geometry Size Effect on Folding Defect

To investigate the influence of geometry size on flow-induced folding defect, two different size scales and the testing materials prepared with different grain sizes are used for forming of the non-axisymmetric part. Figures 5.47 and 5.48 show the folding defect and the comparison between the meso- and microscaled parts by using the material with different grain sizes in S–S and O–O cross sections, respectively. Columns 1 and 2 represent the mesoscaled part and Column 3 is the microscaled part. Rows 1–4 indicate the materials with different grain sizes from the as-received case with the grain size of 10 to other cases with the grain sizes from 20, 30 to 50 μm.

Generally, no matter what grain size the material has, the mesoscaled part has the folding defects in both the inner hole and the flange portion while the microscaled part does not have any folding defect. Taking Row 1 in S–S section, as shown in Fig. 5.47, as an instance, the part using the as-received material has small folding lines. For the O–O section of the mesoscaled part, as shown in Fig. 5.48, there is folding defect in the inner hole, but there is no any folding defect in the inner hole and the flanged portion in the microscaled part. Comparing the formation of folding defect in the meso- and microscaled parts in different cross sections using the material with same grain size, it can be concluded that the geometry size significantly affects the formation of folding defect and the smaller part geometry size is easier to form with defect-free.

To compare the flow behaviors, which lead to the formation of flow-induced defects in meso- and microscaled parts, the deformation flow nature of polycrystals needs to be considered. The polycrystal is generally considered as a composite made out of a continuous network of grain–boundary film with discontinuous grains of bulk material. Discontinuous grain bulk is easier to deform in sliding direction than grain boundary and the grain boundary blocks the material plastic flow in deformation process. For the microscaled part, the number of grains involved in deformation is less than that in mesoscaled part. To maintain the strain continuity, single grain needs to undergo larger and more severe deformation, which means in the deformation region with the smaller geometry size, grains undergo severe and complex plastic deformations inside the grain driven by the external force and a few grains in the die cavity could accommodate the large deformation needed for shaping of the geometry to form the desired shape and geometry of the microscaled part. In this scenario, the deformation not only happens in grain boundary, majority of the deformation happens inside the grain. In this process, the relative change of neighboring position among of the grains is less since there is only a few of grains existing in the deformation zone and it is thus not easy to form the obvious flowlines and deformation texture, and furthermore difficult to form the folding defect, which is the meeting or merging of the different flow streams of grains in opposite directions the deformation zone. In mesoscaled part, however, there are tens or even hundreds of grains in the deformation zone, the deformation of grains not only happens inside the grain, majority of deformation would happen in grain boundaries, which can form flowlines and deformation texture. In addition, plenty of grains involved in deformation would be able to form flow streams in different directions and the meeting of them would form the folding defect. This could be the main reason why the deformation in microscaled part is more uniform than that in mesoscaled part and the latter one has plenty of flowlines, deformation texture, and folding defects, as shown in Figs. 5.47 and 5.48.

5.4.2.2 Effect of Grain Size and Microstructure on Folding Defect

For the non-axisymmetric part with the same geometry size, the grain size effect on the formation of folding defect is also shown in Figs. 5.47 and 5.48. For the mesoscaled part, there are severe folding defects in the inner hole and the flanged portion in all the deformed parts. Since the grain size is quite small compared to the geometry size of the workpiece, the deformation region thus contains quite a lot of grains and individual grains with different properties can be randomly and evenly distributed and do not significantly affect the flow behavior of material, which in turn results in an isotropic deformation behavior. Therefore, the grain size does not have an obvious effect on the formation of folding defect in mesoscaled part.

As for the microscaled part, however, Column 3 in Figs. 5.47 and 5.48 shows that the material flows in the deformed parts with different grain sizes. For the as-received case with the grain size of about 10 μm, the folding defect and flow-lines do exist. The main reason for this is that there are still plenty of grains in the deformation zone. Taking the flange part as instance, it has the height of 0.5 mm and the grain size is 10 μm; there are about 50 grains in the flange part in height direction for the given size scale. Its deformation is thus similar to that in mesoscaled part. For other three cases with the grain size from 20 to 40 μm, there is no folding found in the deformed part, but the flowlines and broken grains do exist and do not obviously change with the increase in grain size.

Similar to the explanation in the previous section, the fact that the grain size effect does not significantly affect the material flow in microscaled part can be explained by the relative grain number in the deformation zone. The deformation zone in microscaled part only contains less grain or only several grains and the individual grain and its boundary have a significant effect on material deformation behavior. In this scenario, the deformation inside the grain is important and in addition part of it also happens in grain boundary. Furthermore, fewer grains involved in the deformation cannot form the flow streams and the relative position change of the neighboring grains is also not significant. Therefore, the grain size does not significantly affect the formation of obvious flowline, deformation texture, or even the folding defect. Therefore, the change of grain size in four scenarios as shown in Figs. 5.47 and 5.48 does not significantly affect the material flow as the change of geometry size does.

Figures 5.47 and 5.48 also present the microstructure and flow behavior of the formed parts in microextrusion process. The formed parts of Columns 1 and 2 have a bigger ratio of geometry size to grain size than that in Column 3. It can be also seen that the flow-induced defects occur in the mesoscaled cases with different grain sizes. However, for the microscaled parts as shown in Column 3 in the both figures, there is a small ratio of the geometry size to grain sizes and there is no flow-induced defect, viz., folding observed. The parts with fine grains have the obvious flow lines which display the flow-induced defects due to the large possibility to form flow-lines, deformation texture, or even folding defect in microextrusion process as explained in the previous section. For the parts with coarse grains, there is no flow-induced defect and even no obvious flow line due to the fact that the material flow line is mostly represented by the shape of grain boundary. The increase in grain size leads to the decrease in grain boundary fraction, which further results in the unclear flow lines. This phenomenon shows that flow-induced defects could be improved or even avoided in microscaled part via reduction of grain boundary and deformation flexibility of individual grain.

In microextrusion process, the grains in free surfaces have irregular shape due to the orientation of single grain and lack of constraints. The grains located in the interior of the part are restricted by neighboring grains, and each grain undergoes a severe deformation in order to keep the strain continuity. Once the plastic deformation exceeds a certain limit, the grains could be broken and become a potential insecurity factor, as shown in Column 3 in Fig. 5.47 with the grain size of 30 and 50 μm. The large deformation portion of the deformed parts became rough with the increase in grain size in microextrusion process.

Based on the microstructure and the flow pattern of the deformed parts, it can be seen that the case with fine grains has flow-induced defects; meanwhile, the case with coarse grains has broken grains as shown in Column 3 in Fig. 5.47. However, under such a severe deformation condition, there is a balance between flow pattern and degree of deformation. Column 2 in Figs. 5.47 and 5.48 with the grain size of 20 μm shows the unshaped flowlines which means the flowline texture will have good mechanical properties and do not have any broken grains. In microextrusion of the complicated parts, the relationship of geometry and grain sizes has a significant effect on flow-induced defects.

For the axisymmetric parts formed by using the materials prepared with different grain sizes, the grain size effect on the formation of folding defect is also shown in Figs. 5.49 and 5.50. For the large mesoscaled part, the obvious folding defects are formed in the inner hole and the flange portion of the deformed part. Since the grain size is extremely small compared with the geometry size, the deformation region thus has a lot of grains, and the single grain with anisotropy can be randomly and evenly distributed, does not affect the material flow behavior, and further leads to the isotropic deformation pattern. Therefore, the grain size does not have a prominent effect on the formation of folding defect in the part with this size scale.

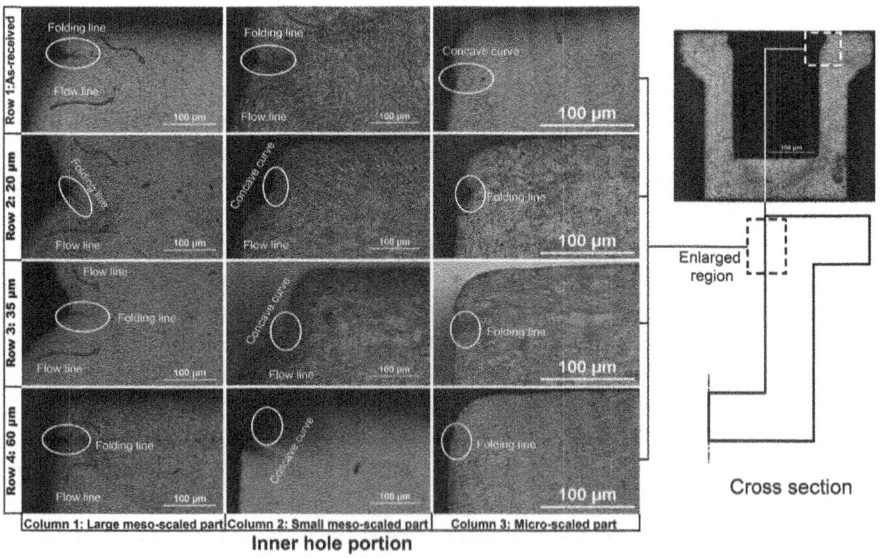

Fig. 5.49 Folding defect formation in forming of the axisymmetric parts with large macro-, small meso-, and microscale and different grain sizes [6]

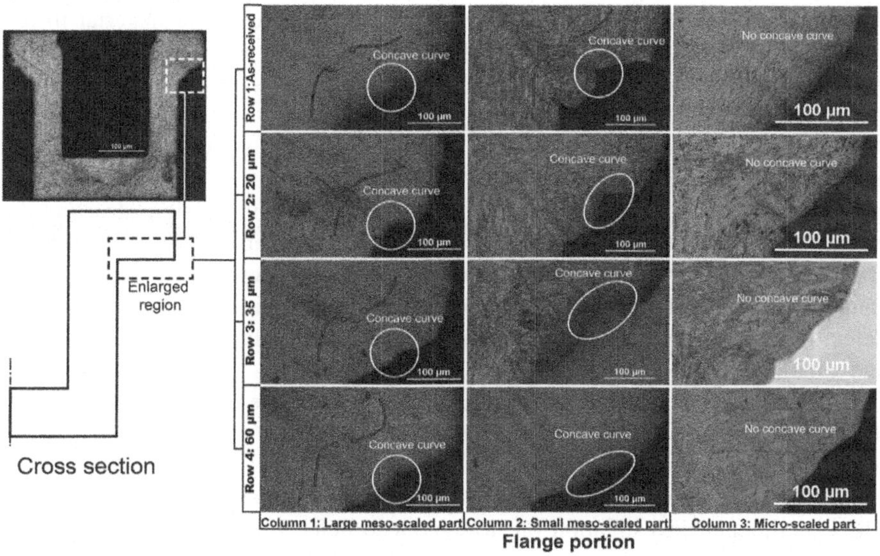

Fig. 5.50 Folding defect formation in forming of the axisymmetric parts with large meso-, small meso-, and microscale and different grain sizes [6]

As for the small mesoscaled part, Column 2 in Figs. 5.49 and 5.50 shows that the flowlines in the workpiece are not so obvious with the increase in grain size. With the scenario of the increasing grain size, the individual grain becomes important in deformation. The boundaries of grains block material flow, which resists the formation of flow-induced defects to a certain extent.

For the microscaled part, the complex deformation region only contains several grains, and the individual grain and its grain boundary significantly affect the material deformation behavior. In the inner hole portion of the part, the flow lines become smaller with the increase in grain size. In the flange portion, however no matter what grain size is, there are no concave curves which can lead to the formation of folding defect, and this reveals that the grain size does not have a significant effect on the material flow behavior. This phenomenon can be explained by the reduction of grain boundaries and the random characteristics and orientation of the individual grains.

5.5 Summary

In this chapter, the experimental investigation and FE simulation of the formation of flow-induced defects in multiscaled plastic deformation are presented. The flow-induced defects are caused by irregular flow pattern in the deformation processes and the formation mechanisms are revealed via examination of the deformation behaviors, flowlines, and deformation texture of the material in real plastic deformation process and the FE simulation results of the deformation processes in different size scales.

The simulation-enabled method and feature-based method are used to aid the design of defect-free forming process, sequence, and die structure. A sliding die insert is introduced into the die structure for avoidance of folding defect. A design framework through introducing spring and die insert into design structure for prediction and avoidance of folding defect via CAE simulation and a feature-based approach for defect-free forming sequence design are developed. Through case studies, the presented design framework and the feature-based approach for defect-free forming sequence design are validated and verified by both experiment and simulation.

In meso- and microscaled forming processes, the occurrence of size effect affects deformation behavior and product quality of the deformed parts. To fabricate high-quality meso- and microparts, the flow-induced defects such as folding defect caused by the irrational material flow in microforming process should be avoided. The experimental investigation on flow-induced defects, microstructure, and flow pattern in meso- and microforming of non-axisymmetric and axisymmetric parts with complicated shape is presented and the findings related to flow-induced defects in terms of formation mechanism and size effect can be leveraged to fabrication of the meso- and microscaled parts with similar geometries and the non-axisymmetric and axisymmetric nature by using meso- and microforming processes.

References

1. Chan WL, Fu MW, Lu J, Chan LC (2009) Simulation-enabled study of folding defect formation and avoidance in axisymmetrical flanged components. J Mater Process Technol 209:5077–5086
2. Scientific Forming Technoloiges Corporation (2014) DEFORM™ 3D Manual, USA
3. Wang JL, Fu MW, Ran JQ (2013) Analysis and avoidance of flow-induced defects in meso-forming process: simulation and experiment. Int J Adv Manuf Technol 68:1551–1564
4. Fu MW, Luo ZJ (1992) The prediction of macro-defects during the isothermal forging process by the rigid viscoplastic finite-element method. J Mater Process Technol 32:599–608
5. Wang JL, Fu MW, Ran JQ (2014) Analysis of size effect on flow-induced defect in micro-scaled forming process. Int J Adv Manuf Technol 73:1475–1484
6. Wang JL, Fu MW, Yu JX, Wang X, Yang WB (2016) Study of the flow behavior and defect formation in forming of axisymmetric flanged and multi-scaled parts. Int J Precis Eng Manuf 17(10):1–9

Chapter 6
Ductile Fracture and Stress-Induced Defects in Multiscaled Plastic Deformation

6.1 Introduction

Fracture failure, as one of the main failure modes in product engineering, has been fully recognized and extensively studied in the past a few decades. Product fracture failure can happen in different stages of product life cycle including manufacturing, service, and maintenance. In manufacturing, especially for the parts and components made via plastic deformation of materials or the so-called metal-forming processes, the existence of fracture and fracture defects in the metal-deformed parts is a critical and tantalized issue to be addressed. Any fracture defects formed in forming process would reduce the load-carrying capacity, deteriorate the performance of product in service, and thus disqualify the produced parts. In the traditional metal-forming arena, the size scale of metal-formed parts is mostly large scale and the plastic deformation process is the so-called macroscaled forming. For the large-scaled parts made via plastic deformation, ductile fracture often occurs in the process and has been extensively studied with the intended goal of defect-free production. The systematic knowledge in terms of fracture formation mechanism, affecting factor, prediction criterion, classification, etc., has also been well established. When the part size is scaled down to microscale, however, the well-established knowledge and the insights into this widely used manufacturing process may not be fully valid in microscaled plastic deformation or microforming, so is the knowledge of ductile fracture and stress-induced defects. The barrier of this knowledge transfer from macro- to microscale plastic deformation is generally believed to be caused by the size effect, which is a ubiquitous and unique phenomenon in many engineering processes. This is a relatively new topic and has not yet been extensively explored and investigated.

In microscaled plastic deformation, the size effects arising from the variations in both geometry and grain sizes and their interactions affect fracture occurrence and

M.W. Fu, *Design and Development of Metal-Forming Processes and Products Aided by Finite Element Simulation*, Engineering Materials and Processes,
DOI 10.1007/978-3-319-46464-0_6

behavior deformation process. It thus becomes a non-trivial issue and needs to be extensively explored and investigated. In this chapter, the ductile fracture in multiscaled plastic deformation is presented, but the focus is more on meso- and microscaled deformations. How the size effect affects ductile fracture behavior and occurrence is investigated via experiments and finite element simulations. From experimental point of view, a series of ductile fracture experiments from macro-, meso-, to microscale are conducted and the experimental results are compared with FE simulation. From simulation point of view, the constitutive modeling-based FE simulation, which considers geometry and grain size effects, microstructure, and phase of materials, is presented. The FE simulations using the constitutive models established based on the widely used surface layer model in microscaled plastic deformation are also conducted and corroborated with experiments. The applicability of ductile fracture criteria (DFC) in microscaled plastic deformation is evaluated via FE simulation and physical experiment. The systematic and in-depth understanding of ductile fracture in microscaled plastic deformation is summarized and elucidated.

6.2 Ductile Fracture and Stress-Induced Defects

Fracture can be considered as the separation of a part in service or a workpiece in manufacturing process in which its internal working stress exceeds the strength limit of part material or the materials being processed. It results in the loss of load-carrying capacity of the part in service or the plasticity of workpiece for further and continuous deformation. Fracture is formed via fracture initiation, propagation, growth, and finally occurrence of fracture in plastic deformation process or fracture failure of the parts and components in service. The initiation of fracture, on the other hand, could cover void nucleation, growth, coalescence, and further formation of microcrack. There are many factors affecting the fracture behavior and occurrence of the parts to be fabricated in manufacturing process or fracture failure of the components in service. These factors include material properties, applied external loading, stress condition, service circumstance, and the design of the parts and components. From engineering materials perspective, there are two major fracture modes, viz. ductile and brittle fractures. The main difference of these two fracture modes is the amount of plastic deformation the material undergoes before fracture occurs. Generally, ductile materials demonstrate large amounts of plastic deformation, while brittle materials show little or no plastic deformation before fracture. Figure 6.1 shows the schematic illustration of ductile and brittle fractures of materials in tensile stress–strain curve. In the stress–strain curve, the area under the curve is associated with the energy needed for plastic deformation to a certain deformation level. From the curve, it is found that the energy needed for ductile fracture of ductile materials is much greater than that for brittle fracture of brittle materials. Since the focus of this book was more on the fracture behavior and occurrence of materials in plastic deformation process, the ductile fracture is thus the focus.

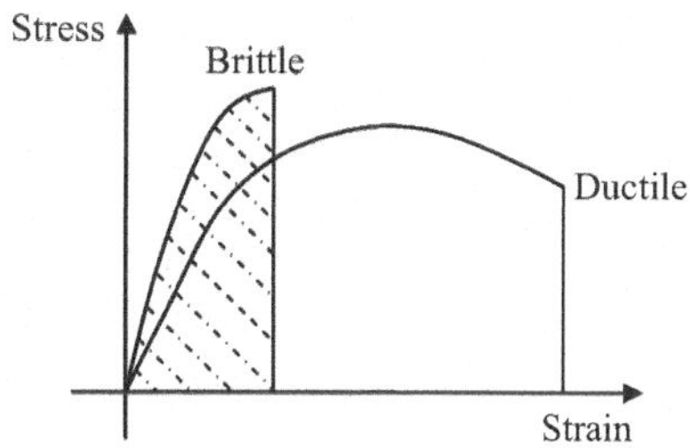

Fig. 6.1 Schematic illustration of ductile and brittle fractures in tensile deformation

Furthermore, fracture, no matter whether it is ductile or brittle, is a permanent deformation behavior of materials under loading. From product service point of view, the occurrence of fracture, even though it is quite small, deteriorates the load-carrying capacity and the desired performance of products, or eventually causes catastrophic consequence. Therefore, it is necessary to avoid the occurrence of cracks in the service of products or the fracture formed in manufacturing process, especially in metal-forming process. To minimize the possibility of fracture failure of products in service caused by ductile or brittle fracture, the root causes of fracture need to be extensively studied and seriously considered in up-front design stage. They include inappropriate design of product, selection of wrong materials, inherent defects generated in prior manufacturing processes, overloading in the service of products, and the inadequate maintenance. From manufacturing point of view, fracture-based defects are easy to occur when the products are made by metal-forming processes, which employ the ductility of materials for plastic deformation via the change in the shape and geometries of workpiece and redistribution of materials in die cavity. In this process, when the amount of deformation exceeds the capacity of deformation represented by the ductility or plasticity of materials, fracture happens and fracture defects are formed in the deformation body. In design of metal-formed part and determination of metal-forming process, how to avoid fracture occurrence in each forming process needs to be carefully investigated and addressed such that the designed metal-formed parts can be successfully fabricated via plastic deformation and the defect-free production can be efficiently realized.

To produce defect-free metal-formed parts, fracture formation mechanisms in forming process need to be fully understood. In addition to different amounts of deformation the ductile and brittle fractures experience, the energy absorption in fracture process is also different. For ductile fracture, high energy is absorbed by microvoid coalescence in the formation of ductile fracture and the fracture usually takes time to grow and propagate. The fracture mode is thus less catastrophic. For brittle fracture, however, it absorbs less energy in transgranular cleavage or intergranular fracture, and therefore, it is a low-energy fracture mode and exhibits more catastrophic. In addition, ductile fracture is generally formed via microvoids initiation, growth, coalescence, and finally the formation of microcracks, which propagate in the direction normal to tensile stress. The cracks then rapidly grow, and eventually, ductile fracture happens. In tensile test, ductile cracks generally propagate through the periphery along the shear plane at 45° to tensile direction, resulting in the cup-and-cone fractures on the fracture surfaces of the testing sample. For brittle fracture, the process of cleavage fracture could have three steps. The first one is the plastic deformation to produce dislocation pileups and then followed by crack initiation. The crack can very quickly propagate to the failure of materials. In this process, there is no obvious plastic deformation and the process is difficult to predict. From the grain deformation point of view, transgranular cleavage and intergranular fracture are the main fracture mechanisms of brittle fracture.

In metal-forming process, ductile fracture, instead of brittle fracture, occurs in forming process as the ductility of materials is employed for realization of this manufacturing process. For plastic deformation of brittle materials, the deformation condition needs to be carefully selected in such a way the brittle-to-ductile transition can be realized and the materials can undergo a sufficient plastic deformation. In addition, if the brittle materials such as BCC structure metallic materials need to undergo plastic deformation, the amount of plastic deformation needs to be minimized since they have limited slip systems at low deformation temperature. Increasing the deformation temperature facilitates more slip systems to be activated and yields a large plastic deformation before failure occurs. This brittle-to-ductile transition behavior needs to be considered in forming process determination and process window design such that the best ductility of materials can be used to plastically deform the materials and the defect-free deformed parts can be produced.

If the amount of deformation in each deformation operation is not properly allocated, fracture may happen in the corresponding forming process. To analyze and predict the ductile fracture in forming process, two analysis approaches have been employed. The first approach uses the deformation kinematics of materials to analyze the material flow in different deformation zones. As shown in Fig. 6.2a, if the plastic deformation zones 1 and 2 do not meet with each other at the axis of symmetry, chevron crack may occur at the central line of the deformation body, as shown in Fig. 6.2b. This approach has been used by Avitzur [1] for his researches on material flow in drawing and extrusion such that the central burst or chevron crack could be predicted and analyzed.

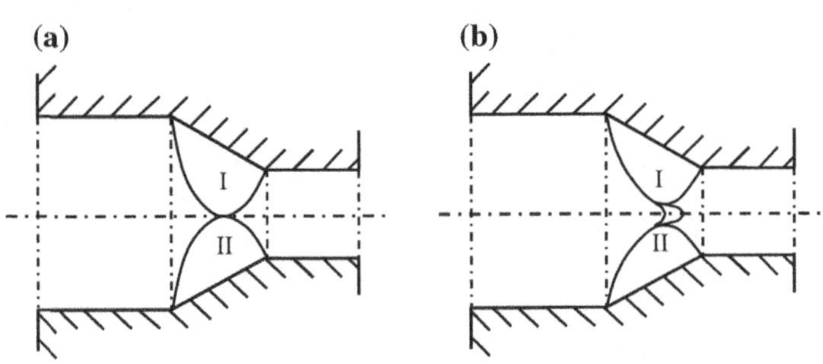

Fig. 6.2 Deformation coordination in deformation zones [1]

The other approach uses fracture criterion to analyze the fracture behavior in plastic deformation of materials. Fracture criteria are proposed based on a specific physical basis and the essence of fracture mechanism such as the deformation energy accumulation or a certain critical stress condition, to predict the occurrence of ductile fracture in plastic deformation process. Table 6.1 gives some widely used DFCs, which are generally classified into two categories, viz. the uncoupled and coupled ones. The uncoupled DFC neglects the effects of damage on the yield surface of materials, while the latter considers it and incorporates the damage accumulation into the constitutive equations to analyze the fracture in deformation process. In the traditional or the so-called macroscaled forming arena, ductile fracture in tandem with DFCs has been extensively explored, investigated, and studied in terms of formation mechanism and the applicability of DFCs. However, the same cannot be said to the ductile fracture in microscaled plastic deformation despite of the fact that some efforts have been provided into the exploration in this niche area recently. In this chapter, the focus is more on the ductile fracture in

Table 6.1 The widely used DFCs [2]

Criterion	Formula	Background	Type
Freudenthal et al.	$\int_0^{\bar{\varepsilon}_f} \bar{\sigma} \mathrm{d}\bar{\varepsilon} = C_1$ (plastic work)	Plastic work	Uncoupled DFC
Cockcroft and Latham (C&L)	$\int_0^{\bar{\varepsilon}_f} \langle\sigma_1\rangle/\bar{\sigma} \mathrm{d}\bar{\varepsilon} = C_2$	Empirical model with underlying pressure effect	
Brozzo et al.	$\int_0^{\bar{\varepsilon}_f} \frac{2\langle\sigma_1\rangle}{3\langle\sigma_1-\sigma_m\rangle} \mathrm{d}\bar{\varepsilon} = C_3$		
Ayada	$\int_0^{\bar{\varepsilon}_f} \sigma_m/\bar{\sigma} \mathrm{d}\bar{\varepsilon} = C_4$	Based on the effect of stress triaxiality on void growth and neglecting void nucleation and coalescence	
Rice and Tracey (R&T)	$\int_0^{\bar{\varepsilon}_f} \exp\left(\frac{3}{2}\frac{\sigma_m}{\bar{\sigma}}\right) \mathrm{d}\bar{\varepsilon} = C_5$ (multiaxial stress state)		
Oyane et al.	$\int_0^{\bar{\varepsilon}_f} \langle 1 + A\frac{\sigma_m}{\bar{\sigma}}\rangle \mathrm{d}\bar{\varepsilon} = C_7$ (based on void growth and porous plasticity theory)		
Tresca	$\tau_{\max} = \frac{\sigma_1-\sigma_3}{2} = C_8$	Shear band	
Xue–Wierzbick model	$\int_0^{\bar{\varepsilon}_f} \frac{\mathrm{d}\bar{\varepsilon}}{F(\eta,\zeta)} \mathrm{d}\bar{\varepsilon} = 1$		
Gurson/Tvergaard (GTN)	$\Phi(\boldsymbol{\sigma}, \bar{\varepsilon}, f) = \left[\frac{\bar{\sigma}}{\sigma_0(\bar{\varepsilon})}\right]^2 + 2q_1 f*(f)\cosh\left[-\frac{3q_2\sigma_m}{2\sigma_0(\bar{\varepsilon})}\right] - (1 + q_3[f*(f)]^2) = 0$		Coupled DFC
CDM-based Lemaitre	$\dot{D} = \frac{\partial F_D(Y,\bar{\varepsilon},D,\ldots)}{\partial Y}\dot{\varepsilon}(1-D)$		

Where $\langle\ldots\rangle$ – Mac Auley bracket $\langle x\rangle = x$ if $x > 0$, $\langle x\rangle = 0$ if $x \le 0$

meso- and microscaled plastic deformation. How the ductile fracture behaviors in these downsized forming processes differ from those in macroscaled forming processes and the applicability of DFCs in meso- and microscaled plastic deformation are discussed and articulated.

The defects induced by ductile fracture in plastic deformation process, on the other hand, are termed as stress-induced defects, which are different from the flow-induced defects caused by the irrational flow described in Chap. 5. The exploration on stress-induced defects generated by ductile fracture in meso- and microscaled deformation provides a basis to understand the ductile fracture formed in the small-scaled plastic deformation and further to avoid the occurrence of this type of defects. It is very crucial for the promising applications of meso- and microscaled plastic deformations in the forms of meso- and microscaled forming processes for mass production of defect-free meso- and microscaled parts and components.

6.3 Size Effect on Ductile Fracture

In macroscaled plastic deformation processes, viz. macroforming, ductile fracture has been extensively studied in terms of physical nature, formation mechanisms, affecting factors, and the prediction criteria. In microscaled plastic deformation, or the so-called microforming in this book, all of these are relatively new and have not yet been extensively investigated. In tandem with this, this section explores how size effect affects the ductile fracture behaviors in microforming via experiments and FE simulations. Using microscaled flanged upsetting as a case study process, the ductile fracture in this microscaled deformation process is studied. The FE simulation is conducted using the constitutive model established based on the widely accepted surface layer model. On the basis of the proposed surface layer model, the fracture in microscaled plastic deformation is predicted by considering the size factor in Cockcroft fracture criterion, and the results and findings are corroborated and verified by experiments. It is found that the ductile fracture affected by size effect is difficult to happen in microforming under the same deformation conditions in which the fracture happens in macroforming scenario. In other words, it needs a larger plastic deformation in microscaled deformation than that in macroscaled one. In addition, the deformation-related behaviors can be different from those in macroscaled domain, which is caused by size effect in the deformation process. All of these provide an in-depth understanding of deformation and fracture in microscaled plastic deformation.

In this book, two types of size effects arising from the scale variation in geometrical size of workpiece and microstructure or grain size of materials are systematically studied from the perspectives of deformation and fracture behaviors by experiments and simulations. From experimental point of view, the central-heading process in different size scales is conducted to investigate the geometrical size effect of testing samples. From microstructure or grain size effect perspective, the materials are prepared with different grain sizes via vacuum heat treatment processes. In terms of modeling, the constitutive models considering size effect and even microstructure phase of materials are developed for efficiently representing the reality of deformation. By employing both the experiment and simulation, the knowledge related to the ductile fracture in microscaled plastic deformation is thus established.

6.3.1 Modeling of Deformation Behaviors Considering Size Effect

In modeling of the deformation behavior in microscaled plastic deformation via considering size effect phenomenon, how to handle the size effect in establishment of constitutive models is critical. Figure 6.3 shows a methodology and the procedure for modeling of the size effect affected deformation behaviors in plastic deformation processes. The stress–strain relationship of the testing material of brass is first established through macro- and microscaled upsetting experiments. The size factor, which represents the percentage of the grains in surface layer among all the grains in the workpiece, is introduced, defined, and calculated according to the specimen dimension and grain size. A microscaled upsetting stress–strain relationship model is then established. The simulations of microscaled flanged upsetting are also conducted using the established model. The corresponding simulation results are compared with the experimental ones to determine whether the deformation behavior in forming process is affected by size effect. Finally, by implementing the stress–strain relationship model into the Cockcroft fracture criterion, the fracture prediction is carried out. The fracture energy of the testing specimen is determined, and the fracture strain is calculated. They are further compared with the experiments; in such a way, the presented methodology is validated and verified; and the simulation results are corroborated by experiments.

In microscaled plastic deformation, there are two main size effects to be considered, viz. grain and geometry size effects. The grain size effect deals with how the size variation in microstructure grain affects the deformation behavior and performance of deformation body, while the geometry size effect is concerned about how the variation in geometry size scale of workpiece and its features affects the deformation behavior and performance of the workpiece in deformation process. On the other hand, the geometry and grain size effects can occur simultaneously with the variation in geometry and grain sizes in different deformation scenarios. When the ratio between the dimensions of workpiece and the grain size of its material is introduced into the constitutive equation, how the geometry and grain size effects affect the deformation behavior and performance in microforming process can be considered and handled simultaneously. The investigation of the two size effects affected deformation behaviors and fracture phenomena in microscaled flanged upsetting process is conducted in this way via experiments and simulations.

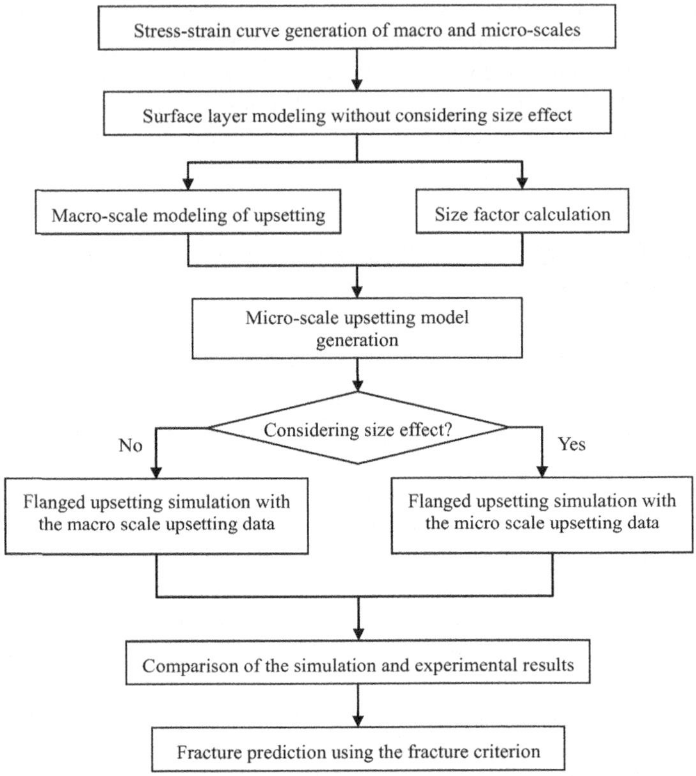

Fig. 6.3 Methodology for modeling of size effect affected deformation behaviors [3]

6.3.2 *Surface Layer Model*

In deformation process, the workpiece can be assumed to be consisting of surface and internal grains. The surface grains have fewer constraints in deformation as part of the grain has free surfaces. It is thus easier for them to deform compared with the internal grains. In the conventional macroforming process, the ratio between the surface and internal grains is very small such that the contribution of surface grains to the entire deformation of workpiece can be ignored. In microforming process, however, this ratio is much larger and the size effect exists. In alignment with this fact, the surface layer model is proposed, which formulates the stress contributions from the grains in surface layer and interior of the workpiece in the following:

$$\sigma = \frac{N_s \sigma_s + N_i \sigma_i}{N} (N = N_s + N_i) \tag{6.1}$$

In Eq. (6.1), σ and N are the flow stress and the total grain number of specimen. N_i and σ_i are the number and the flow stress of internal grains, while N_s and σ_s are the number and the flow stress of surface grains, respectively.

By employing this model, a hybrid material flow stress–strain model is developed by Peng et al. [1] and Lai et al. [5], in which the flow stress of specimen is contributed by the stresses of surface grains and internal grains. The size effect and its existence are thus caused by the ratio change of surface and internal grains. For the surface grains, the grain mechanical property is similar to single crystal. For the internal grains, a polycrystal flow stress model is used to represent their material behavior. By applying Schmid method, Hall–Petch equation, and crystal plasticity theory [6, 7], a hybrid model can be represented as follows:

$$\begin{cases} \sigma_s(\varepsilon) = m\tau_R(\varepsilon) \\ \sigma_i(\varepsilon) = M\tau_R(\varepsilon) + \frac{k(\varepsilon)}{\sqrt{d}} \end{cases} \tag{6.2}$$

In Eq. (6.2), d represents the grain size; m and M are the orientation factors of the surface layer and internal grains; $\tau_R(\varepsilon)$ is the main shear stress; η is the size factor to quantitatively represent the percentage of surface grains in all the grains of the deformation body. Compared with Eq. (6.1), let $N_s = \eta N$. The flow stress model of material can thus be designated in the following:

$$\begin{cases} \sigma(\varepsilon) = \sigma_{\text{ind}} + \sigma_{\text{dep}} \\ \sigma_{\text{ind}} = M\tau_R(\varepsilon) + \frac{k(\varepsilon)}{\sqrt{d}} \\ \sigma_{\text{dep}} = \eta\left(m\tau_R(\varepsilon) - M\tau_R(\varepsilon) + \frac{k(\varepsilon)}{\sqrt{d}}\right) \end{cases} \tag{6.3}$$

In Eq. (6.3), σ_{ind} represents the flow stress of the conventional polycrystal, which is feature size independent. σ_{dep} is the flow stress related to size effect.

In flanged upsetting process, the original specimen is divided into two sections: surface grain and internal grain sections, as shown in Fig. 6.4. In the figure, the surface grain section is shaded, while the internal grain section is not. The height and diameter of the specimen are h and D, and the grain size is d, respectively.

According to Eq. (6.1) and geometrical calculation, the size factor η is determined as follows:

$$\begin{cases} \eta = \frac{N_s}{N} = \dfrac{\frac{\frac{\pi D^2 h}{4} - \left(\frac{\pi}{4}(D-2d)(h-2d)\right)}{\frac{4}{3}\pi d^3}}{\frac{\frac{\pi D^2 h}{4}}{\frac{4}{3}\pi d^3}} = 2d\left(\frac{\left(1-\frac{2d}{D}\right)^2}{h} + \frac{2\left(1-\frac{d}{D}\right)}{D}\right) \\ \eta = 1 \end{cases} \tag{6.4}$$

In Eq. (6.4), the first one represents the size factor when the specimen is still polycrystalline. And when $d = D$, it means the specimen is single crystalline material and the size factor is 1. Compared with Peng's research, the size factor in this chapter considers the influence of the grain volume in both the bottom and top

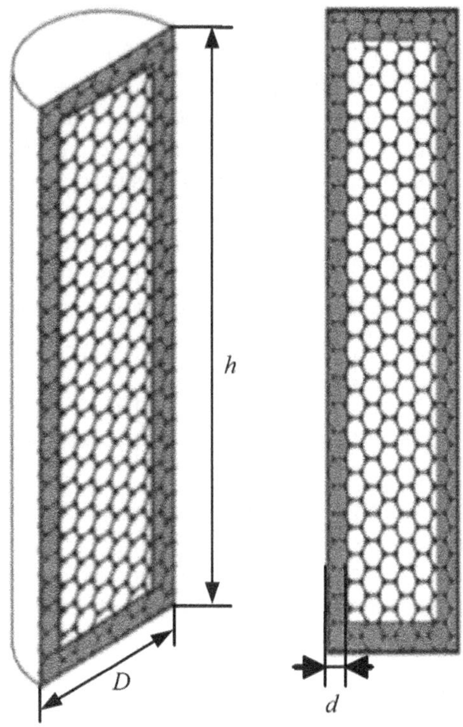

Fig. 6.4 The surface layer model in the microscaled simple upsetting and flanged upsetting [3]

cross sections, which is represented as $\frac{2d\left(1-\frac{2d}{D}\right)^2}{h}$. It is believed that the billet length must also be considered when the grain size is close to the feature diameter.

6.3.3 Calculation and Comparison of Flow Stress Models in Simple Upsetting

To demonstrate the influence of size effect on material flow behavior and compare different models, curve fitting is employed to generate the stress–strain curves based on the experimental results. After compared with several mathematical models, the exponential function is used to establish the flow stress model.

First of all, the curve fitting is conducted by using the experimental data of the macroscaled billet (2 mm × 3 mm, *H/D* = 1.5) to obtain the coefficients of Hall–Petch equation without consideration of size effect. The surface layer model is then built with these coefficients. The results obtained by using the developed model and the conventional bulk forming model without consideration of size effect are compared with the experimental data of microscaled simple upsetting.

To establish the size effect-dependent flow stress model, the experimental data with the small size factor are applied. In this case, the size factor is 0.04 as the grain size of raw material is 18.54 μm. In Fig. 6.5, four color lines are used to describe the stress and strain relationships of the four types of specimens heat-treated with different conditions and with the average grain sizes of 18.54, 24.21, 45.48, and 87.73 μm, respectively. From Armstrong's flow stress model, the following equation is obtained:

$$\sigma_i(\varepsilon) = M\tau_R(\varepsilon) + \frac{k(\varepsilon)}{\sqrt{d}} \tag{6.5}$$

where M is known as 3.06 [8] for BCC material based on Taylor's model. M is calculated at the yielding point of material and varies differently when the strain is larger. When the strain exceeds the yield strain, this parameter needs recalculation. $\tau_R(\varepsilon)$ and $k(\varepsilon)$ can be represented by an exponential function of $y = kx^n$. If the strain is set to a certain value, the equation will converge to the classic Hall–Petch equation. If a series of strains are given, a series of flow stresses can then be obtained via the simple upsetting experiment [9].

Therefore, if the strain is set to a certain value, $\sigma_i(\varepsilon) = M\tau_R(\varepsilon) + \frac{k(\varepsilon)}{\sqrt{d}}$ will converge to the classic Hall–Petch equation $\sigma_i = M\tau_R + \frac{k}{\sqrt{d}}$. In this equation, both $M\tau_R$ and k are constant. Accordingly, when the grain size d changes after heat treatment, the corresponding true stress will also change. Since all the grain sizes and stresses can be obtained from experiment, a linear function between grain size d and true stress σ can thus be established and the $M\tau_R$ for a certain strain can be obtained. Let the strain be 0.1, 0.2, 0.3, and finally 1.0 with an increasing interval of 0.1, the set of $M\tau_R$, which contains the value for these ten strains, can be determined and the relationship between $M\tau_R(\varepsilon)$ and ε is obtained as shown in Fig. 6.6.

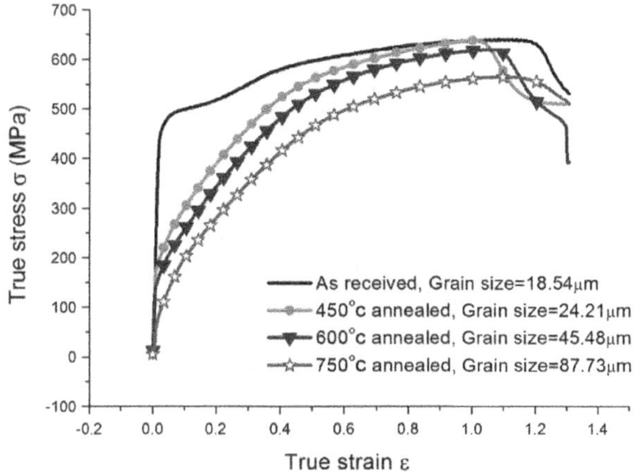

Fig. 6.5 Stress–strain curves of the testing material with different grain sizes [3]

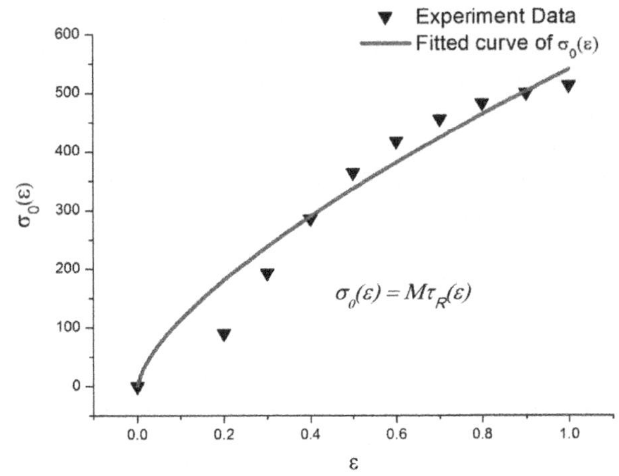

Fig. 6.6 Relationship between $M\tau_R(\varepsilon)$ and ε [3]

Therefore, the hypothetic equation of $\tau_R(\varepsilon)$ can be calculated using the nonlinear curve fitting method and formulated in the following:

$$\tau_R(\varepsilon) = 176.32\varepsilon^{0.68} \tag{6.6}$$

Putting Eq. (6.6) into the surface layer model and letting the grain orientation factor m equal to 2, which is the minimum limit of this parameter, a nonlinear fitting with the size factor of $\eta = 0.6$ and the grain size of $d = 18.54$ μm can be conducted and $k(\varepsilon)$ is thus obtained as follows:

$$k(\varepsilon) = 73.09\varepsilon^{0.06} \tag{6.7}$$

The final model with $M = 3.72$ and $m = 2$ (the lower bound) is formulated as follows:

$$\begin{cases} \sigma(\varepsilon) = \sigma_{\text{ind}} + \sigma_{\text{dep}} \\ \sigma_{\text{ind}} = M\tau_R(\varepsilon) + \frac{k(\varepsilon)}{\sqrt{d}} = 539.53\varepsilon^{0.68} + d^{-\frac{1}{2}}.73.09\varepsilon^{0.06} \\ \sigma_{\text{dep}} = \eta\left(m\tau_R(\varepsilon) - M\tau_R(\varepsilon) - \frac{k(\varepsilon)}{\sqrt{d}}\right) = \eta\left(352.64\varepsilon^{0.68} - 539.53\varepsilon^{0.68} - d^{-\frac{1}{2}}.73.09\varepsilon^{0.06}\right) \end{cases} \tag{6.8}$$

Schematically, the experimental data and the fitted curve are shown in Fig. 6.7.

In addition, Fig. 6.8 shows two sets of the experimental data generated via simple upsetting test with the billet dimensions of 2 × 3 and 0.5 mm × 0.75 mm. From $\varepsilon = 0$ to $\varepsilon = 1.2$, the flow stresses of both the macro- and microscaled upsetting are increased. In the figure, the area between the curve and the coordinate for the case with the dimensions of 2 mm × 3 mm is obviously larger than the area of the case with the dimensions of 0.5 mm × 0.75 mm, and it indicates that the specimen absorbs more energy in macroscaled forming than that in microscaled forming.

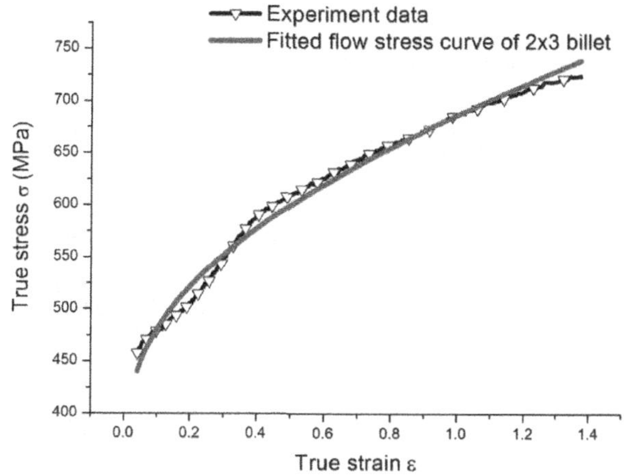

Fig. 6.7 Experimental data and the fitted curve of 2 mm × 3 mm specimens [3]

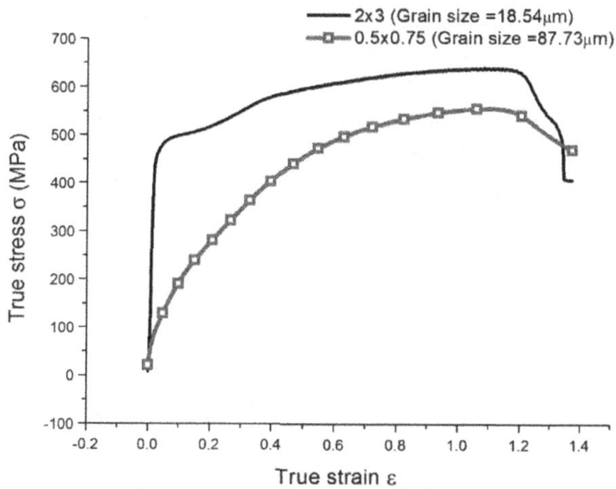

Fig. 6.8 Stress–strain curves generated based on simple upsetting experiment [3]

In order to evaluate the accuracy of the developed surface layer model, the experimental data, the stress and the strain curve generated by Peng's model [4, 5], and the flow stress curve without consideration of size effect are compared with the stress and strain curve constructed based on the developed model. By applying different size factors, the flow stress curves are constructed as shown in Fig. 6.9. The flow stress curve without consideration of size effect is the conventional model with the size factor assumed to be zero, viz. $\eta = 0$. The current surface layer model and Peng's surface layer model are quite similar. The difference in-between is the calculation methodology of $\tau_R(\varepsilon)$ and the definition of size factor η. In addition, Peng's model is mainly used in microsheet metal-forming and simple upsetting processes. The newly proposed surface layer model, however, is more accurate in flanged upsetting process. In addition, the length of billet is an important parameter

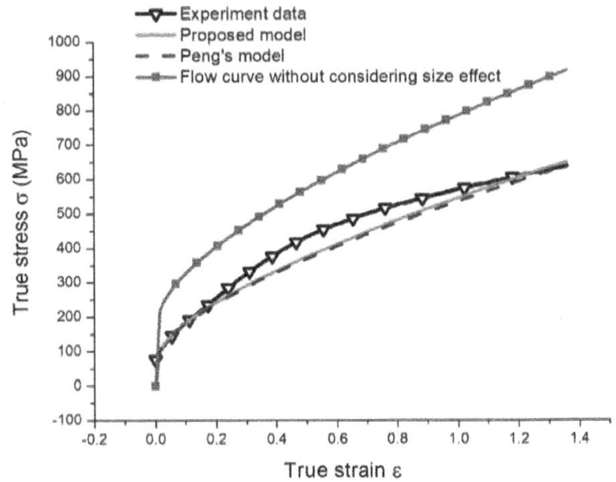

Fig. 6.9 Comparison of the experimental data and different mathematical models [3]

and the volume portion of the grains in the top and bottom surfaces cannot be neglected in the flanged upsetting.

In the curves shown in Fig. 6.9, the flow stress without considering size effect is much higher than the actual experimental results, revealing that the influence of size effect is significant. In addition, it is also greater than the values predicted by other flow stress models. However, the generated results based on the proposed surface layer model and Peng's surface layer model are quite similar in simple upsetting process as the length and diameter ratio is larger than 1, in which the surface volume portion of the grains on the top and bottom surfaces can be neglected.

6.3.4 *Experiments and Simulations*

The purpose of the macroscaled simple upsetting experiment and the development of the surface layer model is to accurately predict and analyze the fracture behavior in microscaled plastic deformation using the microscaled flanged upsetting as a case study process.

To systematically conduct experiments and simulations, an implementation framework is developed as shown in Fig. 6.10. The implementation framework includes three stages, viz. experiment, flow stress and fracture modeling, and FE simulation, as shown in the figure. In FE simulation stage, the data needed for preparation of material properties are determined based on the surface layer model and fracture criterion. As described in the previous section, the stress and strain curves are first generated via macroscaled upsetting experiment. The determined flow stress is then used as the yield stress in plastic deformation and the surface layer model. By using ductile fracture criterion, the fracture strain and the C value of the criterion are determined according to the fracture energy per unit the specimen absorbs before fracture initiation. The results of the expected fracture strain

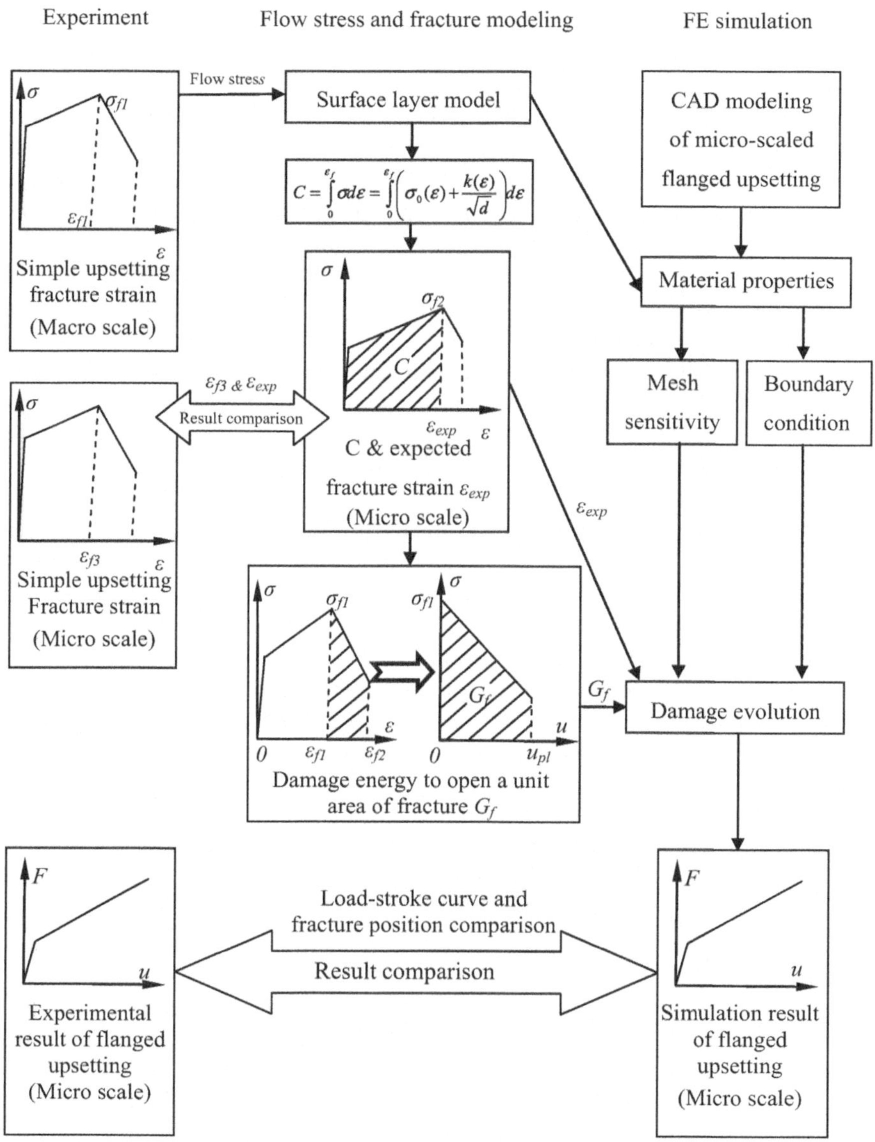

Fig. 6.10 Implementation framework of experiments and simulations [3]

are compared with the actual fracture strain of the microscaled simple upsetting. By using the fracture strain and damage energy as the input in fracture evolution analysis, the final fracture location and the load–stroke curve of the microscaled simple upsetting can thus be predicted.

In simulation stage, meshing is conducted based on the requirement of damage evolution. In compression simulation, which is unlike other microforming processes, the mesh must be generated first before the material properties are specified to the meshed workpiece. This is because in compression simulation, when a mesh element is damaged and considered as an invalid mesh element, the invalid element is deleted and a "blank area" will be formed in the mesh. As the original boundary condition is not applied to this "blank area," other elements around this invalid element would overlap with each other if the simulation continues. In terms of the simulation conducted in this book, a four-node bilinear element CAX4R, which is often used in the simulations with a very large mesh distortion, is employed to model simple upsetting and flanged upsetting processes.

The data needed in damage evolution analysis includes the maximum fracture strain and damage energy, which are determined in modeling stage. Fracture strain is obtained from the surface layer model and can be used to determine the absorbed energy per unit in the hardening portion in the flow stress curve. In the softening part, however, the flow stress curve cannot represent the material property. To address this issue, the damage energy G_f, which is the energy needed to open a unit area of crack, is adopted by Hillerborg [10]. The damage energy to open a unit area can be determined based on the flow stress–displacement curve. By taking into account the mesh sensitivity, the damage evolution law can be specified in terms of the damage energy per unit area. When the material stiffness of a damaged element is fully degraded, this invalid element will be removed and the "blank area" left is considered as the result of fracture formation.

Upon completion of the simulation, the predicted results of the microscale flanged upsetting are compared with the actual experimental ones. The load–stroke curve and the fracture location comparison are also conducted, and the final fracture prediction can thus be made.

To demonstrate the above-described implementation framework, Brass C3602 is used as the testing material for flanged upsetting of different size-scaled specimens. The specimens are annealed to obtain different microstructures. The annealing heat treatment conditions are presented in Table 6.2. The cylinder samples are annealed in vacuum condition with different temperatures and holding times, viz. 750 °C for 3 h and 600 and 450 °C for 2 h to obtain different grain sizes. Furthermore, the metallographic examination is performed with the specimen etched in a solution of 5 g of $FeCl_3$, 15 ml of HCl, and 85 ml of H_2O for 15 s. The grain sizes are then measured correspondingly and also given in Table 6.2.

In terms of experiment realization, the upsetting experiment is conducted in a MTS testing machine. The tools and specimen are lubricated with machine oil to

Table 6.2 Heat treatment parameters of Brass C3602 for cold upsetting [3]

	Heating temperature	Dwelling time	Grain size (μm)
Group 1	As-received	As-received	18.54
Group 2	450 °C	2 h	24.21
Group 3	600 °C	2 h	45.48
Group 4	750 °C	3 h	87.73

Fig. 6.11 Microstructures of the billet materials heat-treated by different conditions [3]

reduce the friction in between. The punch speed is 0.01 mm/s to eliminate the strain rate effect. The original specimens with the dimensions of 2 × 3, 1 × 1.5, and 0.5 mm × 0.75 mm are used, and their microstructures are shown in Fig. 6.11. These samples are compressed to the height reduction of 75 %, which is equivalent to the effective strain of 1.2. Upon completion of the flanged upsetting process, the scanning electron microscope (SEM) technique is used to investigate the potential fracture location in the barreling surface predicted by FE simulation. The specimen is sectioned perpendicular to the bottom surface to study the formation of interior fracture in the workpiece.

Furthermore, the die structure and assembly used in the experiments are shown Fig. 6.12. The tooling components can be changed to do the upsetting experiments using the specimens with different dimensions and scales. In other words, by changing the die components with different geometries and dimensional sizes, the macro-, meso-, and microscaled flanged upsetting can be realized in the same experimental platform and measurement system and in such a way to ensure the consistency and repeatability of the experiments. In addition, the comparison among the experiments can have the same basis.

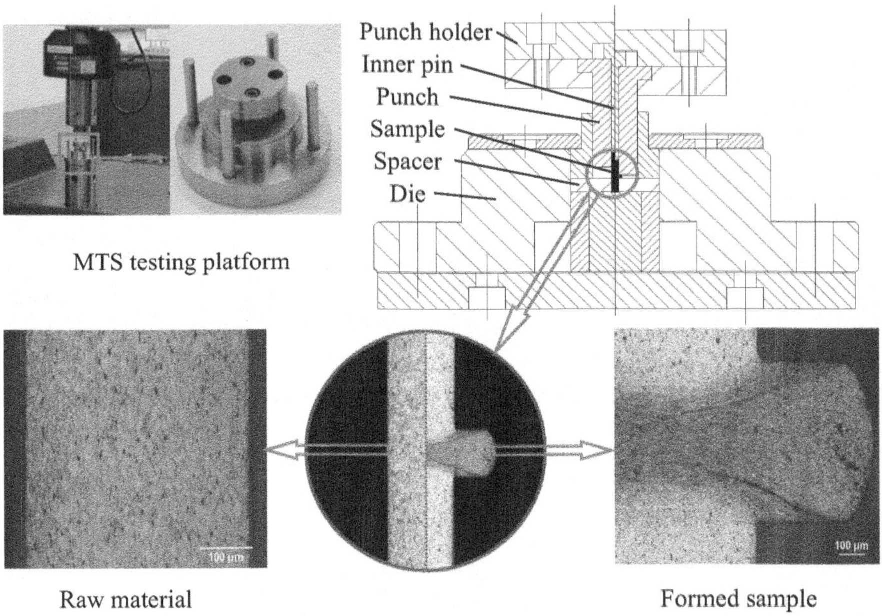

Fig. 6.12 Die assembly for flanged upsetting of different scaled samples [3]

6.3.5 Size Effect on Ductile Fracture

Figure 6.13 shows the deformed parts via macro- and microscaled flanged upsetting. From Fig. 6.13a, it is found that there are two fractures located at two different places when the stroke reaches the deformation limit with the height reduction of 75 %. The first one is located at the circumference of the flanged feature and on the plane with 45° to the axial direction or compression direction. The second one occurs at the interior cross-shear band, as shown in Fig. 6.15, at which the maximum strain is located. In addition, there is no fracture found in the small-scaled parts as shown in Fig. 6.13c–f. Figure 6.13b is the top view of the flanged part with the billet dimension of 2.0 mm × 9.0 mm. To examine the size effect in microscaled flanged upsetting, simulation of the deformation process is conducted based on the surface layer model and the size-independent model. The corresponding results are compared with the experiments. From the load–stroke curves shown in Fig. 6.14, it is found that the proposed surface layer model is more accurate than the traditional constitutive model, which does not consider the size effect, in deformation loading prediction.

Figure 6.15 shows the mechanism of fracture formation in microscaled flanged forming process. The picture in the left top corner shows the expected fracture location and the actual defect location in microscaled flanged upsetting. When the damage energy of the mesh exceeds the fracture energy, which can be determined

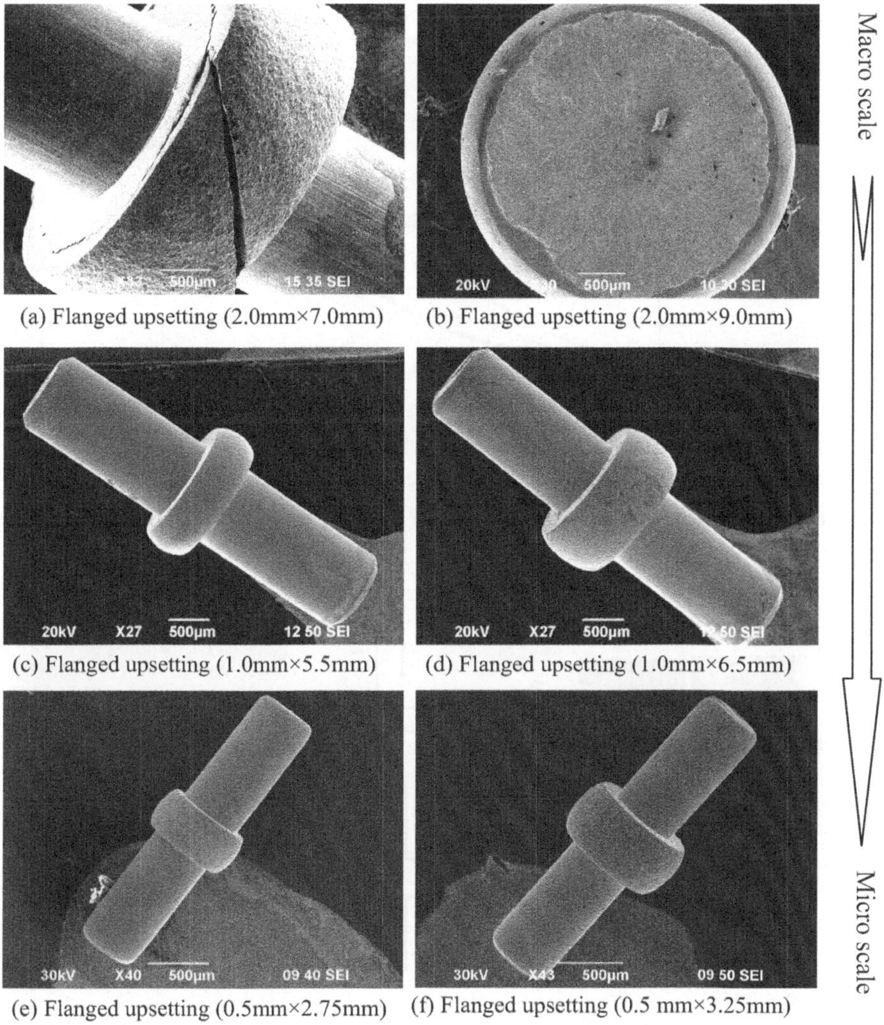

(a) Flanged upsetting (2.0mm×7.0mm) (b) Flanged upsetting (2.0mm×9.0mm)

(c) Flanged upsetting (1.0mm×5.5mm) (d) Flanged upsetting (1.0mm×6.5mm)

(e) Flanged upsetting (0.5mm×2.75mm) (f) Flanged upsetting (0.5 mm×3.25mm)

H/D=1.5:1 H/D=2.5:1

Fig. 6.13 SEM photographs of the flanged upsetting with different scales and *H/D* ratios (the ratio of original height to diameter of the sample) in flanged upsetting. Each sample is compressed with the height reduction of 75 %. Samples (**a**) and (**b**) are considered as macroscaled flanged upsetting, while (**e**) and (**f**) are considered as microscaled flanged upsetting [3]

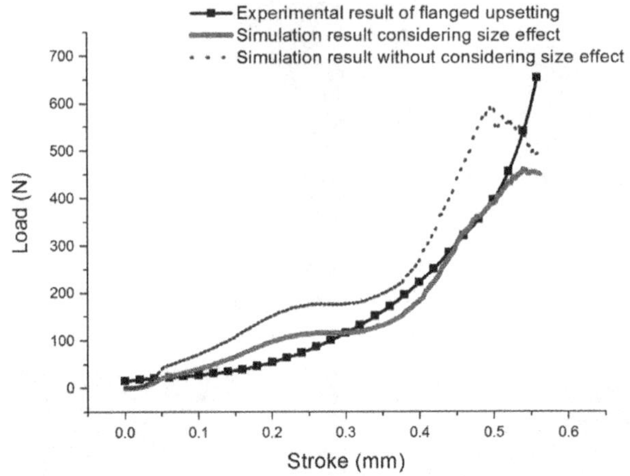

Fig. 6.14 Comparison of the FE simulation results using the proposed surface layer model and the conventional model [3]

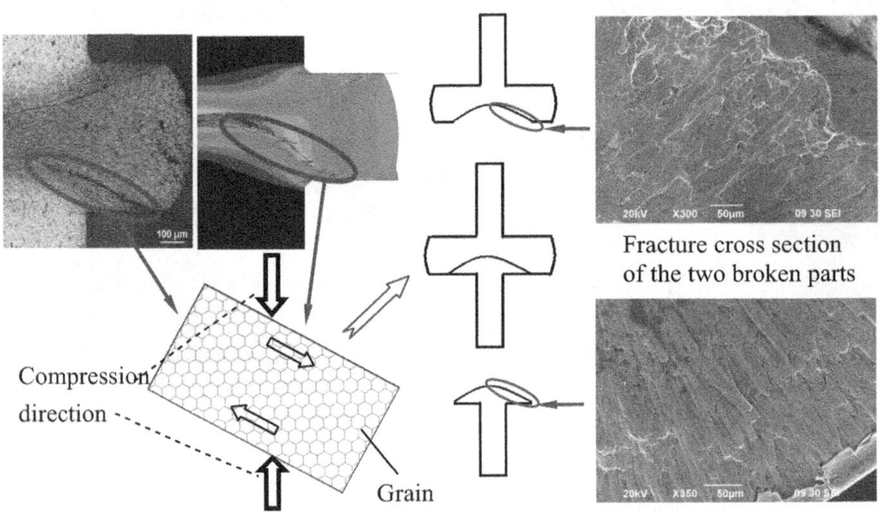

Fig. 6.15 Shear fracture in the microscale flanged upsetting process [3]

based on the stress–strain curve generated from the actual experiment, this mesh will be considered as an invalid one. The invalid mesh is generated by the large mesh distortion and shape change in deformation and will be automatically deleted in simulation. The blank area is then used to represent the invalid elements. In the figure, a slight difference of the fracture in shear band between the actual experiment and simulation is found. This is mainly caused by the mesh distortion of the model. Furthermore, the typical shear dimple is found at the fracture section of the specimen. When the punch contacts the top surface of the specimen, the material in the main deformation area, viz. the flanged part, flows in two opposite directions along the shear band, as shown in Fig. 6.15, which tallies with the research done by Saanouni et al. [11]. The step-like elongated dimples are formed in the two fracture surfaces in the broken specimen.

In macroscaled flanged upsetting process, the specimen dimension is much larger than grain size and the size factor is approximately equal to zero. In microscaled flanged upsetting process, the deformation is much easier to initiate and the ductility of the specimen decreases accordingly with the increase in grain size, which is caused by the reduction in grain boundary constraint as the total amount of grain boundaries is reduced accordingly.

To predict the fracture in macroforming, damage evolution is judged by the fracture energy, which is considered as constant for the same material. Cockcroft and Latham defined this constant with fracture stress and equivalent strain [12–14]. Using Hall–Petch equation and Armstrong's methodology, the damage evolution equation can be redesignated in the following equation:

$$C = \int_{0}^{\varepsilon_f} \sigma \mathrm{d}\varepsilon = \int_{0}^{\varepsilon_f} \left(\sigma_0(\varepsilon) + \frac{k(\varepsilon)}{\sqrt{d}} \right) \mathrm{d}\varepsilon \tag{6.9}$$

As discussed above, when the size effect is considered in microforming process, the established surface layer model is more accurate compared with the conventional macroforming model. Therefore, the damage evolution equation can be represented in the following form:

$$C = \int_{0}^{\varepsilon_f'} (\sigma_{\text{ind}} + \sigma_{\text{dep}}) \mathrm{d}\varepsilon = \int_{0}^{\varepsilon_f'} \left(M\tau_R(\varepsilon) + \frac{k(\varepsilon)}{\sqrt{d}} + \eta \left(m\tau_R(\varepsilon) - M\tau_R(\varepsilon) - \frac{k(\varepsilon)}{\sqrt{d}} \right) \right) \mathrm{d}\varepsilon \tag{6.10}$$

For the material annealed with the same heat treatment process and having a strain gradient in deformation process, it should follow the fracture criterion in Eq. (6.11) with the same C. Figures 6.16 and 6.17 show the relationship between the expected strain when the maximum stress is achieved and the actual strain under the same heat treatment condition. In these figures, the vertical lines represent the existence of the maximum flow stress in deformation process. The area, which is formed by the vertical lines indicated in the figures, the stress–strain curve, and the coordinate axis, represents the energy the specimen absorbed in the hardening process.

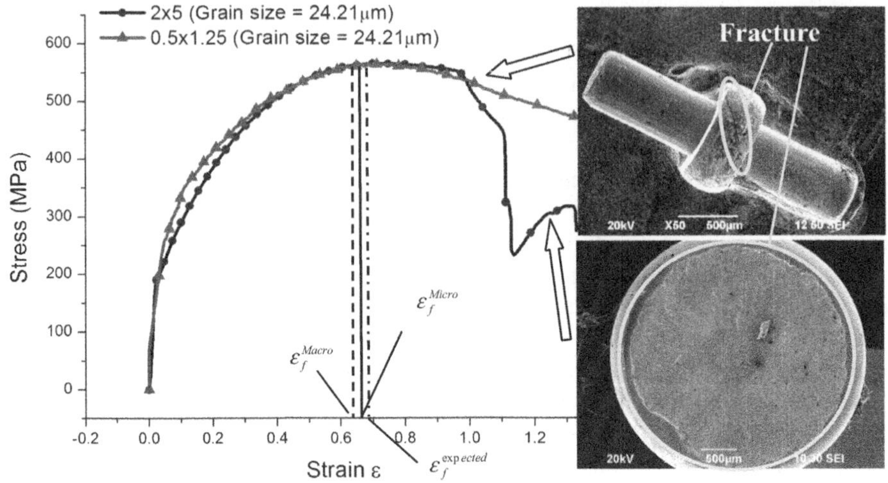

Fig. 6.16 Expected and actual maximum fracture strains of the brass annealed at 450 °C [3]

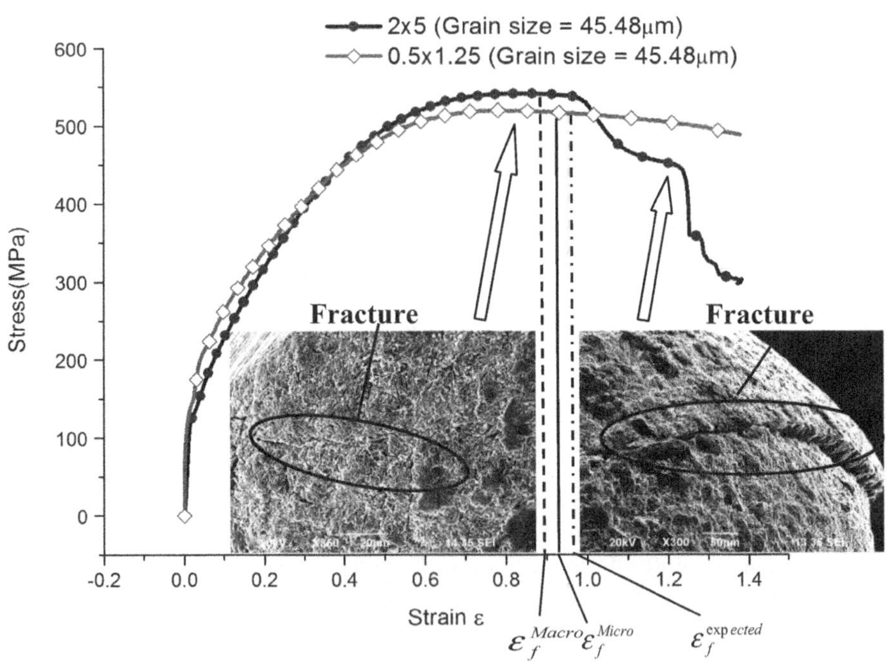

Fig. 6.17 Expected and actual maximum fracture strain of brass annealed at 600 °C [3]

$$\begin{cases} C = \int_0^{\varepsilon'_{f1}} (\sigma_{\text{ind}} + \sigma_{\text{dep}}) \text{d}\varepsilon = \int_0^{\varepsilon'_{f1}} \left(539.53\varepsilon^{0.68} + d^{-\frac{1}{2}}.73.09\varepsilon^{0.06} + \eta_1 \left(352.64\varepsilon^{0.68} - 539.53\varepsilon^{0.68} - d^{-\frac{1}{2}}.73.09\varepsilon^{0.06}\right)\right) \text{d}\varepsilon \\ C = \int_0^{\varepsilon_{f2}} (\sigma_{\text{ind}} + \sigma_{\text{dep}}) \text{d}\varepsilon = \int_0^{\varepsilon_{f2}} \left(539.53\varepsilon^{0.68} + d^{-\frac{1}{2}}.73.09\varepsilon^{0.06} + \eta_2 \left(352.64\varepsilon^{0.68} - 539.53\varepsilon^{0.68} - d^{-\frac{1}{2}}.73.09\varepsilon^{0.06}\right)\right) \text{d}\varepsilon \end{cases} \tag{6.11}$$

In Eq. (6.11), the first equation represents the fracture criterion of macroscaled forming, while the second one is for microscaled forming. The fracture factor C and the grain size d in both the macro- and microscaled specimens are all the material constants with the same value. By solving Eq. (6.11), the relationship of the fracture strain in macro- and microscales is obtained and the prediction of microscaled fracture strain can be made. As shown in Fig. 6.16, when the grain size equals to 24.21 μm, the microscaled fracture strain is given as $\varepsilon_f^{\text{expected}} = 1.07\varepsilon_f^{\text{Macro}}$. In Fig. 6.17, it can be predicted that the microscaled fracture strain $\varepsilon_f^{\text{expected}} = 1.09\varepsilon_f^{\text{Macro}}$. This explains the reason why the fracture is difficult to form in microscaled plastic deformation process.

In Fig. 6.18, the maximum fracture strain is determined as $\varepsilon_f^{\text{expected}} = 1.28\varepsilon_f^{\text{Macro}}$ using Eq. (6.11). With the height reduction of 75 % in upsetting deformation, the microscaled specimen is difficult to break due to the contribution of size effect. However, it is also obvious that the actual maximum fracture strain of the microscaled specimens is much larger than the expected fracture strain. This can be illustrated by the SEM photographs as shown in Fig. 6.18. In the SEM

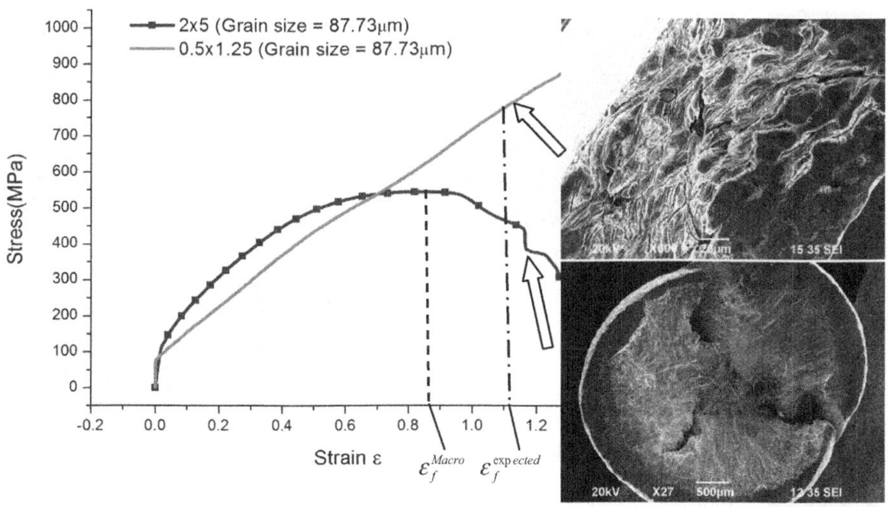

Fig. 6.18 Expected and actual maximum fracture strain of brass annealed at 750 °C [3]

photographs, although the microfractures are observed on the barreling surface of specimen, the continuous fracture has not yet formed when the strain reaches the maximum fracture strain. This may be caused by the friction force between the specimen and the die.

With the support of experimental results, the stress without consideration of size effect is much higher than the size dependent one. As the damage energy is considered as the same, the fracture strain predicted by the size dependent model is believed to be larger than that of the size-independent model [10]. Therefore, a larger deformation is needed in microscaled flange upsetting to reach its damage evolution energy.

Figures 6.19 and 6.20 illustrate the simulation and experimental results of the macro- and microscaled flanged upsetting. In Fig. 6.19, two fractures are found on the surface of the macroscale specimen. The fracture located at the barreling surface of the flanged feature with 45° to the axis direction of the sample is considered as the main fracture. The other is the second fracture located at the cross-shear bands at the flanged part of the central-heading deformation. The cross-shear bands are the unique feature for this central-heading deformation as fracture is easy to occur in the shear band and the maximum effective strain is also located at the shear bands. The simulation results show that when the main fracture is initiated and grows to a certain level, the second fracture will happen.

As discussed before, fracture is more difficult to form in microscaled specimen. The pictures in Fig. 6.20 are the SEM photograph and the FE simulation of the microscaled specimen with H/D of 1.5 and the original dimensions of 0.5 mm

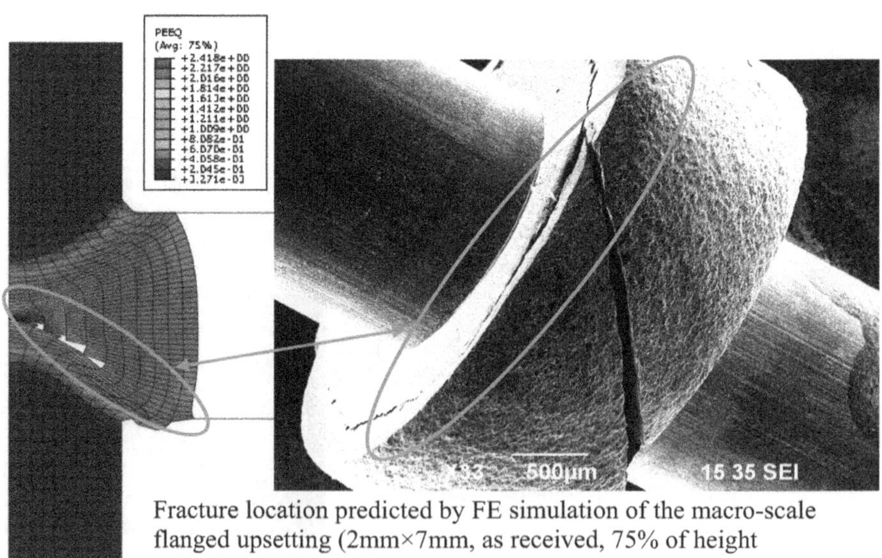

Fig. 6.19 Experimental and simulation results of the macroscale flanged upsetting [3]

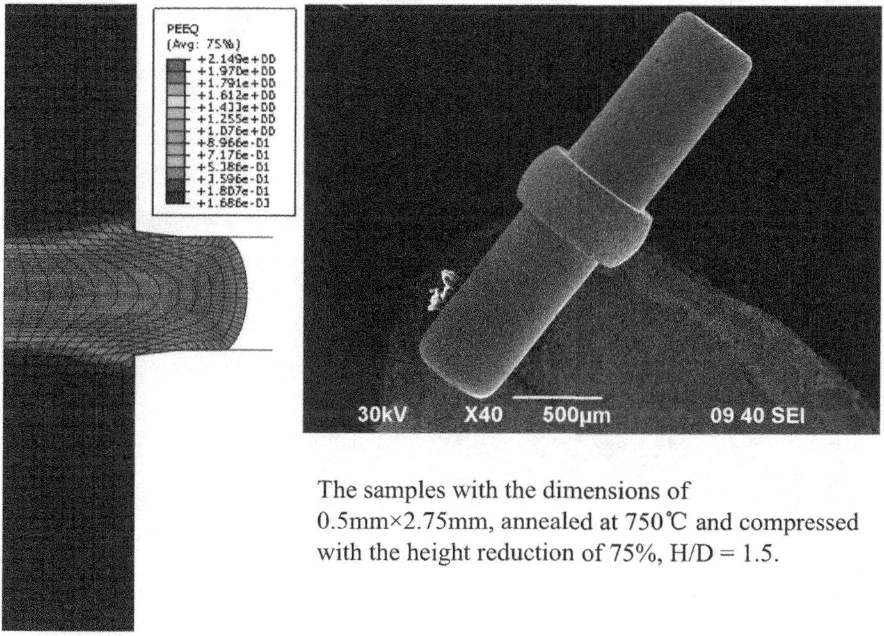

Fig. 6.20 Experimental and simulation results of the microscale flanged upsetting [3]

2.75 mm. In this microscaled central-heading process, the sample is deformed to the height reduction of 75 %. From the picture, it is found that there is no continuous ductile fracture on the barreling surface of the flanged feature and the cross-shear band in the flanged part. Compared with the simulation result of the macroscaled flanged part, the effective strain of the microscaled flanged part in the shear band is smaller than that in macroscaled case with the same height reduction percentage. The fracture initiation in microscaled part would happen only if the microscaled part undergoes more deformation. This also tallies with the phenomenon revealed in Fig. 6.18.

Figure 6.21 shows the fractography of the specimens with different grain sizes. It thus has different scaling factors as the specimens have the same geometry and dimensions. Compared with the result of the previous literature [15], the fracture type is considered as transgranular fracture and the dimple size is quite different, which is caused by different grain size. The image on the left illustrates the fractography of specimen without heat treatment. The transgranular fracture is found, and the dimples with different sizes exist. The image on the right is a shear fracture showing the elongated shear dimples with uniform distribution.

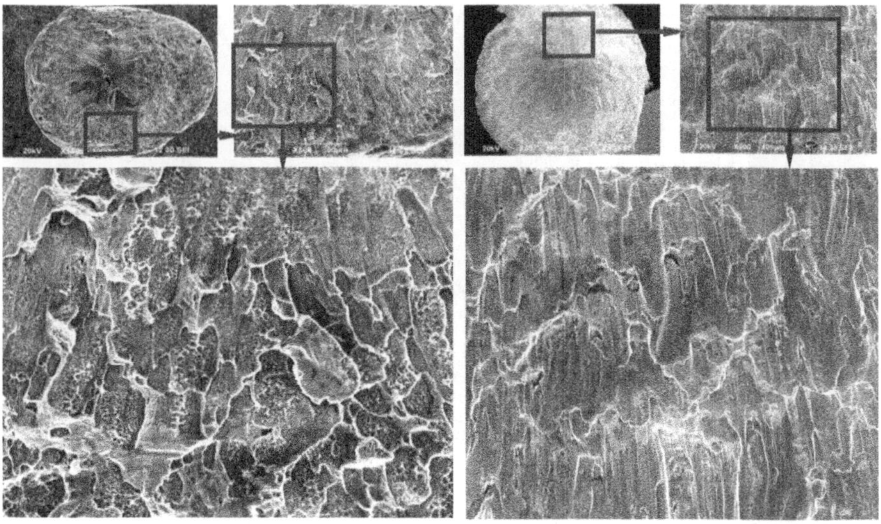

Grain size = 18.54μm Grain size = 87.73μm

Fig. 6.21 Fractography of the samples with different grain sizes [3]

6.4 Hybrid Constitutive Modeling of Fracture in Microscaled Plastic Deformation

In the last section, the size effects of workpiece geometry and material microstructure grain on the ductile fracture behavior in microscaled plastic deformation are summarized. However, this prior art does not consider the multiphase nature of materials, which ubiquitously exists in many engineering materials. For multiphase materials, the detailed properties such as the flow stress of each phase may affect the ductile fracture and deformation behaviors of materials in microscaled deformation process. This is also an eluded and tantalized issue, and there is no compelling and epistemological understanding so far. In-depth exploration and study are thus needed. In this section, this issue will be explored. To model and represent the contribution of individual phases to fracture and deformation behaviors in microscaled plastic deformation, a hybrid model, which considers the individual phase property and percentage in the entire phase composition of material, is proposed. Based on the proposed model, the fracture energy of the alloys with single or multiphase is calculated and the fracture strain is then determined. This hybrid model is more accurate in fracture prediction as it considers the influences of size effect and individual phase on material fracture energy. Using Brass C3602 with different grain sizes obtained via heat treatment as the testing material, the grain and geometry size effects are investigated. Through FE simulation by using the developed hybrid model and the physical experiment, a proposed methodology to represent and model the influence of individual phases on the deformation and fracture behaviors in microscaled plastic deformation of multiphase alloys is validated and verified. Furthermore, the stress-induced fracture map, which articulates the relationship of size effect, fracture energy, and the expected fracture strain in microscaled plastic deformation, is proposed and constructed for easy analysis and prediction of fracture in microforming processes.

6.4.1 *Hybrid Flow Stress Modeling*

Flow stress modeling is important in fracture prediction in microscaled plastic deformation. In stress modeling, the most critical thing is to represent and distinguish the size effect in macro- and microscaled scenarios. For hybrid flow stress modeling, the size factor η is introduced to the conventional stress model, and therefore, the flow stress of the material in microscaled deformation can be more accurately described.

6.4.1.1 Macroscaled Flow Stress Modeling

In deformation process, the flow stress σ can be represented by the microscaled shear stress τ in the following [16]:

$$\sigma = M\tau \tag{6.12}$$

where M is the Taylor factor, which is not related to the grain size of materials [17]. Generally, the flow stress model needs to consider two parts of contribution: One is the grain size-independent part and the other is the dependent part [16]. The well-known Hall–Petch equation is employed by considering both the two parts:

$$\sigma = \sigma_0 + kd^{-\frac{1}{2}} \tag{6.13}$$

where σ_0 is a material constant representing the stress of single crystalline, k is also a material constant, and d is the grain size [18].

For the Armstrong's model designated in the following, it considers the flow stress for a given strain and has the form of:

$$\tau(\varepsilon) = \tau_0(\varepsilon) + k(\varepsilon)d^{-\frac{1}{2}} \tag{6.14}$$

In Eq. (6.14), $\tau_0(\varepsilon)$ and k are constant for a given strain. Equation (6.12) is then formulated as follows:

$$\sigma(\varepsilon) = M\left(\tau_0(\varepsilon) + k(\varepsilon)d^{-\frac{1}{2}}\right) \tag{6.15}$$

Equation (6.15) is used to calculate the macroscaled flow stress in Peng's research [4] and Ran's work [3]. Based on the crystal plasticity theory [19], shear stress $\tau(\varepsilon)$ can be represented by the lattice friction stress $\tau_0(\varepsilon)$ and the interaction among dislocations in the following:

$$\tau(\varepsilon) = \tau_0(\varepsilon) + \alpha\mu b\sqrt{\rho_{\mathrm{T}}} \tag{6.16}$$

In Eq. (6.16), α is a particular constant for the phase of alloy and describes the dislocation interaction; μ is the corresponding shear module of the phase of alloy; b is the Burgers vector and it is different for FCC and BCC phases; and ρ_{T} is the total dislocation density. To explain the strengthening effect, which is related to the grain size change, Ashby [20] proposed a model which classified the dislocations into statistically stored dislocation $\rho_{\mathrm{S}}(\varepsilon)$ and geometrically necessary dislocation $\rho_G(\varepsilon)$. The statistically stored dislocation can be represented by a monotonic function formulated in the following:

$$\rho_{\mathrm{S}}(\varepsilon) = \frac{C_{\mathrm{S}}\varepsilon}{bL^{s}} \tag{6.17}$$

where C_{S}, ε, b, and L^s are material constant, strain, Burgers vector, and slip length, respectively [16]. The geometrically necessary dislocation is generally formulated as follows:

$$\rho_G(\varepsilon) = \frac{C_G\varepsilon}{bd} \tag{6.18}$$

where C_G is material constant. Therefore, the total dislocation density takes the following form:

$$\rho_{\mathrm{T}} = \rho_{\mathrm{S}}(\varepsilon) + \rho_G(\varepsilon) = \frac{C_{\mathrm{S}}\varepsilon}{bL^{s}} + \frac{C_G\varepsilon}{bd} \tag{6.19}$$

When there is almost no coarse grain existing in the deformation body, the grain boundary strengthening becomes critical in the deformation. This means the influence of the statistically stored dislocation can be neglected, viz. $\rho_{\mathrm{S}}(\varepsilon) \approx 0$. In Eq. (6.15), it is known that $\sigma_0(\varepsilon)$ is equivalent to $M\tau_0(\varepsilon)$ and can be formulated as an exponential function of $y = k\varepsilon^n$. Thus, the constitutive model for single phase metal in macroscaled plastic deformation is proposed in Eq. (6.20).

$$\sigma(\varepsilon) = M\left(\tau_0(\varepsilon) + \alpha\mu b\sqrt{\rho_{\mathrm{S}}(\varepsilon) + \frac{C_G\varepsilon}{bd}}\right) = Mk_1\varepsilon^{n_1} + \alpha M\mu b\sqrt{\frac{C_G\varepsilon}{bd}} \tag{6.20}$$

Compared with the previous model developed in Ref. [3], the present constitutive model specifically articulates the form of $k(\varepsilon)d^{-\frac{1}{2}}$ instead of using the curve fitting approach. The model in Eq. (6.20) has been proven to be valid for single phase alloys. For multiphase alloys, the coefficients μ, b, and C_G in Eq. (6.20) are different for each phase. The flow stress consists of the contribution from the stress of each phase by introducing the volume fraction of each phase and expressed as follows:

$$\sigma = \sum \sigma_i f_j \tag{6.21}$$

Equation (6.21) is the multiphase mixture rule in which σ_i and f_j are the corresponding stress and volume fraction of each phase, respectively.

6.4.1.2 Hybrid Constitutive Model

Based on the surface layer model [5], the flow stress of the deformation material is formulated as follows:

$$\sigma = \frac{N_s\sigma_s + N_i\sigma_i}{N} = \eta\sigma_s + (1-\eta)\sigma_i \quad \left(N = N_s + N_i, \eta = \frac{N_s}{N}\right) \tag{6.22}$$

where σ and N are the total flow stress and the grain number; the flow stress and grain number of surface grains are N_s and σ_s; and N_i and σ_i are for the internal grains. The independent stress–strain relationship is as follows:

$$\begin{cases} \sigma_s(\varepsilon) = m\tau_0(\varepsilon) = mk_1\varepsilon^{n_1} \\ \sigma_i(\varepsilon) = \sigma_\alpha f_\alpha + \sigma_\beta(1-f_a) \\ \quad = \left(Mk_2\varepsilon^{n_2} + \alpha M\mu_\alpha b_\alpha\sqrt{\frac{C_{G\alpha}\varepsilon}{b_\alpha d}}\right)f_\alpha + \left(Mk_2\varepsilon^{n_2} + \alpha M\mu_\beta b_\beta\sqrt{\frac{C_{G\beta}\varepsilon}{b_\beta d}}\right)(1-f_\alpha) \\ \quad = Mk_2\varepsilon^{n_2} + f_\alpha\alpha M\mu_\alpha b_\alpha\sqrt{\frac{C_{G\alpha}\varepsilon}{b_\alpha d}} + (1-f_\alpha)\alpha M\mu_\beta b_\beta\sqrt{\frac{C_{G\beta}\varepsilon}{b_\beta d}} \end{cases} \tag{6.23}$$

where σ_α, σ_β, f_α, and f_β are the stress and volume fraction of α and β phases; m is the grain orientation factor. By combining Eqs. (6.22) and (6.23), the final formulation of the hybrid model is thus obtained as follows:

$$\begin{aligned} \sigma_{\text{total}}(\varepsilon) = {} & Mk_2\varepsilon^{n_2} + f_\alpha\alpha M\mu_\alpha b_\alpha\sqrt{\frac{C_{G\alpha}\varepsilon}{b_\alpha d}} + (1-f_\alpha)\alpha M\mu_\beta b_\beta\sqrt{\frac{C_{G\beta}\varepsilon}{b_\beta d}} \\ & + \eta\left(mk_1\varepsilon^{n_1} - \left(Mk_2\varepsilon^{n_2} + f_\alpha\alpha M\mu_\alpha b_\alpha\sqrt{\frac{C_{G\alpha}\varepsilon}{b_\alpha d}} + (1-f_\alpha)\alpha M\mu_\beta b_\beta\sqrt{\frac{C_{G\beta}\varepsilon}{b_\beta d}}\right)\right) \end{aligned} \tag{6.24}$$

Equation (6.24) is suitable for both the macro- and microscaled deformation when there are few coarse grains in the specimen. For the heat treatment processed billet, the contribution of grain boundary to dislocation density is small as the grain becomes coarse and the volume percentage of grain boundary in the billet is smaller. Thus, the effect of statistically stored dislocation cannot be ignored and Eq. (6.24) is further formulated as follows:

$$\sigma_{\text{total}}(\varepsilon) = Mk_2\varepsilon^{n_2} + f_\alpha \alpha M\mu_\alpha b_\alpha \sqrt{\frac{C_{S\alpha}\varepsilon}{b_\alpha L^s} + \frac{C_{G\alpha}\varepsilon}{b_\alpha d}} + (1-f_\alpha)\alpha M\mu_\beta b_\beta \sqrt{\frac{C_{S\beta}\varepsilon}{b_\beta L^s} + \frac{C_{G\beta}\varepsilon}{b_\beta d}}$$
$$+\eta\left(mk_1\varepsilon^{n_1} - \left(Mk_2\varepsilon^{n_2} + f_\alpha \alpha M\mu_\alpha b_\alpha \sqrt{\frac{C_{S\alpha}\varepsilon}{b_\alpha L^s} + \frac{C_{G\alpha}\varepsilon}{b_\alpha d}} + (1-f_\alpha)\alpha M\mu_\beta b_\beta \sqrt{\frac{C_{S\beta}\varepsilon}{b_\beta L^s} + \frac{C_{G\beta}\varepsilon}{b_\beta d}}\right)\right) \tag{6.25}$$

where $C_{S\alpha}$ and $C_{S\beta}$ are all material constants. As discussed in the previous paragraph, total dislocation density can be described as the combination of statistically stored dislocation density and geometrically necessary dislocation density. Equation (6.25) is the general form of the hybrid constitutive model considering the influence of both the size effect and the dislocation density. In this equation, $C_{G\alpha}$, $C_{G\beta}$, $C_{S\alpha}$, and $C_{S\beta}$ are the undetermined material constants, and the conduction of curve fitting is easier if $C_{G\alpha}$ and $C_{G\beta}$ are determined first by ignoring the influence of statistically dislocation density when there is almost no coarse grain in the billet. After $C_{G\alpha}$ and $C_{G\beta}$ are calculated according to Eq. (6.24), $C_{S\alpha}$ and $C_{S\beta}$ can be eventually determined using Eq. (6.25). In this book, $C_{S\alpha}$ and $C_{S\beta}$ are assumed to be the same for simplicity and they are designated as C_S in the subsequent curve fitting.

6.4.2 Methodology and Calculation

Figure 6.22 schematically illustrates the methodology to establish the hybrid model for analysis of ductile fracture of multiphase alloys. In the figure, the macroscaled material properties are determined based on the load–stroke curve via simple upsetting experiment. The flow stress curve of the material is then calculated. By observing the microstructure and using energy-dispersive X-ray spectroscopy (EDX), the volume fraction of each phase in the multiphase alloys is determined. The coefficient of the macroscaled constitutive model of the multiphase materials is calculated by curve fitting using the data of volume fraction and flow stress curve. To identify the influence of size effect, the macroscaled constitutive model is embedded into the surface layer model, which introduces the size factor η to describe the influence of size effect in microscaled plastic deformation. The final hybrid model can be used to calculate the flow stress of all the scenarios of multiphase materials.

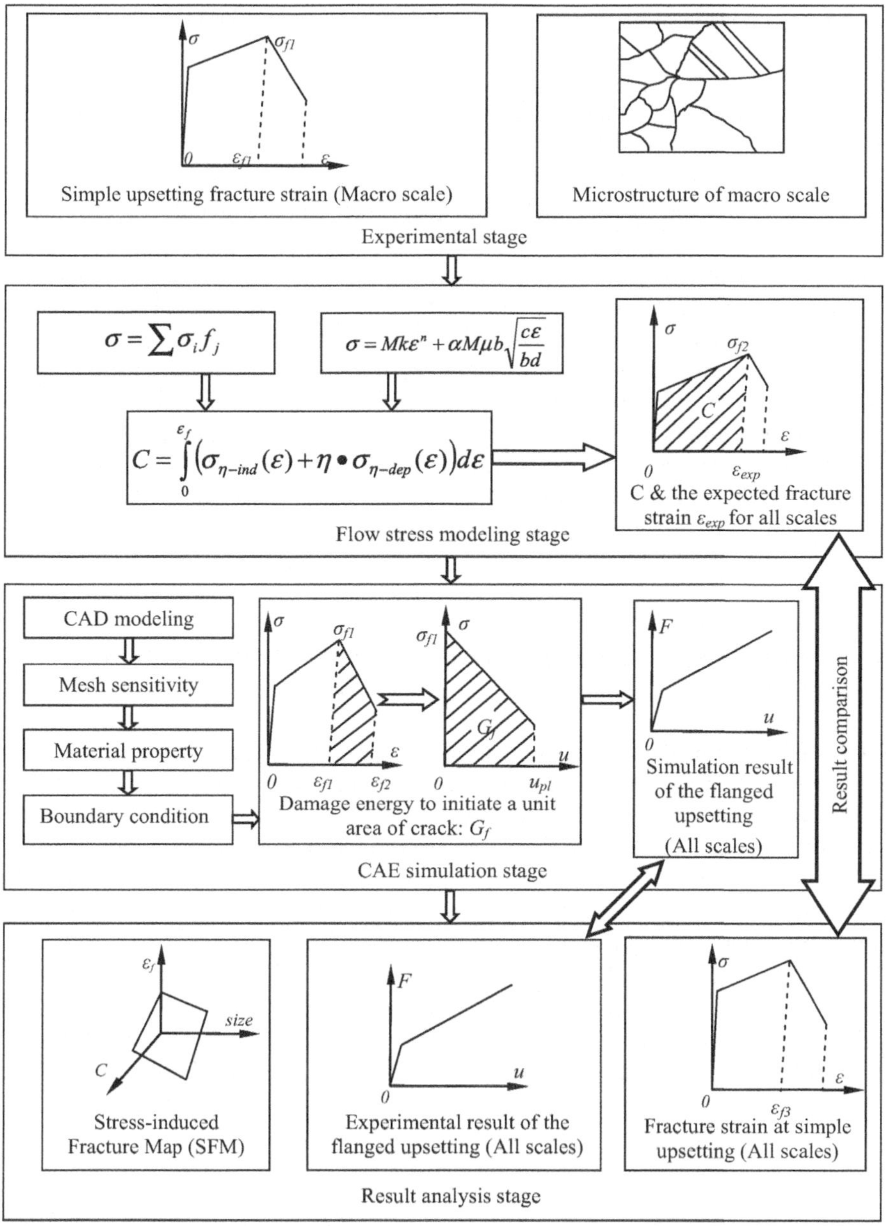

Fig. 6.22 Methodology for hybrid constitutive modeling and analysis of fracture [21]

For ductile fracture prediction, the fracture energy C is calculated based on the hybrid constitutive model. After that, the FE simulation can be carried out. The meshing of computer-aided design (CAD) model is then conducted, and the boundary condition is set up. The damage energy G_f determined based on the flow stress–displacement curve is applied to the material property so that the damage evolution behavior of the material can be identified in deformation process. Finally, the load–stroke curve of the flanged upsetting simulation is extracted and constructed in comparison with the experimental result.

In most prior researches, the influence of geometry and grain size effects is considered and handled separately, and their interaction and interplay in microscaled ductile fracture from analysis and prediction perspective have not yet been fully explored and investigated. To address this issue, the so-called stress-induced fracture map (SFM) is proposed and constructed for microscaled fracture prediction. SFM schematically articulates the interaction among size effect, fracture energy, and the expected fracture strain in microscaled plastic deformation of multiphase alloys. It can be used to reveal the influence level of different size effects with the same size factor. The expected fracture strain of multiphase alloys in different scenarios can thus be predicted. Finally, the load–stroke curve of the microscaled flanged upsetting generated by FE simulation based on the established hybrid model is compared with the experimental results to verify the efficiency of the developed hybrid model.

6.4.2.1 Coefficient Calibration Procedure of the Hybrid Constitutive Model

Equation (6.25) is the final formulation of the hybrid constitutive model. The unknown coefficients of the equation are determined by curve fitting based on the flow stress of the macroscaled upsetting experiment. How the experimental data are used for coefficient calibration is presented in Table 6.3. The stress–strain curve of the sample with the dimensions of 2 × 3 mm is used for curve fitting, while the experimental data of the samples with other three dimensions are used for result comparison with simulation.

Equation (6.25) is applicable for all scenarios as it is suitable for both the fine and coarse grains. The equation contains five unknown coefficients, viz. k_2, n_2, $C_{G\alpha}$, $C_{G\beta}$, and C_S ($C_S = C_{S\alpha} = C_{S\beta}$). Figure 6.23 shows the curve fitting procedure used to determine the five unknown coefficients. In the figure, the coefficients marked

Table 6.3 Experimental data processing [21]

Size	Heat treatment			
	As-received	Annealed at 450 °C	Annealed at 600 °C	Annealed at 750 °C
2 × 3 mm	Macroscaled experimental result, used for curve fitting			
1 × 1.5 mm	Microscaled experimental result, used for result comparison			
0.5 × 0.75 mm	Microscaled experimental result, used for result comparison			
0.25 × 0.375 mm	Microscaled experimental result, used for result comparison			

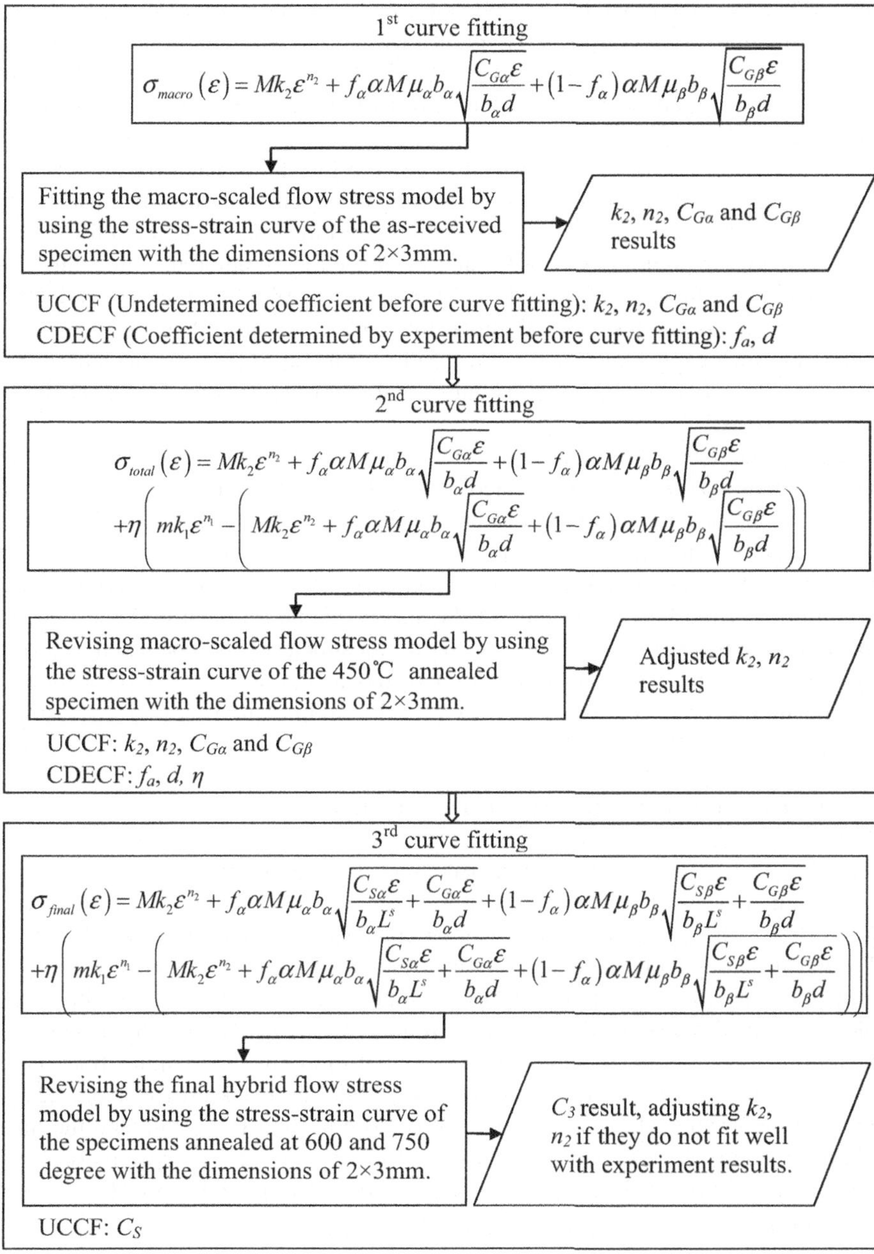

Fig. 6.23 Curve fitting procedure [21]. ($C_{S\alpha} = C_{S\beta} = C_S$)

with red color are unknown to be decided in the shown curve fitting stage, while the volume fraction of alpha phase f_α, grain size d, and size factor η are directly determined using the experimental results before fitting. The remaining coefficients are determined based on the available information and the established available knowledge.

There are three curve fitting stages as shown in Fig. 6.23. The first curve fitting uses the flow stress data of the as-received specimen with the dimensions of 2×3 mm. The size factor η of this scenario determined based on the information of metallographic photograph is 4 %, and its size effect is ignored. Thus, only the independent part in Eq. (6.24) needs to be considered. With the flow stress data of the as-received specimens with the dimensions of 2×3 mm and annealed at 450 ° C, the unknown coefficients k_2, n_2, $C_{G\alpha}$, and $C_{G\beta}$ in Eq. (6.24) are calculated and determined by curve fitting, which is only suitable for the deformation of specimen with fine grains.

The second curve fitting uses the flow stress data of the specimen with the dimensions of 2×3 mm and annealed at 450 °C. The coefficients k_2, n_2, $C_{G\alpha}$, and $C_{G\beta}$ are recalculated and adjusted to ensure the hybrid model fits both the experimental results of the as-received samples and those annealed at 450 °C. After the two rounds of fitting, the remaining unknown coefficient C_S can be determined via curve fitting of the flow stress data of the specimens annealed at 600 and 750 °C with the dimensions of 2×3 mm. A slight adjustment of the coefficients k_2, n_2, $C_{G\alpha}$, and $C_{G\beta}$ is necessary if the fitted curve has a significant deviation from any experimental result of the four scenarios with the sample dimension of 2×3 mm.

6.4.2.2 Calculation of Microscaled Deformation

According to Eq. (6.21), the macroscaled stress obtained from the simple upsetting experiment consists of the stresses of different phases. As there is only 2 % lead in Brass C3602, the deformation body can be considered to majorly being composed of α and β phases without consideration of the stress influence of lead phase.

In Fig. 6.24, there are about 5 % of the total pixels of EDX picture with black color. It indicates that there is no spectrum to be detected. By sharpening the metallurgical photograph as shown in Fig. 6.25, it shows that the volume fraction of the white area is 77 %, representing α phase, while the black area is about 22 %, representing β and lead phases. This result is close to the EDX result, and the final volume fractions of α, β, and lead phases are 81, 17, and 2 %, respectively.

In Eq. (6.24), $\alpha = 0.34$ [22]; $\mu_\alpha = 78{,}500$ MPa (shear modulus for FCC phase); $\mu_\beta = 72{,}000$ MPa (shear modulus for BCC phase); $b_\alpha = \frac{\sqrt{2}}{2} \times 3.69 \times 10^{-10} = 2.608 \times 10^{-10}$ m (Burgers vector for FCC phase); $b_\beta = \frac{\sqrt{3}}{2} \times 2.94 \times 10^{-10} = 2.546 \times 10^{-10}$ m (Burgers vector for BCC phase); and the volume fraction of each phase is obtained based on the EDX results, which is 83 % for α phase and 17 % for β phase. The unknown coefficient in Eq. (6.24) is the constant of dislocation

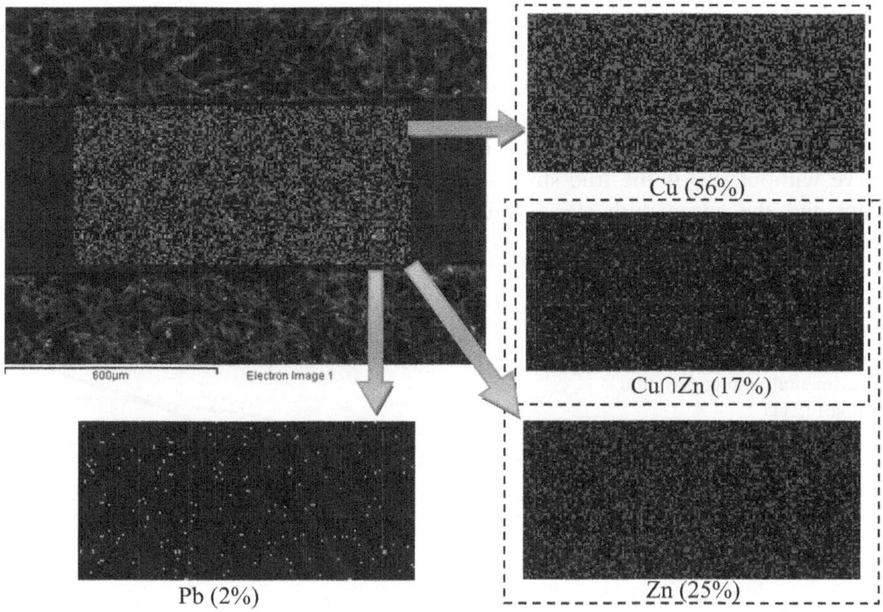

Fig. 6.24 Determination of the phase volume fraction by EDX [21]

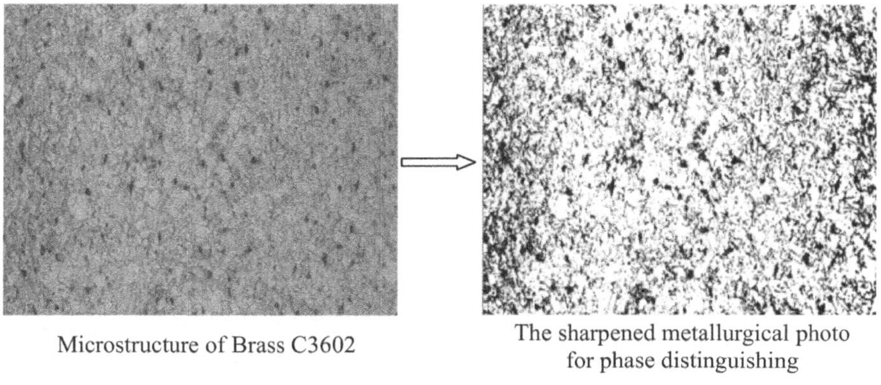

Fig. 6.25 Determination of the phase volume fraction via metallurgical photograph [21]

density in α and β phases. By using the curve fitting approach, the constants $C_{G\alpha}$ and $C_{G\beta}$ can be determined and Eq. (6.24) thus becomes:

$$\sigma_{\text{total}}(\varepsilon) = 587.52\varepsilon^{0.23} + 0.498\varepsilon^{0.5} \cdot d^{-\frac{1}{2}} - \eta\left(203.52\varepsilon^{0.23} + 0.498\varepsilon^{0.5} \cdot d^{-\frac{1}{2}}\right) \tag{6.26}$$

where $C_{G\alpha} = 0.14$, $C_{G\beta} = 0.18$, $k_1 = k_2 = 192$, $n_1 = n_2 = 0.23$, $M = 3.06$, $m = 2$. k_1 and k_2, n_1, and n_2 here are actually the same coefficient in the equation of $\tau_0(\varepsilon) = k\varepsilon^n$. The k_1 and k_2, and n_1 and n_2 are the same as those in Eq. (6.23).

The comparison between the calculation result and the true stress–strain curve from the actual experiment is shown in Figs. 6.26 and 6.27. In Fig. 6.26a, the solid curve without star is the true stress–strain curve obtained from the actual experiment and the curve with star is the curve fitting result of macroscaled simple upsetting. By using the data in Eq. (6.26), the unknown coefficient in Eq. (6.25) can

Fig. 6.26 Comparison of the curve fitting and the experimental results (original sample) [21]

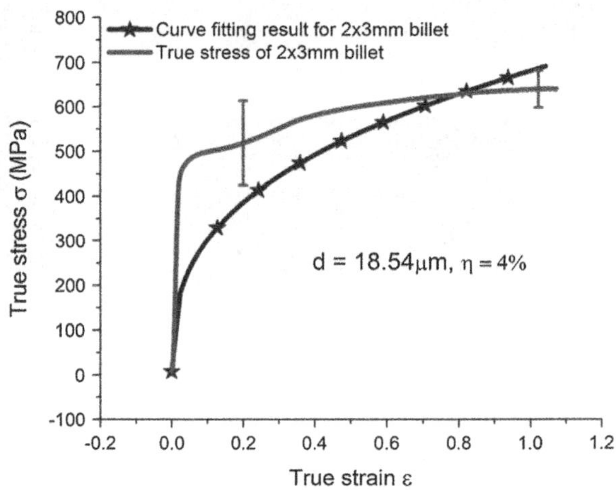

(a) 2 ×3 mm billet, As-received.

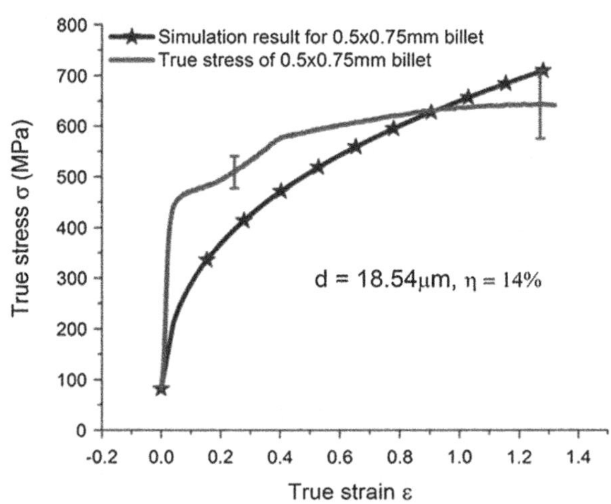

(b) 0.5 ×0.75 mm billet, As-received.

Fig. 6.27 Comparison of the simulation and experimental results of the material annealed at 750 °C [21]

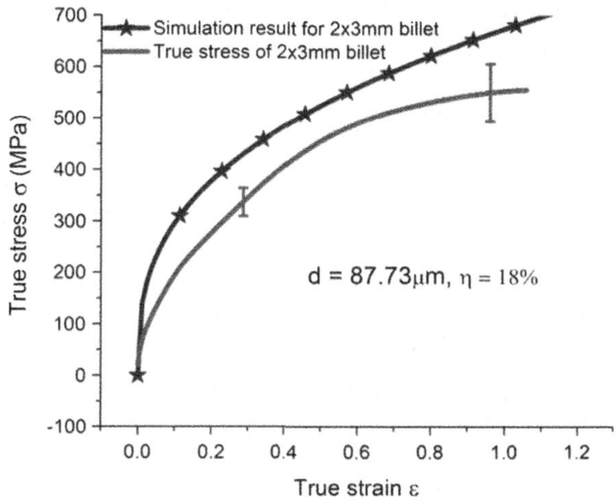

(a) 2 × 3 mm billet, Annealed at 750°C.

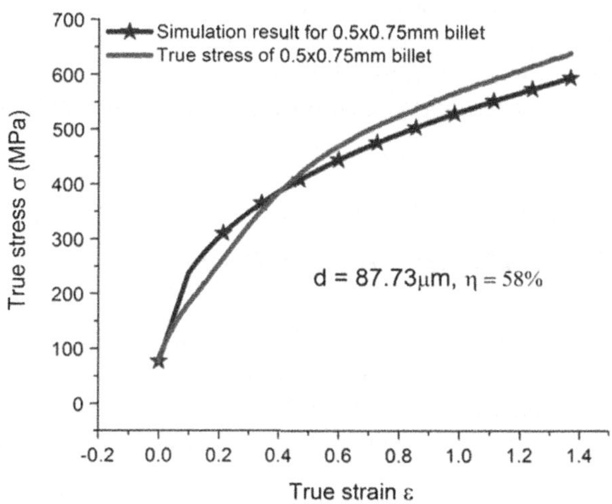

(b) 0.5 × 0.75 mm billet, Annealed at 750°C.

be determined. The final description of flow stress and size factor η is obtained and presented in Eq. (6.27):

$$\sigma_{\text{total}}(\varepsilon) = 587.52\varepsilon^{0.23} + 0.681\varepsilon^{0.5} \cdot d^{-\frac{1}{2}} - \eta\left(203.52\varepsilon^{0.23} + 0.681\varepsilon^{0.5} \cdot d^{-\frac{1}{2}}\right) \tag{6.27}$$

Figure 6.27 shows the simulation results using the flow stress model formulated by Eq. (6.27). It indicates that when the size factor η is higher, in other words, it is the scenario of the small workpiece size, viz. the billet size is 0.5 × 0.75 mm, the simulation result is closer to the actual experimental result.

Although Figs. 6.26 and 6.27 show a consistent variation trend between the simulation results and the experimental ones, some part of the simulated stress curve has a big deviation from the experimental result. This is caused by the fact that the unknown coefficients of Eq. (6.25) are calibrated with the flow stress data of all the four scenarios with the sample dimension of 2 × 3 mm. If the four unknown coefficients k_2, n_2, $C_{G\alpha}$, and $C_{G\beta}$ are determined without adjustment after the first time of curve fitting, the simulation result of the as-received scenario is quite close to the experimental result. However, the simulation results of other three scenarios, including the samples annealed at 450, 600, and 750 °C, have a deviation from the actual experimental results. Thus, the coefficients k_2 and n_2 need to be calibrated and adjusted after the two rounds of curve fitting to ensure all the simulation results are as close as possible to the corresponding experimental results.

When the experimental data of macroscaled upsetting is used, the unknown coefficients of Eqs. (6.24) and (6.25) can be determined via curve fitting and Eqs. (6.26) and (6.27) are thus obtained. Equation (6.26) is suitable for macro- and microscaled deformation when there are a few coarse grains in the specimen, while Eq. (6.27) is applicable for both macro- and microscaled deformations of all kinds of scenarios. Therefore, the stress–strain curve of the microscaled upsetting process can be determined based on Eqs. (6.26) and (6.27) and compared with the experimental results.

6.4.3 Ductile Fracture Prediction

In order to predict the ductile fracture in microscaled deformation, the fracture energy of the deformation material, which is assumed to be material-dependent constant, must be determined first. Based on Freudenthal's criterion [23], the damage evolution equation can be expressed as follows:

$$C = \int_0^{\varepsilon_f} \sigma(\varepsilon)\mathrm{d}\varepsilon \tag{6.28}$$

To use the fracture energy C to predict the ductile fracture, the size effect needs to be considered. By combining Eqs. (6.27) and (6.28), the fracture energy C is formulated in the following:

$$C = \int_0^{\varepsilon_f} \left(587.52\varepsilon^{0.23} + 0.681\varepsilon^{0.5} \cdot d^{-\frac{1}{2}} - \eta\left(203.52\varepsilon^{0.23} + 0.681\varepsilon^{0.5} \cdot d^{-\frac{1}{2}}\right)\right)\mathrm{d}\varepsilon \tag{6.29}$$

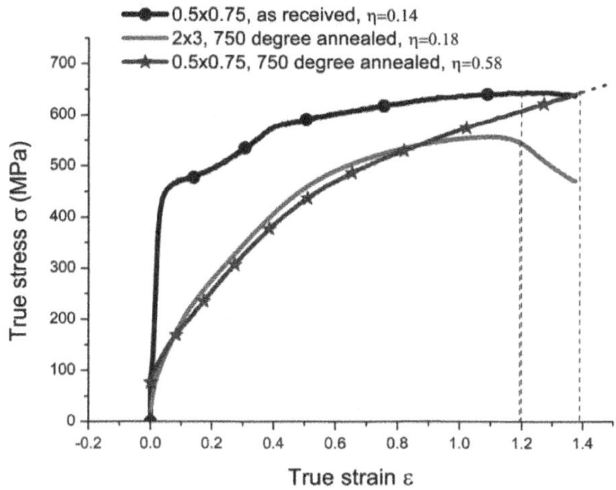

Fig. 6.28 Fracture strain prediction for different size factors [21]

Using the original sample with the dimensions of 2 × 3 mm and without heat treatment, the fracture strain of $\varepsilon = 1.15$ is obtained. With the corresponding grain size $d = 18.54$ μm and the size factor $\eta = 0.04$, C is determined as follows:

$$C = \int_0^{1.15} \left(587.52\varepsilon^{0.23} + 0.681 \times \varepsilon^{0.5} \times \left(18.54 \times 10^{-6}\right)^{-\frac{1}{2}} - 0.04 \times \left(203.52\varepsilon^{0.23} + 0.681 \times \varepsilon^{0.5} \times \left(18.54 \times 10^{-6}\right)^{-\frac{1}{2}}\right)\right) d\varepsilon = 684.09 \tag{6.30}$$

With the hybrid fracture model formulated in Eq. (6.29) and the fracture energy C, the fracture strain and grain size of other scenarios can thus be determined. In Fig. 6.28, the corresponding expected fracture strains in different scenarios are $\varepsilon_{0.5\times0.75,\mathrm{AS}} = 1.19$, $\varepsilon_{2\times3,750^\circ\mathrm{C}} = 1.19$, and $\varepsilon_{0.5\times0.75,750^\circ\mathrm{C}} = 1.39$, which are quite close to the actual experimental results.

6.4.4 *Stress-Induced Fracture Map*

The FE simulation of microscale flanged upsetting can be implemented after the C value and the fracture strain are determined. Table 6.4 shows the expected fracture strain and the size factor for each case. In the table, the fracture energy C for each scenario of the materials heat-treated with different conditions is calculated from Eq. (6.30). The size factor η is determined based on Eq. (6.22). The expected fracture strain is figured out using Eq. (6.29) when the fracture energy C in Eq. (6.29) is determined in each scenario.

Table 6.4 The expected fracture strain and size factor of Brass C3602 [21]

C	Size			
	2 × 3 mm	1 × 1.5 mm	0.5 × 0.75 mm	0.25 × 0.375 mm
684 (as-received)	$\varepsilon_f = 1.15$, $\eta = 4\,\%$	$\varepsilon_f = 1.16$, $\eta = 7\,\%$	$\varepsilon_f = 1.19$, $\eta = 14\,\%$	$\varepsilon_f = 1.26$, $\eta = 27\,\%$
578 (annealed at 450 °C)	$\varepsilon_f = 1.03$, $\eta = 5\,\%$	$\varepsilon_f = 1.05$, $\eta = 9\,\%$	$\varepsilon_f = 1.08$, $\eta = 18\,\%$	$\varepsilon_f = 1.16$, $\eta = 35\,\%$
598 (annealed at 600 °C)	$\varepsilon_f = 1.11$, $\eta = 9\,\%$	$\varepsilon_f = 1.15$, $\eta = 18\,\%$	$\varepsilon_f = 1.22$, $\eta = 33\,\%$	$\varepsilon_f = 1.36$, $\eta = 59\,\%$
606 (annealed at 750 °C)	$\varepsilon_f = 1.19$, $\eta = 18\,\%$	$\varepsilon_f = 1.25$, $\eta = 32\,\%$	$\varepsilon_f = 1.39$, $\eta = 58\,\%$	$\varepsilon_f = 1.63$, $\eta = 91\,\%$

Upon acquisition of the data from simulation, the stress-induced fracture map (SFM) for multiphase alloys is proposed. In Fig. 6.29, all the expected fracture strains calculated by the hybrid ductile fracture model are presented in the map. For the multiphase alloys such as C3602, the SFM is established by using the data shown in Table 6.4. In the SFM, the axes stand for specimen size, fracture energy, and the expected fracture strain of the materials. In this case, the diameters of the samples are 0.25, 0.50, 1.00, and 2.00 mm, and they are shown in the size axis in the figure. When the grain size effect is considered, it is assumed that the grain size changes while the specimen size remains constant. The fracture energy C represents this kind of size effect as C is a material constant for the material with the same microstructure. For the geometry size effect, the dimension of specimen represents the influence of geometry size change. Therefore, the SFM demonstrates the influence of grain and geometry size effects in microscaled ductile fracture deformation.

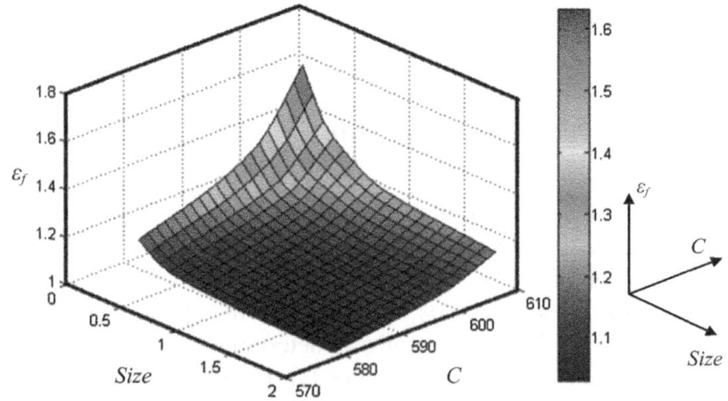

Fig. 6.29 Stress-induced fracture map [21]

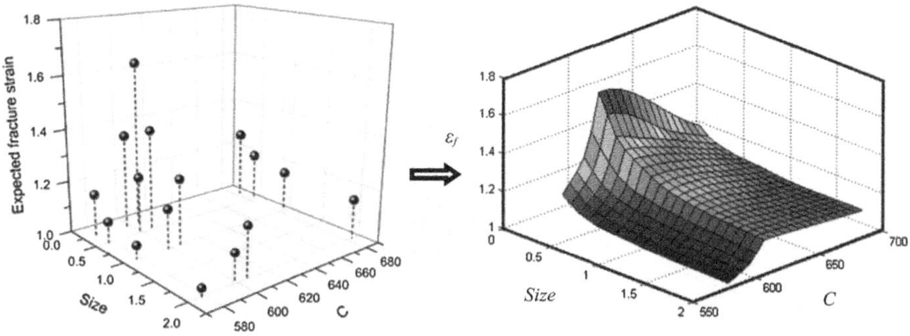

Fig. 6.30 Preliminary SFM of Brass C3602 [21]

By using the data in Table 6.4, a preliminary SFM is first established and shown in Fig. 6.30. From the figure, it is clear that the original material without heat treatment has the highest fracture energy of 684, but its fracture strain is greater than that of the material annealed at 450 °C. This reveals that the "original samples" may have received some kind of heat treatment and it has the complex and unknown preprocessed history. The final of SFM is constructed after the "as-received" case is taken away from the preliminary SFM as shown in Fig. 6.31.

When the SFM is completed, each scenario in the range of size scale and fracture energy can have an identified fracture strain in the SFM. The fracture strain is the fracture initiation condition for the ideally isotropic metal alloys.

In Fig. 6.31a, lines 1, 2, 3, and 4 show the variation in the expected fracture strain of the samples with the same sample size, but heat-treated with different conditions, while in Fig. 6.31b, lines 5, 6, and 7 show the change in the expected fracture strain of the samples heat-treated with the same condition, but with different geometry sizes. From the figures, it is found that the fracture strain is increased rapidly for the sample with smaller size, as shown in line 1 in Fig. 6.31a with the sample diameter of 0.25 mm. For the scenarios in Fig. 6.31b, line 5 has a significant increase in fracture strain, which is the scenario of the sample heat-treated at 750 °C. The variation gradient of the curves of 6 and 5 as shown in Fig. 6.31b increases rapidly, especially for the specimen diameter smaller than 0.5 mm.

On the other hand, the size factor, η, which is defined as the percentage ratio between the surface layer grains and the total grains in the workpiece, is used to analyze the SFM. As shown in Fig. 6.32, the lines, which link different scenarios in the figure, have the same size factor η. How the size factor η affects the fracture strain can be elaborated and discussed using the following equation Eq. (6.31):

$$C = 77.1\varepsilon_f^{1.5} + 477.7\varepsilon_f^{1.23} - \eta \cdot \left(77.1\varepsilon_f^{1.5} + 165.5\varepsilon_f^{1.23}\right) \tag{6.31}$$

Equation (6.31) is the solution of Eq. (6.29). In discussing the geometry size effect, the microstructure of the material is assumed to be the same, and thus, the fracture

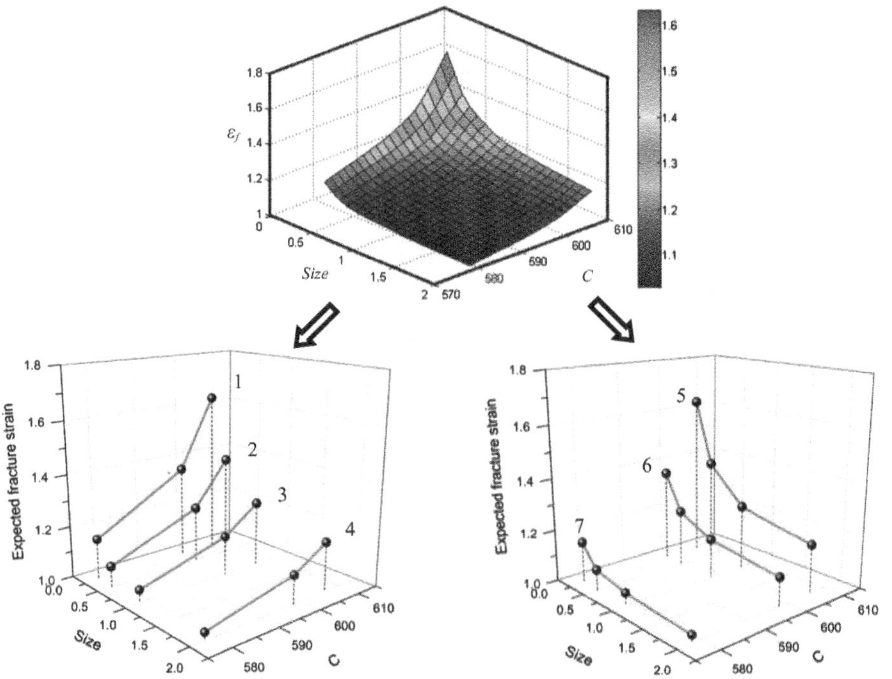

(a) Grain size effect

(b) Feature size effect

Fig. 6.31 Variation in the fracture strain in different size effect scenarios [21]. **a** Grain size effect and **b** feature size effect

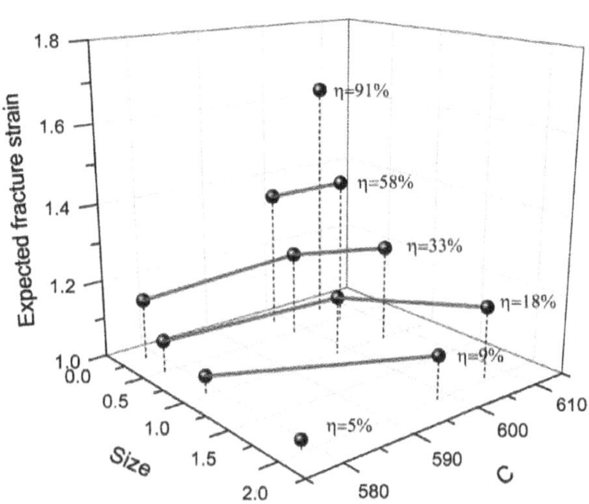

Fig. 6.32 The expected fracture strain with the same size factor η [21]

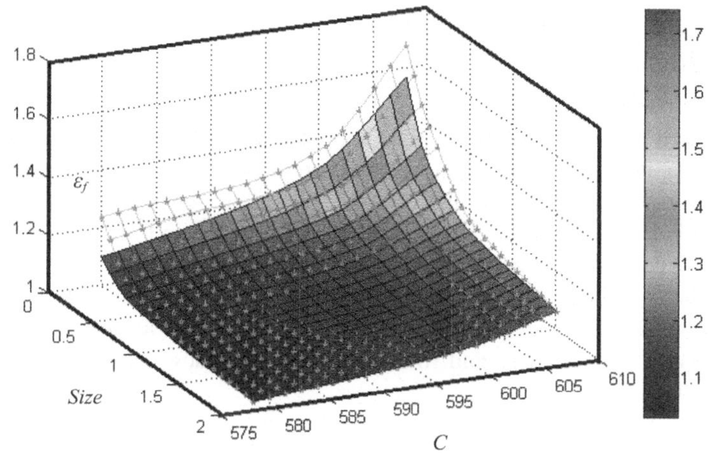

Fig. 6.33 Comparison of the fracture maps constructed based on simulation and experiment [21]

energy C remains constant for each scenario as the geometry size effect does not affect grain size or fracture energy. When the size factor η increases, the geometry scale of the sample is decreased and the expected fracture strain must increase to maintain the same C based on Eq. (6.31) as the increase in η leads to the increase in the absolute value of the negative terms in the equation and the fracture strain must increase in the first two positive terms to cancel out the increase in the absolute value in the negative terms in Eq. (6.31). In analysis of the grain size effect, the geometry size of sample is assumed to be the same. With the increase in size factor η, it is found from the figure that the fracture energy C also increases with the size factor η. Therefore, the expected fracture strain needs to increase much more than such that it can cancel out the increase from both C and the absolute value in the negative terms in Eq. (6.31) to maintain the equality of the equation.

The SFM can not only predict the fracture strain of certain multiphase alloys, but also reveal the in-depth relationship of grain and geometry size effects in microscaled plastic deformation, viz. microforming, from ductile fracture perspective. It can be considered as a useful tool to predict the stress-induced fracture in microforming processes in such a way to reduce physical experiments and numerical simulations needed to address this issue.

Figure 6.33 demonstrates the simulation results and the actual experimental ones by SFM. In this figure, the surface marked with red stars represents the real fracture map constructed based on the actual experimental results, which actually are the real fracture strains in different scenarios summarized in Table 6.4. To make an ease comparison, the two fracture maps generated based on the actual experimental results and the simulation ones are shown in the same figure, viz. Figure 6.33. It is found that there is a good agreement in between. When the deformation is large and with the large fracture strain, as shown in Fig. 6.33, the deviation in between is smaller, but it becomes greater for the samples with the smaller size scale.

6.5 Applicability of DFCs in Microscaled Plastic Deformation

As mentioned before, a number of DFCs proposed for macroscaled plastic deformation have been used for analysis and prediction of fracture behaviors and occurrence in plastic deformation processes. When the geometry size of workpiece is scaled down to microscale, however, the applicability and accuracy of DFCs in microscaled deformation could be different from that in macroscale, which thus needs further exploration and systematic study. In this section, a few uncoupled DFCs are employed for evaluation and assessment of their applicability in microscaled plastic deformation and their detailed accuracy and applicability are summarized.

6.5.1 The Uncoupled DFCs

Table 6.5 presents six widely used uncoupled DFCs to be evaluated in this section. Each criterion has its own fracture critical value and designated as C, which is exactly the same as the fracture energy for the fracture energy-based DFC in the previous section. To analyze and predict the fracture occurrence in deformation process, the representation of the uncoupled DFCs can be generally formulated in the following Eq. (6.32). When the integral of the following criterion is larger than or equal to the critical value C of fracture, the fracture is assumed to happen in the deformation body.

Table 6.5 A few uncoupled fracture criteria used in this chapter [29]

Macroscaled fracture criteria	Criterion formulation
Cockcroft and Latham model	$\int_0^{\varepsilon_f} \bar{\sigma}\left(\frac{\sigma_1}{\bar{\sigma}}\right)\mathrm{d}\bar{\varepsilon} = \int_0^{\varepsilon_f} \sigma_1 \mathrm{d}\varepsilon = C$
Oyane model	$\int_0^{\bar{\varepsilon}_f} \left(\frac{1}{\alpha_0}\cdot\frac{\sigma_m}{\bar{\sigma}} + 1\right)\mathrm{d}\bar{\varepsilon} = C$
Ayada model	$\int_0^{\bar{\varepsilon}_f} \left(\frac{\sigma_m}{\bar{\sigma}}\right)d\bar{\varepsilon} = C$
Brozzo model	$\int_0^{\bar{\varepsilon}_f} \frac{2\sigma_1}{3(\sigma_1-\sigma_m)}\mathrm{d}\bar{\varepsilon} = C$
Rice and Tracey model	$\int_0^{\bar{\varepsilon}_f} \mathrm{e}^{\alpha\frac{\sigma_m}{\bar{\sigma}}} = C$
Freudenthal model	$\int_0^{\bar{\varepsilon}_f} \bar{\sigma}\mathrm{d}\bar{\varepsilon} = C$

$$\int_{0}^{\bar{\varepsilon}_f} f(\bar{\sigma}, \bar{\varepsilon})d\bar{\varepsilon} - C < 0 \tag{6.32}$$

In the table, $\bar{\varepsilon}_f$ is the fracture strain, C is the fracture critical value, $\bar{\varepsilon}$ is the equivalent strain, $\bar{\sigma}$ is the equivalent stress, and σ_1 is the maximum principal stress.

In Eq. (6.32), the fracture critical value C is a material constant and usually determined by experiment such as upsetting and tensile tests of materials. The value of C is assumed to be the same for the same material with the same microstructure. For the materials processed at different heat treatment conditions, their microstructures are generally different, and thus, the fracture critical value C is also different.

Among the six DFCs, Cockcroft and Latham criteria is one of the early-developed criteria in ductile fracture arena [24]. It was proposed based on the Freudenthal criterion, which is actually the first proposed DFC [23]. It is found that the yield stress at fracture point is not affected by the shape of the necked region in tensile test, which is different from the actual experimental scenario. Cockcroft criterion is thus developed for bulk forming and applicable to the deformation with a low stress triaxiality. The reduced form of the criterion is represented as follows:

$$\int_{0}^{\bar{\varepsilon}_f} \bar{\sigma} \cdot \frac{\sigma_1}{\bar{\sigma}} d\varepsilon = \int_{0}^{\bar{\varepsilon}_f} \sigma_1 d\varepsilon = C \tag{6.33}$$

In Eq. (6.33), σ_1 is the maximum principal stress and $\left(\frac{\sigma_1}{\bar{\sigma}}\right)$ is a non-dimensional stress concentration factor.

For the Oyane fracture model, it articulates the concept of ductile fracture with four development stages, viz. microscaled void formation caused by dislocation pileup, void distance becoming closer due to void growth, plastic deformation concentration, and the dimple initiation on the surfaces of material [25]. For porous materials, by employing the relationship between the equivalent strain $\bar{\varepsilon}$ and the volumetric strain ε_v, the stress–strain relationship can be represented by Eq. (6.34) as follows:

$$d\varepsilon_v = \frac{d\bar{\varepsilon}}{\gamma f^2}\left(\frac{\sigma_m}{\bar{\sigma}} + \alpha_0\right) \tag{6.34}$$

where γ is the ratio between the nominal density ρ and the constituent metal density of porous material ρ_0. f is a function of γ and expressed as $f = \frac{1}{3}\left(1 + \sqrt{\frac{\gamma}{1-\gamma}}\right)$. In addition, σ_m is the hydrostatic stress and α_0 is a material constant. The volumetric strain ε_v represents the volume change of the porous materials and can be represented as the ratio between the volume of the porous material v and the volume of the constituent metal with the same weight v_0 and formulated as follows:

$$\varepsilon_v = \ln\frac{v}{v_0} = -\ln\frac{\rho}{\rho_0} = -\ln\gamma \tag{6.35}$$

In Eq. (6.34), the fracture strain can be determined if the fracture occurs at a particular volumetric strain. Equation (6.34) is further formulated as follows:

$$\int_0^{\bar{\varepsilon}_f} \frac{\gamma f^2}{\alpha_0} d\varepsilon_v = \int_0^{\bar{\varepsilon}_f} \left(\frac{1}{\alpha_0} \cdot \frac{\sigma_m}{\bar{\sigma}} + 1\right) d\bar{\varepsilon} \tag{6.36}$$

As α_0, γ, and f are all material constant, Eq. (6.36) can be written as follows:

$$\int_0^{\bar{\varepsilon}_f} \left(\frac{1}{\alpha_0} \cdot \frac{\sigma_m}{\bar{\sigma}} + 1\right) d\bar{\varepsilon} = C \tag{6.37}$$

where C is the fracture critical value of this criterion.

The Ayada criterion is proposed based on the Cockcroft and Oyane criteria to provide an evaluation of fracture in compression-dominative deformation, since the result predicted by the Cockcroft criterion is very unsatisfactory when the strain is large and the tensile stress is small [26]. The detailed formulation of the Ayada criterion is in the following Eq. (6.38).

$$\int_0^{\bar{\varepsilon}_f} \left(\frac{\sigma_m}{\bar{\sigma}}\right) d\bar{\varepsilon} = C \tag{6.38}$$

where σ_m is the mean stress and C is the fracture critical value of the criterion and considered to be inversely proportional to the hardness of materials according to experiment. $\frac{\sigma_m}{\bar{\sigma}}$ is called stress triviality.

In addition, the Brozzo criterion in Eq. (6.39) is established based on the Cockcroft criterion and articulates the relationship between the maximum stress and mean stress when fracture happens [27].

$$\int_0^{\bar{\varepsilon}_f} \frac{2\sigma_1}{3(\sigma_1 - \sigma_m)} d\bar{\varepsilon} = C \tag{6.39}$$

where C is the fracture critical value of the criterion.

The Rice and Tracey model is more focused on the modeling of the growth of microvoid with sphere shape in fracture formation process [28]. The model assumes that the fracture growth rate is affected by stress triaxiality $\frac{\sigma_m}{\bar{\sigma}}$ and is designated as follows:

$$\int_{0}^{\bar{\varepsilon}_f} e^{\alpha \frac{\sigma_m}{\bar{\sigma}}} = C \tag{6.40}$$

where α is material constant and C is the fracture critical value of the criterion.

The last criterion to be investigated in this research is the Freudenthal criterion, which describes the influence of damage accumulation in plastic deformation process [23]. The Freudenthal criterion is formulated as follows:

$$\int_{0}^{\bar{\varepsilon}_f} \bar{\sigma} d\bar{\varepsilon} = C \tag{6.41}$$

The integral of the equivalent stress provides the physical meaning of the energy required to initiate a crack tip per unit area. In terms of the prediction of ductile fracture in microscaled plastic deformation, Freudenthal criterion has been used in microscaled flanged upsetting process by considering size effect and proven to be able to provide a satisfactory result [21].

In this chapter, the applicability of the above-described uncoupled DFCs in microscaled plastic deformation for fracture prediction and analysis is systematically studied and summarized. The related findings and concluding comments are also presented.

6.5.2 Applicability of the DFCs in Microscaled Plastic Deformation

The applicability of the above-described DFCc in microscaled plastic deformation is studied to predict and analyze the fracture occurrence in the deformation process. The prediction results are compared with the outcomes by using the conventional and the hybrid flow stress models developed in prior researches [3, 21]. The size effect on the applicability of the six DFCs in ductile fracture prediction in microscaled deformation is considered. The prediction results by using each DFC and the conventional and hybrid flow stress models are analyzed and compared. From the experimental point of view, the flanged upsetting in both macro- and microscales were conducted to validate the predictions using the DFCs. The flanged upsetting selected for experimental corroboration is based on the fact that there exists a cross-shear band in the flanged or the centrally heading portion in the upsetting part and the ductile fracture is easy to occur in the shear band in this unique plastic deformation process.

For the experiment, the multiphase alloy Brass C3602 is used as the testing material. The composition of the alloy is Cu: 59.0–63.0, Pb: 1.8–3.7, Fe (max): 0.5, and S (max): 1.2, and the balance is Zn. To obtain different microstructures of the

Table 6.6 Heat treatment conditions for annealing of the testing material [29]

	Heating temperature (°C)	Dwelling time (h)	Grain size (μm)
Group 1	As-received	As-received	18.54
Group 2	450	2	24.21
Group 3	600	2	45.48
Group 4	750	3	87.73

test material, annealing heat treatment of the testing material is conducted. The heat treatment conditions and the average grain sizes of the material after annealing are presented in Table 6.6. For realization of the flanged upsetting in macro- and microscales, the deformation velocity is set to 0.01 mm/s in such a way to ensure the strain rate does not affect the experimental result. In addition, all the specimens are compressed to 75 % of the original sample height to facilitate the occurrence of ductile fracture. Meanwhile, the height and diameter ratio H/D of the sample is 1.5.

6.5.2.1 Comparison of the Predicted Fracture Strain

To explore the applicability of different DFCs, a hybrid model, which considers size effect and the contribution of each phase in the multiphase alloy to microscaled plastic deformation, presented in Eq. (6.24) is used. The model describes the stress contribution of each phase by using the dislocation density and the size effect via introducing the size factor η into the model.

In addition, all the unknown coefficients in Eq. (6.24) are determined based on the known coefficients and using the curve fitting approach. The final form of the hybrid constitutive model formulated in Eq. (6.27) is used to predict the flow stress in both the macro- and microscaled deformation scenarios via considering the size effect. Since this model is established using the results of microscaled upsetting deformation, it can thus be used to determine the flow stress in other microscaled deformation scenarios.

By using Eq. (6.29), the fracture critical value C for the testing material can be determined. The physical meaning of the hybrid constitutive model is the threshold energy required to initiate the stress-induced fracture with the fracture strain. When other uncoupled DFCs are implemented, the mean stress and the maximum principal stress are needed for the determination of the fracture critical value C in each DFC. The upsetting simulation of each scenario with a specific heat treatment condition to obtain the principal stress is conducted by using DEFORM 3D. All the upsetting simulation results are extracted, and the stress–strain relationship in the form of $\sigma = k\varepsilon^n$ is generated via curve fitting based on the experimental and simulation results. The summarized results of the mean stress and the maximum principal stress are listed in Table 6.7.

By incorporating the corresponding parameters into each DFC, the fracture critical value C is thus determined. The predicted fracture strain in simple upsetting

Table 6.7 Mean stress and the maximum principal stress for each scenario [29]

	2 × 3 mm		1 × 1.5 mm		0.5 × 0.25 mm		0.25 × 0.375 mm	
	σ_m	σ^*	σ_m	σ^*	σ_m	σ^*	σ_m	σ^*
As-received	$368\varepsilon^{0.21}$	$725\varepsilon^{0.17}$	$716\varepsilon^{0.85}$	$1679\varepsilon^{0.93}$	$186\varepsilon^{0.21}$	$359\varepsilon^{0.14}$	$572\varepsilon^{0.24}$	$907\varepsilon^{0.09}$
Annealed at 450 °C	$397\varepsilon^{0.51}$	$733\varepsilon^{0.38}$	$373\varepsilon^{0.45}$	$586\varepsilon^{0.29}$	$630\varepsilon^{0.95}$	$970\varepsilon^{0.64}$	$471\varepsilon^{0.23}$	$903\varepsilon^{0.15}$
Annealed at 600 °C	$405\varepsilon^{0.65}$	$719\varepsilon^{0.46}$	$287\varepsilon^{0.49}$	$705\varepsilon^{0.52}$	$305\varepsilon^{0.53}$	$714\varepsilon^{0.54}$	$341\varepsilon^{0.43}$	$663\varepsilon^{0.31}$
Annealed at 750 °C	$406\varepsilon^{1.1}$	$733\varepsilon^{0.72}$	$424\varepsilon^{0.92}$	$708\varepsilon^{0.69}$	$297\varepsilon^{0.66}$	$673\varepsilon^{0.64}$	$286\varepsilon^{0.7}$	$569\varepsilon^{0.56}$

process for each case is calculated and presented in Table 6.8, which gives the predicted fracture strain in each scenario of the testing materials with different heat treatment conditions and dimensions.

As the size factor is introduced in the hybrid constitutive model-based fracture model (HFM), the conventional constitutive model-based fracture model (CFM) is thus needed to reveal the size effect by comparing with the prediction accuracy of HFM. The size factor-dependent part of Eq. (6.25) is taken away, and the flow stress model without considering size effect is formulated in the following [21]:

$$\sigma_{\text{total}}(\varepsilon) = Mk_2\varepsilon^{n_2} + f_\alpha \alpha M\mu_\alpha b_\alpha\sqrt{\frac{C_S\varepsilon}{b_\alpha L^s} + \frac{C_{G\alpha}\varepsilon}{b_\alpha d}} + (1-f_\alpha)\alpha M\mu_\beta b_\beta\sqrt{\frac{C_S\varepsilon}{b_\beta L^s} + \frac{C_{G\beta}\varepsilon}{b_\beta d}} \tag{6.42}$$

By applying Eq. (6.42) to each DFC, the predicted fracture strain by CFM is obtained and presented in Table 6.9.

Upon determination of the predicted fracture strains, they can be compared with the actual experimental results. Figure 6.34 shows the actual and the predicted fracture strains using different DFCs in upsetting process. These fracture strains are calculated by using Eq. (6.29) once the fracture critical value is determined. In Fig. 6.34a, it is found that the prediction results using the Brozzo [27] and Ayada criteria [26] are the closest to the experimental results with the deviation of 4.9 %. Meanwhile, the Freudenthal model [23] gives the worst performance with the deviation of 12.6 % from the experiment. This could be caused by the different grain sizes in macro- and microscaled specimens. To distinguish the different stress contribution arising from the grain and geometry size effects, the macro- and microscaled specimens were annealed with the same heat treatment condition. Figure 6.34b shows the result comparison in microscale. In the simple upsetting process with the specimen dimension of 0.5 × 0.75 mm, the specimen does not have any visible macrocrack when the specimen is compressed with the reduction of 75 %. The long dash line in the picture is the predicted flow stress curve if there is no fracture in the deformation process. For the microscaled simple upsetting, however, the Freudenthal criterion provides the best result with the deviation of 3.7 % between the calculation and experiment. The Brozzo criterion, on the other

Table 6.8 The predicted fracture strains using different DFCs [29]

Fracture criteria	Critical value C (annealing temperature)	Specimen dimensions			
		2 × 3 mm	1 × 1.5 mm	0.5 × 0.25 mm	0.25 × 0.375 mm
Freudenthal model	$C = 684$ (as-received)	$\varepsilon_f = 1.15$	$\varepsilon_f = 1.16$	$\varepsilon_f = 1.19$	$\varepsilon_f = 1.26$
	$C = 578$ (450 °C)	$\varepsilon_f = 1.03$	$\varepsilon_f = 1.05$	$\varepsilon_f = 1.08$	$\varepsilon_f = 1.16$
	$C = 698$ (600 °C)	$\varepsilon_f = 1.11$	$\varepsilon_f = 1.15$	$\varepsilon_f = 1.22$	$\varepsilon_f = 1.36$
	$C = 606$ (750 °C)	$\varepsilon_f = 1.19$	$\varepsilon_f = 1.25$	$\varepsilon_f = 1.39$	$\varepsilon_f = 1.63$
Cockcroft model	$C = 730$ (as-received)	$\varepsilon_f = 1.15$	$\varepsilon_f = 0.91$	$\varepsilon_f = 2.1$	$\varepsilon_f = 0.89$
	$C = 553$ (450 °C)	$\varepsilon_f = 1.03$	$\varepsilon_f = 1.17$	$\varepsilon_f = 0.96$	$\varepsilon_f = 0.74$
	$C = 574$ (600 °C)	$\varepsilon_f = 1.11$	$\varepsilon_f = 1.12$	$\varepsilon_f = 1.15$	$\varepsilon_f = 1.1$
	$C = 575$ (750 °C)	$\varepsilon_f = 1.19$	$\varepsilon_f = 1.21$	$\varepsilon_f = 1.23$	$\varepsilon_f = 1.34$
Oyane model	$C = 2.68$ (as-received)	$\varepsilon_f = 1.15$	$\varepsilon_f = 1.03$	$\varepsilon_f = 1.58$	$\varepsilon_f = 0.82$
	$C = 2.2$ (450 °C)	$\varepsilon_f = 1.03$	$\varepsilon_f = 1.03$	$\varepsilon_f = 0.93$	$\varepsilon_f = 0.73$
	$C = 2.39$ (600 °C)	$\varepsilon_f = 1.11$	$\varepsilon_f = 1.22$	$\varepsilon_f = 1.16$	$\varepsilon_f = 0.99$
	$C = 2.43$ (750 °C)	$\varepsilon_f = 1.19$	$\varepsilon_f = 1.11$	$\varepsilon_f = 1.15$	$\varepsilon_f = 1.07$
Ayada model	$C = 0.61$ (as-received)	$\varepsilon_f = 1.15$	$\varepsilon_f = 0.98$	$\varepsilon_f = 2.25$	$\varepsilon_f = 0.67$
	$C = 0.47$ (450 °C)	$\varepsilon_f = 1.03$	$\varepsilon_f = 1.03$	$\varepsilon_f = 0.9$	$\varepsilon_f = 0.57$
	$C = 0.5$ (600 °C)	$\varepsilon_f = 1.11$	$\varepsilon_f = 1.3$	$\varepsilon_f = 1.19$	$\varepsilon_f = 0.89$
	$C = 0.46$ (750 °C)	$\varepsilon_f = 1.19$	$\varepsilon_f = 1.07$	$\varepsilon_f = 1.11$	$\varepsilon_f = 0.96$
Brozzo model	$C = 1.51$ (as-received)	$\varepsilon_f = 1.15$	$\varepsilon_f = 1.15$	$\varepsilon_f = 1.15$	$\varepsilon_f = 1$
	$C = 1.34$ (450 °C)	$\varepsilon_f = 1.03$	$\varepsilon_f = 0.9$	$\varepsilon_f = 0.97$	$\varepsilon_f = 1.03$
	$C = 1.46$ (600 °C)	$\varepsilon_f = 1.11$	$\varepsilon_f = 1.28$	$\varepsilon_f = 1.88$	$\varepsilon_f = 1.16$
	$C = 1.45$ (750 °C)	$\varepsilon_f = 1.19$	$\varepsilon_f = 1.07$	$\varepsilon_f = 1.86$	$\varepsilon_f = 1.18$

(continued)

Table 6.8 (continued)

Fracture criteria	Critical value C (annealing temperature)	Specimen dimensions			
		2 × 3 mm	1 × 1.5 mm	0.5 × 0.25 mm	0.25 × 0.375 mm
Rice and Tracey model	C = 2.55 (as-received)	$\varepsilon_f = 1.15$	$\varepsilon_f = 0.96$	$\varepsilon_f = 1.69$	$\varepsilon_f = 0.58$
	C = 2.05 (450 °C)	$\varepsilon_f = 1.03$	$\varepsilon_f = 1.02$	$\varepsilon_f = 0.89$	$\varepsilon_f = 0.61$
	C = 2.04 (600 °C)	$\varepsilon_f = 1.11$	$\varepsilon_f = 1.15$	$\varepsilon_f = 1.09$	$\varepsilon_f = 0.88$
	C = 2.34 (750 °C)	$\varepsilon_f = 1.19$	$\varepsilon_f = 1.1$	$\varepsilon_f = 1.17$	$\varepsilon_f = 1.08$

hand, has the worst performance with the deviation of 38.2 %. Therefore, the energy-based criterion seems to have a better performance in analysis and prediction of fracture behaviors in microscaled deformation.

6.5.2.2 Stress-Induced Fracture Map Using Different Fracture Criteria

By using the predicted fracture strain presented in Tables 6.8 and 6.9, the SFM based on the hybrid model and the conventional model without considering size effect can thus be constructed. Figure 6.35 shows the SFM comparison between the experiment indicated by red stars and the simulation. In Fig. 6.35a, the experimental results represented by red star have the similar shape with the SFM constructed based on the Freudenthal criterion. It indicates that the Freudenthal criterion can be used for both the macro- and microscaled fracture prediction and provides relatively accurate results.

In addition, Fig. 6.35b–i shows the SFMs generated based on other five uncoupled DFCs. Although the values of the actual fracture strain are the same, the SFM shapes of the actual fracture strain are different based on the Oyane, Ayada, and Rice and Tracey DFCs. The figure also shows the big deviation of the DFCs in microscaled deformation, and thus, their applicability in this domain scale is not desirable.

Figures 6.36, 6.37, and 6.38 show the deviation between the predicted and the actual fracture strains by each DFC. These three figures are used to evaluate the influence of size factor in fracture prediction. In Fig. 6.36, both the hybrid flow stress model (HFM) and the conventional flow stress model (CFM) show a good result in macroscaled plastic deformation. However, the deviation of HFM is less than 11 % in microscaled plastic deformation process, while the deviation of CFM is 20–30 % for the scenarios with different heat treatment conditions. The different deviation indicates that the influence of size effect must be considered if the Freudenthal fracture criterion is used in microscaled scenario.

Table 6.9 The predicted fracture strains using the CFM (without considering size effect) [29]

Fracture criteria	Critical value C (annealing temperature)	Specimen dimensions			
		2 × 3 mm	1 × 1.5 mm	0.5 × 0.25 mm	0.25 × 0.375 mm
Freudenthal model	$C = 697$ (as-received)	$\varepsilon_f = 1.15$	$\varepsilon_f = 1.15$	$\varepsilon_f = 1.15$	$\varepsilon_f = 1.15$
	$C = 592$ (450 °C)	$\varepsilon_f = 1.03$	$\varepsilon_f = 1.03$	$\varepsilon_f = 1.03$	$\varepsilon_f = 1.03$
	$C = 621$ (600 °C)	$\varepsilon_f = 1.11$	$\varepsilon_f = 1.11$	$\varepsilon_f = 1.11$	$\varepsilon_f = 1.11$
	$C = 654$ (750 °C)	$\varepsilon_f = 1.19$	$\varepsilon_f = 1.19$	$\varepsilon_f = 1.19$	$\varepsilon_f = 1.19$
Cockcroft model	$C = 730$ (as-received)	$\varepsilon_f = 1.15$	$\varepsilon_f = 0.91$	$\varepsilon_f = 2.1$	$\varepsilon_f = 0.89$
	$C = 553$ (450 °C)	$\varepsilon_f = 1.03$	$\varepsilon_f = 1.17$	$\varepsilon_f = 0.96$	$\varepsilon_f = 0.74$
	$C = 574$ (600 °C)	$\varepsilon_f = 1.11$	$\varepsilon_f = 1.12$	$\varepsilon_f = 1.15$	$\varepsilon_f = 1.1$
	$C = 575$ (750 °C)	$\varepsilon_f = 1.19$	$\varepsilon_f = 1.21$	$\varepsilon_f = 1.23$	$\varepsilon_f = 1.34$
Oyane model	$C = 2.66$ (as-received)	$\varepsilon_f = 1.15$	$\varepsilon_f = 1.01$	$\varepsilon_f = 1.55$	$\varepsilon_f = 0.89$
	$C = 2.18$ (450 °C)	$\varepsilon_f = 1.03$	$\varepsilon_f = 1.04$	$\varepsilon_f = 0.96$	$\varepsilon_f = 0.81$
	$C = 2.33$ (600 °C)	$\varepsilon_f = 1.11$	$\varepsilon_f = 1.24$	$\varepsilon_f = 1.22$	$\varepsilon_f = 1.13$
	$C = 2.34$ (750 °C)	$\varepsilon_f = 1.19$	$\varepsilon_f = 1.14$	$\varepsilon_f = 1.25$	$\varepsilon_f = 1.28$
Ayada model	$C = 0.6$ (as-received)	$\varepsilon_f = 1.15$	$\varepsilon_f = 0.99$	$\varepsilon_f = 2.41$	$\varepsilon_f = 0.75$
	$C = 0.46$ (450 °C)	$\varepsilon_f = 1.03$	$\varepsilon_f = 1.05$	$\varepsilon_f = 0.93$	$\varepsilon_f = 0.67$
	$C = 0.49$ (600 °C)	$\varepsilon_f = 1.11$	$\varepsilon_f = 1.35$	$\varepsilon_f = 1.31$	$\varepsilon_f = 1.13$
	$C = 0.46$ (750 °C)	$\varepsilon_f = 1.19$	$\varepsilon_f = 1.11$	$\varepsilon_f = 1.29$	$\varepsilon_f = 1.34$
Brozzo model	$C = 1.51$ (as-received)	$\varepsilon_f = 1.15$	$\varepsilon_f = 1.15$	$\varepsilon_f = 1.15$	$\varepsilon_f = 1$
	$C = 1.34$ (450 °C)	$\varepsilon_f = 1.03$	$\varepsilon_f = 0.9$	$\varepsilon_f = 0.97$	$\varepsilon_f = 1.03$
	$C = 1.46$ (600 °C)	$\varepsilon_f = 1.11$	$\varepsilon_f = 1.28$	$\varepsilon_f = 1.88$	$\varepsilon_f = 1.16$
	$C = 1.45$ (750 °C)	$\varepsilon_f = 1.19$	$\varepsilon_f = 1.07$	$\varepsilon_f = 1.86$	$\varepsilon_f = 1.18$

(continued)

Table 6.9 (continued)

Fracture criteria	Critical value C (annealing temperature)	Specimen dimensions			
		2 × 3 mm	1 × 1.5 mm	0.5 × 0.25 mm	0.25 × 0.375 mm
Rice and Tracey model	C = 2.52 (as-received)	$\varepsilon_f = 1.15$	$\varepsilon_f = 0.97$	$\varepsilon_f = 1.71$	$\varepsilon_f = 0.76$
	C = 2.41 (450 °C)	$\varepsilon_f = 1.03$	$\varepsilon_f = 1.22$	$\varepsilon_f = 1.03$	$\varepsilon_f = 0.87$
	C = 2.19 (600 °C)	$\varepsilon_f = 1.11$	$\varepsilon_f = 1.27$	$\varepsilon_f = 1.25$	$\varepsilon_f = 1.14$
	C = 2.21 (750 °C)	$\varepsilon_f = 1.19$	$\varepsilon_f = 1.14$	$\varepsilon_f = 1.28$	$\varepsilon_f = 1.32$

In addition, Fig. 6.37 shows the deviation between the experimental and the calculation results based on Brozzo and C&L criteria. The predicted results with and without considering size effect is the same, which means the size effect does not directly affect the result of the predicted fracture strain. When the specimen dimension is less than 0.5 mm, the predicted result has an over 40 % deviation compared with the experimental results. Thus, these two DFCs are not suitable for analysis and prediction of ductile fracture in microscale.

Furthermore, the deviation percentage of the rest three fracture criteria including Ayada, Oyane, and Rice and Tracey DFCs is shown in Fig. 6.38. These three criteria have one thing in common: The fracture prediction result by CFM is even more accurate than the result predicted by HFM.

6.5.2.3 The Generalized Fracture Model Formulation

Among different DFCs, the uncoupled DFCs neglect the effect of damage on the yield surface of materials in deformation process, while the coupled DFCs incorporate damage accumulation into the constitutive equation of the materials, and thus, its effect on the evolution of yield surface of materials is considered. The coupled DFCs assumes most ductile fractures undergo the process of voids initiation, coalescence, and growth, further to the formation of microscaled fracture. From criterion formulation and representation perspective, the coupled DFCs introduce the damage factor D in their formulation to model and represent the void growth in plastic deformation. As most of these DFCs use tensile test to determine the fracture critical value, their application in prediction of tensile-dominative deformation is acceptable. Their accuracy and efficiency in compression-dominative deformation, however, are beyond satisfactory.

The DFCs developed based on the mechanism of the initiation, growth, and coalescence of voids such as GTN and McClintock DFCs are not well applicable to compression-dominative deformation. The main reason is that the fracture initiation and growth in tensile test is caused by both the voids growth and shear stress

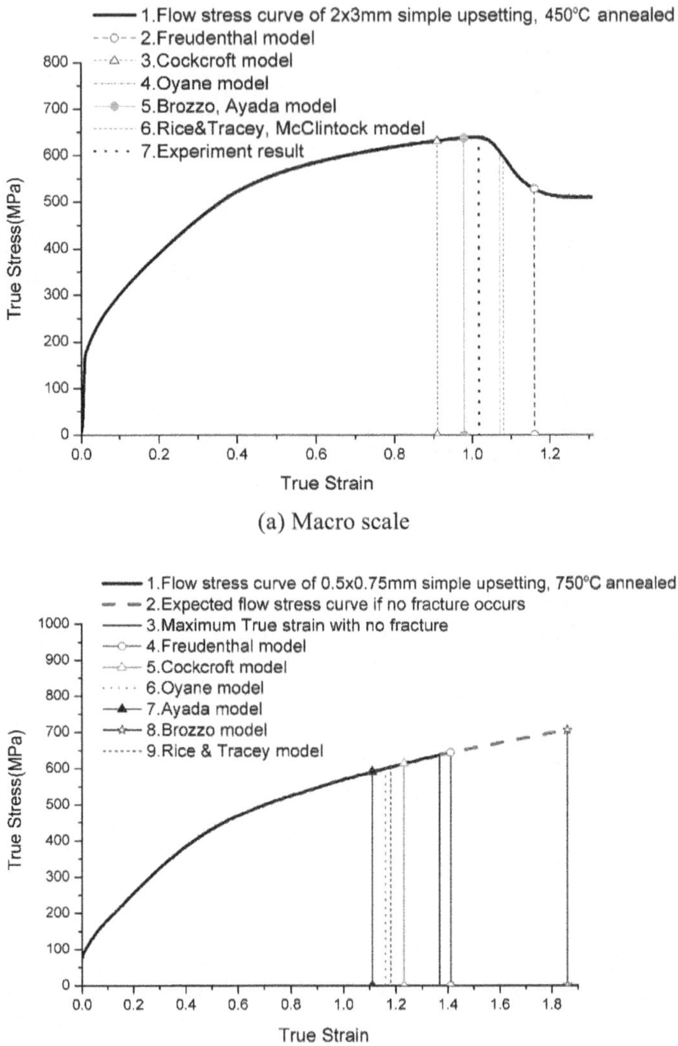

(a) Macro scale

(b) Micro scale

Fig. 6.34 Calculation and experimental results of the fracture strain in macro- and microscales [29]

concentration. In compression-dominative deformation, voids hardly exist inside the specimen. Unlike the tensile-dominative deformation, when the brittle phase or impurity in the multiphase metals is broken down under compression stress, the formed voids can immediately be filled up by its surrounding material. The ductile fracture in the compression-dominative deformation is mainly caused by shear stress concentration.

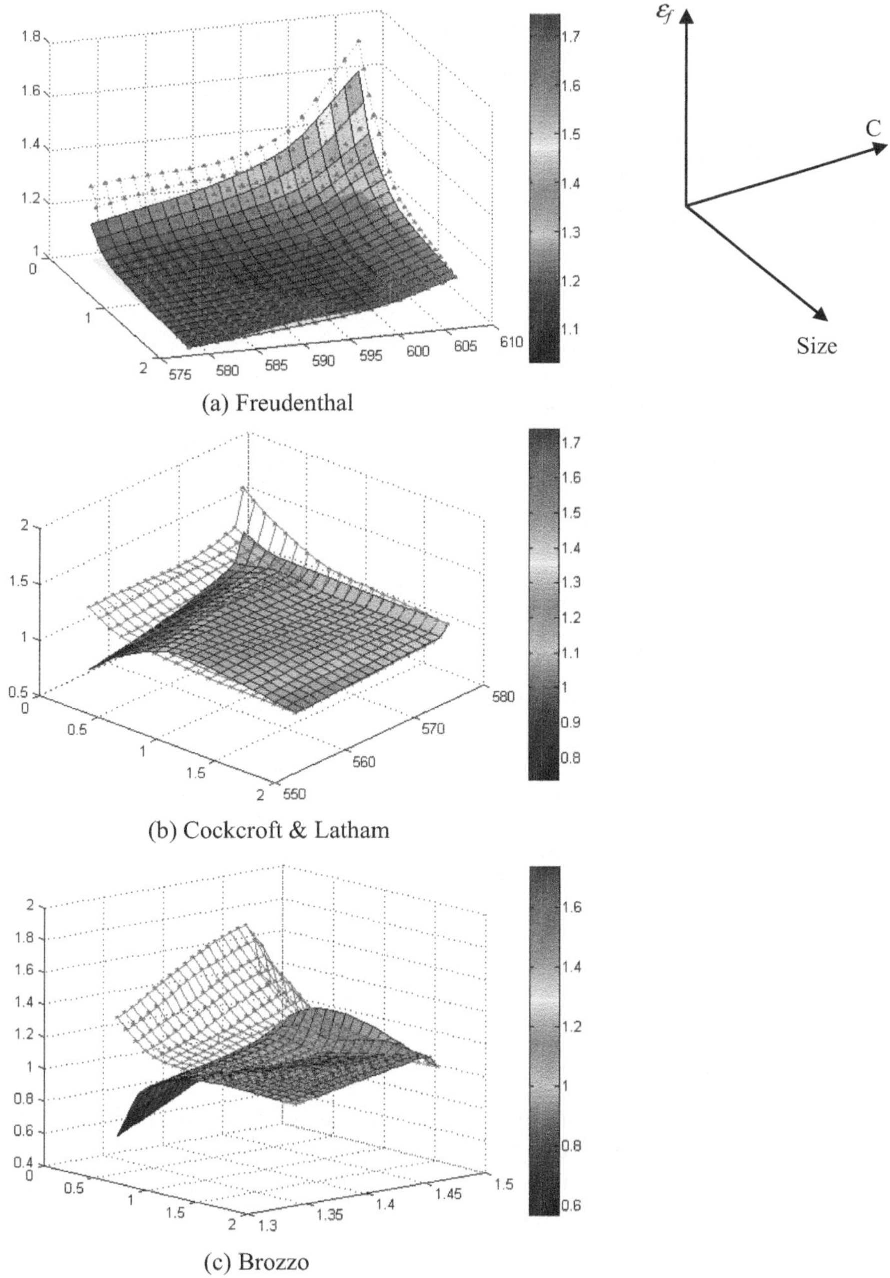

Fig. 6.35 The SFMs constructed based on the uncoupled DFCs [29]

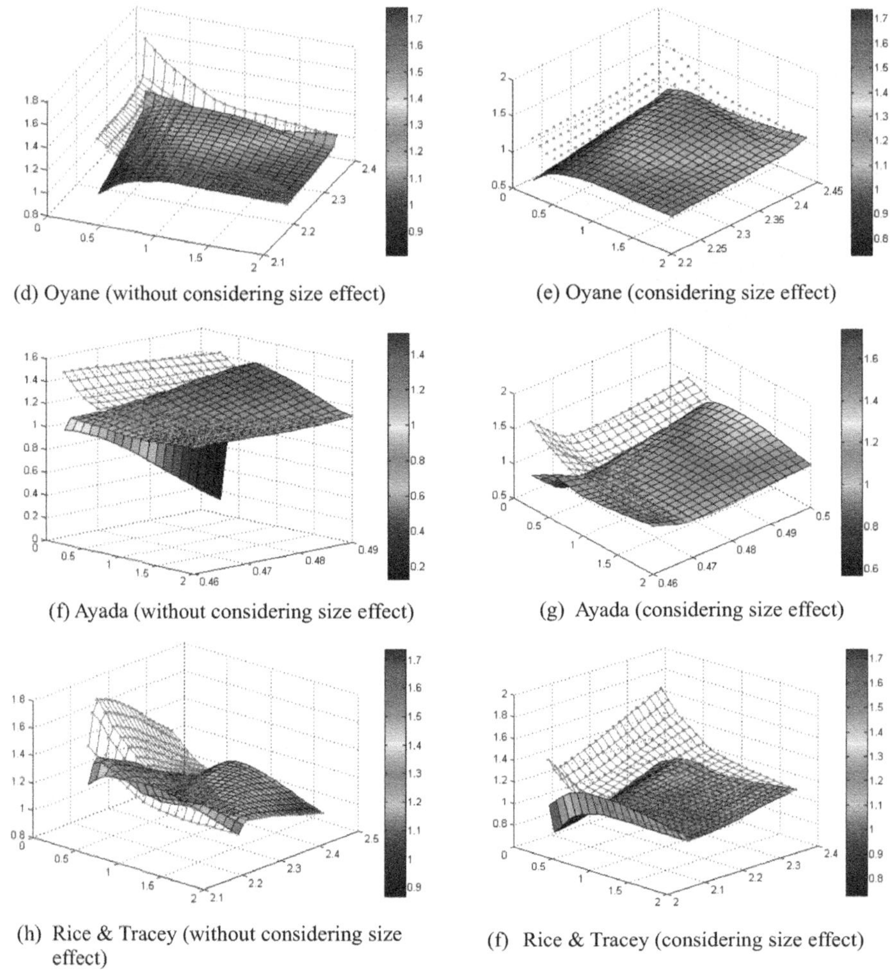

(d) Oyane (without considering size effect)

(e) Oyane (considering size effect)

(f) Ayada (without considering size effect)

(g) Ayada (considering size effect)

(h) Rice & Tracey (without considering size effect)

(f) Rice & Tracey (considering size effect)

Fig. 6.35 (continued)

For the conventional DFCs listed in Table 6.5, the fracture critical value is a parameter used to evaluate the existence of ductile fracture. As represented by Eq. (6.32), the integral of the strain-related function changes in deformation process and is equal to the fracture critical value C when the strain reaches the fracture strain and, in other words, fracture occurs, as shown in Eq. (6.43).

$$F(\varepsilon_f) = C \tag{6.43}$$

In Eq. (6.43), $F(\varepsilon)$ is the damage value function, ε_f is the fracture strain, and C is the fracture critical value. For each DFC, its simplified form can be represented as a strain-related exponent function.

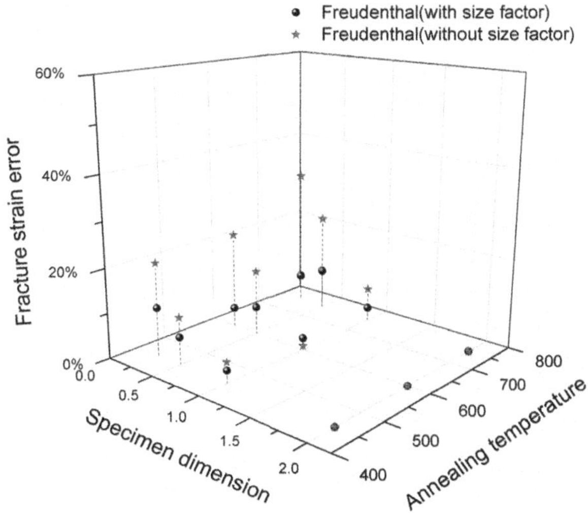

Fig. 6.36 Deviation evaluation of the Freudenthal criterion [29]

In tensile-dominative deformation, stress triaxiality is critical to determine whether void growth or shear stress concentration has a major contribution to fracture initiation. In compression-based deformation, stress triaxiality and non-dimensional stress concentration factor are considered as the two important factors which affect the damage value function $F(\varepsilon)$. Table 6.7 gives the generalized formulations of different DFCs based on the stress triaxiality, non-dimensional stress concentration factor, and mean stress. The damage value function can also be represented in the following general format in Eq. (6.44)

$$F = \int_{0}^{\bar{\varepsilon}_f} \mu(\varepsilon)^{n_1} \cdot \eta(\varepsilon)^{n_2} \cdot \sigma_m^{n_3} \mathrm{d}\varepsilon \quad (6.44)$$

In Eq. (6.44), $\mu(\varepsilon)$ is the stress concentration factor designated as $\frac{\sigma_1}{\bar{\sigma}}$; $\eta(\varepsilon)$ is the stress triaxiality and denoted as $\frac{\sigma_m}{\bar{\sigma}}$. The physical meaning of the damage value function is that the fracture initiation is driven by the stress concentration of the maximum principal stress and affected by stress triaxiality in deformation process.

Regarding the deviation of DFCs in Sect. 6.5.2.2, it can be explained by the general formulations presented in Table 6.7. In Table 6.7, the stress triaxiality in the Freudenthal criterion is inversely proportional to the damage value function. In the deformation of multiphase alloys, the decrease in stress triaxiality means that the main contribution to fracture initiation comes from shear stress concentration. In upsetting experiment, it is found that the stress triaxiality in macroscaled deformation is smaller than that in microscaled deformation, as shown in Fig. 6.39a. The Freudenthal criterion in Table 6.7 can thus be reformulated as follows:

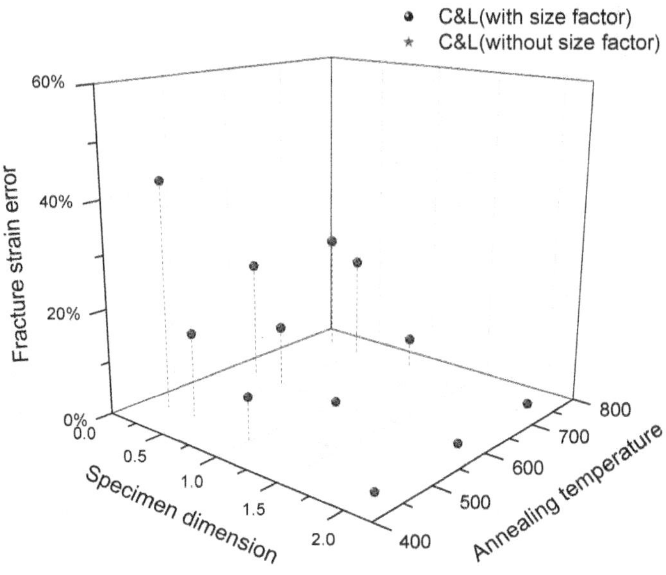

(a) Cockcroft & Latham

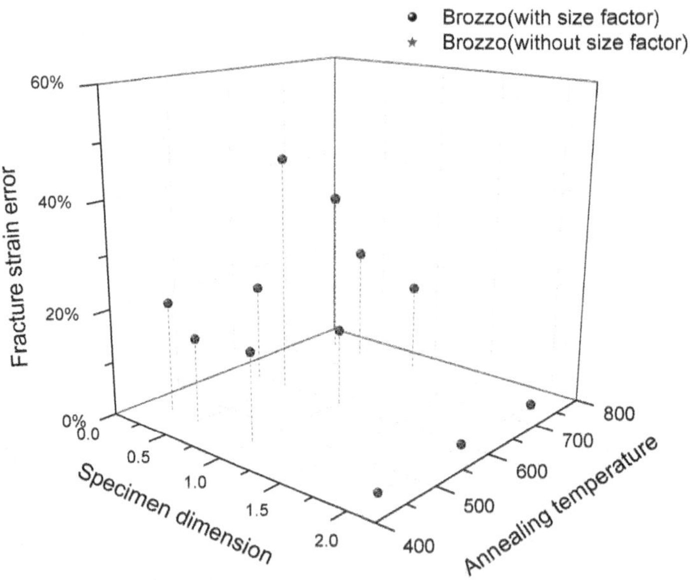

(b) Brozzo

Fig. 6.37 Deviation evaluation of the Cockcroft and Latham and Brozzo criteria [29]. (The *black dots* and the *red stars* are overlapped in the figures (**a**) and (**b**))

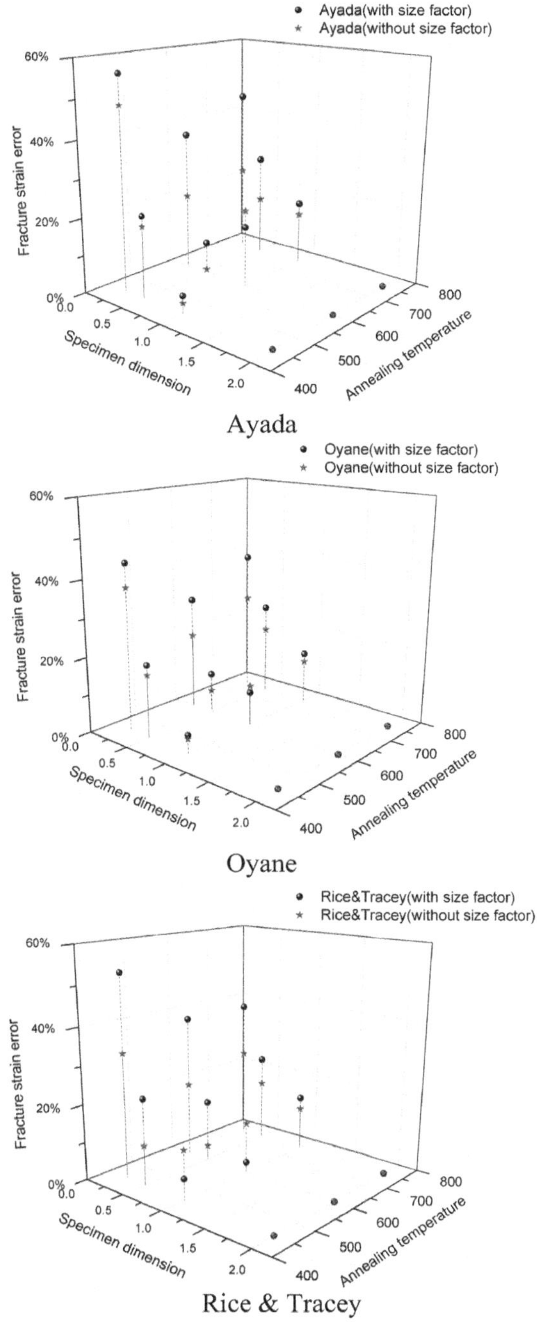

Fig. 6.38 Deviation evaluation of the Ayada, Oyane, and Rice and Tracey DFCs [29]

Fig. 6.39 Relationship between damage value and stress triaxiality [29]

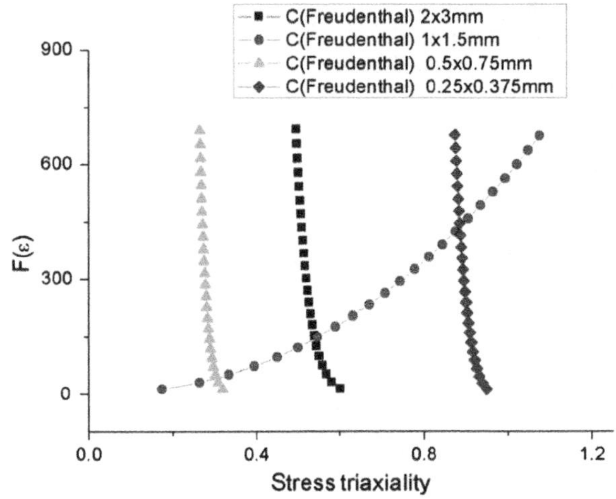

(a) As-receveid

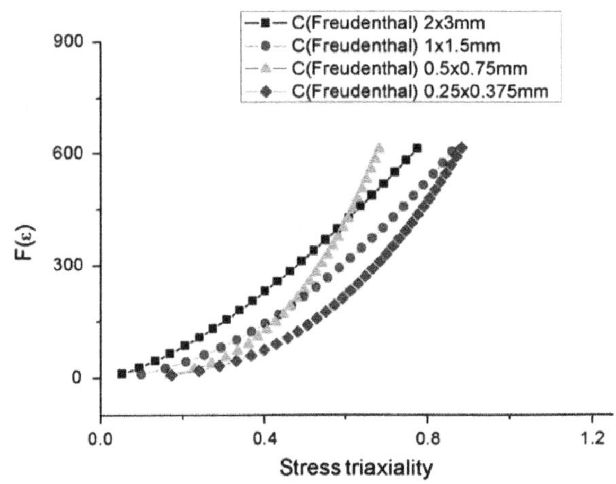

(b) Annealed at 750 degrees

$$F = \int_0^{\bar{\varepsilon}_f} \frac{1}{\eta_{\lambda-\text{dep}}} \cdot \sigma_m d\varepsilon_{\lambda-\text{dep}} = \int_0^{\bar{\varepsilon}_f} \frac{1}{\frac{\sigma_m}{587.52\varepsilon^{0.23} + 0.681\varepsilon^{0.5} \cdot d^{-\frac{1}{2}} - \lambda \cdot \left(203.52\varepsilon^{0.23} + 0.681\varepsilon^{0.5} \cdot d^{-\frac{1}{2}}\right)}} \cdot \sigma_m d\varepsilon_{\lambda-\text{dep}}$$

$$F = \int_0^{\bar{\varepsilon}_f} \frac{1}{\eta_{\lambda-\text{ind}}} \cdot \sigma_m d\varepsilon_{\lambda\text{-ind}} = \int_0^{\bar{\varepsilon}_f} \frac{1}{\frac{\sigma_m}{587.52\varepsilon^{0.23} + 0.681\varepsilon^{0.5} \cdot d^{-\frac{1}{2}}}} \cdot \sigma_m d\varepsilon_{\lambda-\text{ind}} \tag{6.45}$$

Table 6.10 Generalization of the six uncoupled fracture criteria [29]

Freudenthal	$F=\int_0^{\bar{\varepsilon}_f}\bar{\sigma}\mathrm{d}\varepsilon=\int_0^{\bar{\varepsilon}_f}\left(\frac{1}{\frac{\sigma_m}{\bar{\sigma}}}\cdot\sigma_m\right)\mathrm{d}\varepsilon=\int_0^{\bar{\varepsilon}_f}\frac{1}{\eta}\cdot\sigma_m\mathrm{d}\varepsilon$
Cockcroft and Latham	$F=\int_0^{\bar{\varepsilon}_f}\sigma_1\mathrm{d}\varepsilon=\int_0^{\bar{\varepsilon}_f}\sigma_m\cdot\frac{\bar{\sigma}}{\sigma_m}\cdot\frac{\sigma_1}{\bar{\sigma}}\mathrm{d}\varepsilon=\int_0^{\bar{\varepsilon}_f}\frac{\mu}{\eta}\cdot\sigma_m\mathrm{d}\varepsilon$
Ayada	$F=\int_0^{\bar{\varepsilon}_f}\frac{\sigma_m}{\bar{\sigma}}\mathrm{d}\varepsilon=\int_0^{\bar{\varepsilon}_f}\eta\mathrm{d}\varepsilon$
Oyane	$F=\int_0^{\bar{\varepsilon}_f}\left(1+A\frac{\sigma_m}{\bar{\sigma}}\right)\mathrm{d}\varepsilon=\int_0^{\bar{\varepsilon}_f}(1+A\cdot\eta)\mathrm{d}\varepsilon$
Brozzo	$F=\int_0^{\bar{\varepsilon}_f}\frac{2\sigma_1}{3(\sigma_1-\sigma_m)}\mathrm{d}\varepsilon=\int_0^{\bar{\varepsilon}_f}\frac{2\frac{\sigma_1}{\bar{\sigma}}}{3\left(\frac{\sigma_1}{\bar{\sigma}}-\frac{\sigma_m}{\bar{\sigma}}\right)}\mathrm{d}\varepsilon=\int_0^{\bar{\varepsilon}_f}\frac{2}{3\left(1-\frac{\eta}{\mu}\right)}\mathrm{d}\varepsilon$
Rice and Tracey	$F=\int_0^{\bar{\varepsilon}_f}\mathrm{e}^{1.5\frac{\sigma_m}{\bar{\sigma}}}\mathrm{d}\varepsilon=\int_0^{\bar{\varepsilon}_f}\mathrm{e}^{1.5\eta}\mathrm{d}\varepsilon$

where $\varepsilon_{\lambda-\text{dep}}$ and $\eta_{\lambda-\text{dep}}$ are the equivalent strain and stress triaxiality which consider the influence of size effects, while $\varepsilon_{\lambda-\text{ind}}$ and $\eta_{\lambda-\text{ind}}$ are the equivalent strain and stress triaxiality without considering size effect.

The first equation of Eq. (6.45) is the damage value function considering size effect, but the second one does not consider this effect. In microscaled deformation, it is obvious that the stress triaxiality considering size effect $\eta_{\lambda-\text{dep}}$ is larger than the one which does not consider size effect $\eta_{\lambda-\text{ind}}$. To obtain the same damage value F, the predicted fracture strain $\varepsilon_{\lambda-\text{dep}}$, which considers size effect, needs to be larger than $\varepsilon_{\lambda-\text{ind}}$. The predicted fracture strain is thus closer to the experimental results (Table 6.10).

The stress triaxiality in the Ayada [26], Oyane [25], and Rice and Tracey [28] criteria, on the other hand, is all proportional to the damage value function. The influence of size factor makes the calculation result even more deviated from the experimental result. Taking Ayada criterion as an instance, which can be redesignated in Eq. (6.46) as follows:

$$\begin{gathered}F=\int_0^{\bar{\varepsilon}_f}\eta_{\lambda-\text{dep}}d\varepsilon_{\lambda-\text{dep}}=\int_0^{\bar{\varepsilon}_f}\frac{\sigma_m}{587.52\varepsilon^{0.23}+0.681\varepsilon^{0.5}\cdot d^{-\frac{1}{2}}-\lambda\cdot\left(203.52\varepsilon^{0.23}+0.681\varepsilon^{0.5}\cdot d^{-\frac{1}{2}}\right)}d\varepsilon_{\lambda-\text{dep}}\\ F=\int_0^{\bar{\varepsilon}_f}\eta_{\lambda-\text{ind}}d\varepsilon_{\lambda-\text{ind}}=\int_0^{\bar{\varepsilon}_f}\frac{\sigma_m}{587.52\varepsilon^{0.23}+0.681\varepsilon^{0.5}\cdot d^{-\frac{1}{2}}}d\varepsilon_{\lambda-\text{ind}}\end{gathered} \tag{6.46}$$

As mentioned above, $\eta_{\lambda-\text{dep}}$ is larger than $\eta_{\lambda-\text{ind}}$, and $\varepsilon_{\lambda-\text{dep}}$ needs to be smaller than $\varepsilon_{\lambda-\text{ind}}$ to obtain the same damage value and thus makes $\varepsilon_{\lambda-\text{dep}}$ more deviation from the experimental result. This conclusion reveals why the fracture prediction results

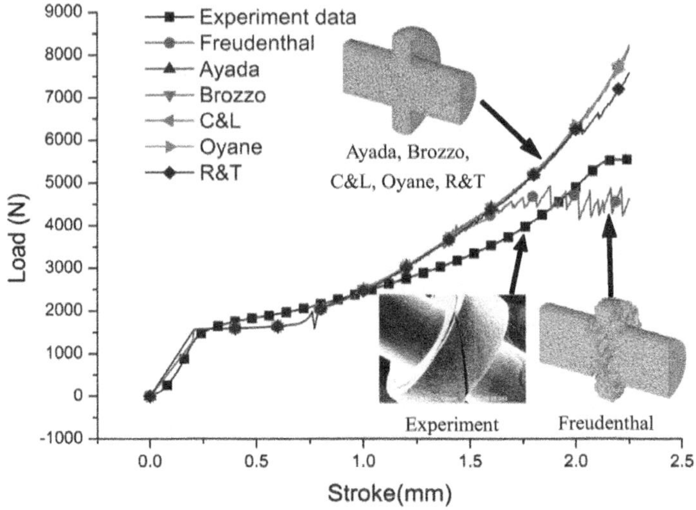

Fig. 6.40 Simulation and experimental verification for macroscaled flanged upsetting [29]

by using the conventional constitutive model is better than the ones using the hybrid constitutive model when Ayada, Oyane, and Rice and Tracey DFCs are used.

Based on HFM and the data in the previous section, the FE simulations for microscaled flanged upsetting and backward extrusion are conducted in a finite element software called DEFORM 3D, and the results are shown in Figs. 6.40, 6.41, and 6.42.

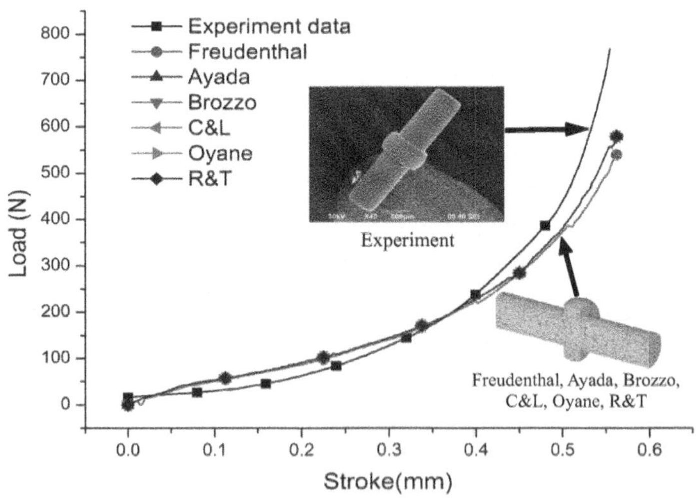

Fig. 6.41 Simulation and experimental verification for microscaled flanged upsetting [29]

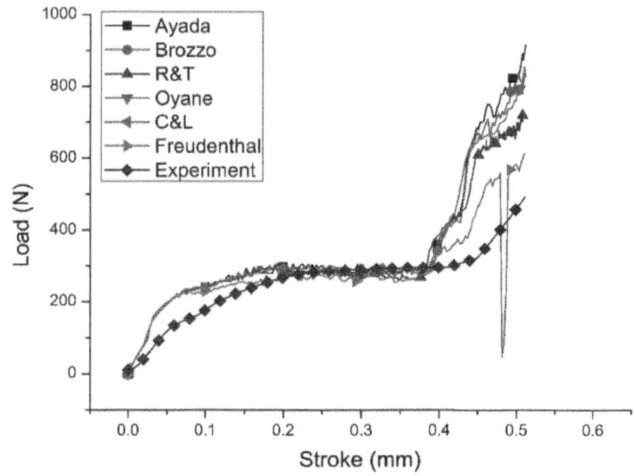

Fig. 6.42 Simulation and experimental verification for microscaled backward extrusion using the samples with the as-received materials [29]

In Fig. 6.40, it is found that the Freudenthal DFC is the only applicable one that can predict fracture in macroscaled flanged upsetting. For other DFCs, as the damage value does not reach its own fracture critical value, no fracture occurs in these scenarios, and their load–stroke curves are almost the same. In Fig. 6.41, the simulation result of the microscaled flanged upsetting shows that no fracture exists by applying all the DFCs. As the load–stroke curve is pretty close to the experimental result, Freudenthal criterion is thus considered to be the most suitable DFC for analysis of microscaled flanged upsetting process.

In Figs. 6.42 and 6.43, all the load–stroke curves of backward extrusion using different DFCs have the same trend as the actual experimental results. When the stroke is higher than 0.4 mm, the deviation between the load obtained by simulation and the one obtained from the actual experiment begins to increase. For both the

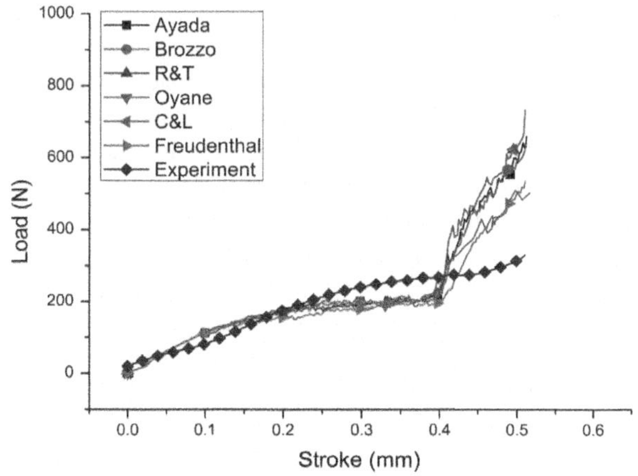

Fig. 6.43 Simulation and experimental verification for microscaled backward extrusion using the samples annealed at 750 °C [29]

as-received samples and the ones annealed at 750 °C, the load–stroke curves based on the Freudenthal and Rice and Tracey DFCs are close to the experimental results.

6.6 Summary

Fracture is a permanent damage of products, which can happen in the service of products or in the course of making the products via plastic deformation of materials. There are two types of fractures, viz. ductile and brittle fractures, which occur depending on the characteristics of materials undergoing plastic deformation and deformation condition. In the service of products, both the fractures can happen, but generally only ductile fracture may occur in the plastic deformation process as the ductility of materials is employed to make the products via the plastic deformation of materials. For brittle materials, the plastic deformation is realized via the brittle-to-ductile transition of materials. In the past a few decades, ductile fracture has been extensively studied in both academia and industry, and the epistemological understanding and systematic knowledge have been established. However, the traditional fracture-related knowledge is established within the domain of macroscale, and the geometry size of the deformed parts and components is basically quite large. With product miniaturization, the meso- and microscaled parts and components made by plastic deformation of materials are getting more and more popular. Whether the established fracture knowledge in macroscaled domain is valid and efficient in small-scaled domain thus becomes a tantalized and eluded issue, which needs to be addressed. In the past decade, many physical experiments and prior arts have shown that the data, information, knowledge, and the insight established in the former domain are not fully accurate and valid in the latter, and thus, it became a hot topic in metal-forming arena.

In this chapter, the ductile fracture in multiscaled plastic deformation is articulated, and the prediction and analysis of fracture behaviors and occurrence by using the constitutive models with and without considering size effect are presented. The ductile fracture criteria developed in macroscaled domain are used in microscaled deformation, and their applicability is investigated. The ductile fracture criteria are classified into uncoupled and coupled ductile fracture criteria, and the focus of this chapter was more on the uncoupled ones. To explore the deformation and fracture behaviors in microscaled plastic deformation by using the developed ductile fracture criteria, the size effects from the geometry and microstructure perspectives of materials are considered. Using the physical experiments in multiscaled deformation, the prediction and analysis of ductile fracture are corroborated by experiments, and the developed methodologies for ductile fracture prediction and analysis are verified and validated. All of these provide a fundamental for prediction, analysis, and avoidance of stress-induced defects in plastic deformation processes. This chapter thus provides a basis for ductile fracture prediction and avoidance in terms of forming process and tooling design in tandem with the improvement in the formability and the enhancement of product quality.

References

1. Avitzur B (1968) Metal forming: processes and analysis. McGraw-Hill, New York
2. Li H, Fu MW, Lu J, Yang H (2011) Ductile fracture: experiments and computations. Int J Plast 27(2):147–180
3. Ran JQ, Fu MW, Chan WL (2013) The influence of size effect on the ductile fracture in micro-scaled plastic deformation. Int J Plast 41:65–81
4. Peng LF, Lai XM, Lee HJ, Song JH, Ni J (2010) Friction behavior modeling and analysis in micro/meso scale metal forming process. Mater Des 31:1953–1961
5. Lai XM, Peng LF, Hu P, Lan SH, Ni J (2008) Material behavior modelling in micro/meso-scale forming process with considering size/scale effects. Comput Mater Sci 43:1003–1009
6. Armstrong RW (1962) The plastic deformation of polycrystalline aggregates. Philos Mag 7:45–48
7. Armstrong RW (1983) The yield and flow stress dependence on polycrystal grain size. In: Baker TN, Petch NJ (eds) Yield, flow and fracture polycrystals, pp 1–31. Appl. Sci. Publ., London
8. Clausen B, Lorentzen T, Leffers T (1988) Self-consistent modeling of the plastic deformation of F.C.C. polycrystals and its implications for diffraction measurements of internal stresses. Acta Mater 46:3087–3098
9. Keller C, Hug E, Habraken AM, Duchene L (2012) Finite element analysis of the free surface effects on the mechanical behavior of thin nickel polycrystals. Int J Plast 29:155–172
10. Hillerborg A, Modeer M, Petersson PE (1976) Analysis of crack formation and crack growth by means of fracture mechanics and finite elements. Mag Concr Res 6:773–781
11. Saanouni K, Mariage JF, Cherouat A, Lestriez P (2004) Numerical prediction of discontinuous central bursting in axisymmetric forward extrusion by continuum damage mechanics. Comput Struct 82:2309–2332
12. Atkins AG, Mai YW (1985) Elastic and plastic fracture: metals, polymers, ceramics, composites, biological materials. Ellis Horwood, New York
13. Cockroft MG, Latham DJ (1968) Ductile and workability of metals. J Inst Metals 96:33–39
14. Gospodinov G, Kerelezova I (2002) Boundary elements modeling of plain concrete notched beams by means of crack propagation formula. Danish Soc Struct Sci Eng 73:1–38
15. Brooks CR, Choudhury C (1993) Metallurgical: failure analysis. McGraw-Hill, New York
16. Hansen N (1985) Polycrystalline strengthening. Metall Mater Trans A 16:2167–2190
17. Bouquerel J, Verbeken K, De Cooman BC (2006) Microstructure-based model for the static mechanical behavior of multiphase steels. Acta Mater 54:1443–1456
18. Petch NJ (1953) The cleavage strength of polycrystals. J Iron Steel Inst 174:25–28
19. Argon AS (2008) Strengthening Mechanisms in Crystal Plasticity. Oxford University Press, Oxford
20. Ashby MF (1970) The deformation of plastically non-homogeneous materials. Philos Mag 21:399–424
21. Ran JQ, Fu MW (2014) A hybrid model for analysis of ductile fracture in micro-scaled plastic deformation of multiphase alloys. Int J Plast 61:1–16
22. Rodriguez R, Gutierrez I (2003) Correlation between nanoindentation and tensile properties. Influence of the indentation size effect. Mater Sci Eng A 361:377–384
23. Freudenthal FA (1950) The inelastic behavior of solids. Wiley, New York
24. Cockroft MG, Latham DJ (1968) Ductile and workability of metals. J Inst Metall 96:33–39
25. Oyane M (1972) Criteria of DF strain. Bull JSME 15:1507–1513
26. Ayada, M, Higashino T, Mori K (1984) Central bursting in extrusion of inhomogeneous materials. In: Proceedings of the first ICTP. Advanced technologies in plastic, vol 1, pp 553–558

27. Brozzo P, Deluca B, Rendina R (1972) A new method for the prediction of formability limits of metal sheets. In: Sheet metal forming and formability. Proceedings of the seventh biennial congress of international deep drawing research group, Amsterdam, Netherlands
28. Rice JR, Tracey DM (1969) On the ductile enlargement of voids in triaxial stress fields. J Mech Phys Solids 17:201–217
29. Ran JQ, Fu MW (2016) Applicability of the uncoupled ductile fracture criteria in micro-scaled plastic deformation. Int J Damage Mech 25:289–314

The manufacturer's authorised representative in the EU is Springer Nature Customer Service Centre GmbH, Europaplatz 3, 69115 Heidelberg, Germany. If you have any concerns regarding our products, please contact ProductSafety@springernature.com

Printed and bound by CPI Group (UK) Ltd, Croydon, CR0 4YY
29/07/2026
02180589-0002